Springer Proceedings in Business and Economics

More information about this series at http://www.springer.com/series/11960

Arnab Kumar Laha
Editor

Advances in Analytics and Applications

Springer

Editor
Arnab Kumar Laha
Indian Institute of Management Ahmedabad
Ahmedabad, Gujarat, India

ISSN 2198-7246 ISSN 2198-7254 (electronic)
Springer Proceedings in Business and Economics
ISBN 978-981-13-1207-6 ISBN 978-981-13-1208-3 (eBook)
https://doi.org/10.1007/978-981-13-1208-3

Library of Congress Control Number: 2018946576

This Springer imprint is published by the registered company Springer Nature Singapore Pte Ltd.
The registered company address is: 152 Beach Road, #21-01/04 Gateway East, Singapore 189721, Singapore

Preface

The Indian Institute of Management Ahmedabad (IIMA) was set up by the Government of India in 1961 as an autonomous institution under the aegis of the Ministry of Human Resource Development of the Government of India. Its creation can be viewed as an early example of public–private partnership where the Government of India, the Government of Gujarat and the leading lights of the local industry bodies collaborated to create an institution of excellence in the field of management education. The legendary scientist, late Dr. Vikram Sarabhai, played a critical role in establishing the mission and vision of the institute and laid the foundation for its eventual success. During the early years, the institute greatly benefitted from the mentorship of the Harvard Business School, which was facilitated by a grant from the Ford Foundation. Over the last 56 years, IIMA has gone from strength to strength to become one of the most coveted places to study management in India. It is regularly ranked at the top of the charts amongst management institutes in India for the last two decades. It is well regarded globally and is featured amongst the top institutes in the field of management by several global ranking agencies.

In anticipation of the major impact that business analytics has on the management of organisations, IIMA started the ICADABAI[1] series of biennial conferences in 2009 to discuss the recent advances in the areas of advanced data analysis, business analytics and business intelligence. From the very beginning, the aim of the conference was to bring together both academicians and practitioners and facilitate open exchange of ideas.

From its very inception, the papers were selected for inclusion in the conference only after a peer review of their extended abstracts.

ICADABAI-2017 was the fifth edition in this series of conferences. This conference too, like its earlier editions, had been fortunate to receive good-quality paper submissions discussing elegant theoretical research and highly innovative applications, some of which have been selected to be included in this volume. The

[1] International Conference on Advanced Data Analysis, Business Analytics and Intelligence.

book has been divided into eight parts, each containing one or more chapters. In the first part, the chapters provide overviews on some topics of current interest, namely machine learning, linear regression, directional data analysis and branching processes. The second part of the book discusses innovative predictive analytics applications, and the chapters in this part discuss click-through rate estimation, predicting success in professional tennis tournaments, predicting person movements in retail spaces, improving email marketing campaigns through personalisation, predicting customer churn for DTH companies and an application to continuous process industry. The third part of the book discusses applications of machine learning, and the chapters in this part discuss automatic detection of tuberculosis and risk-based insurance of connected cars. The fourth part of the book discusses human resource analytics, and the chapters in this part discuss analytics-led talent acquisition and assessment of student employability. The fifth part of the book discusses an innovative application of analytics in oil and gas industry operations. The sixth part of the book discusses applications of analytical models in finance, and the chapters in this part discuss loan loss provisioning practices of banks and models of commodity market returns. The seventh part of the book discusses new methodological developments, and the chapters in this part examine the usefulness of OLS regression when dealing with categorical data, estimation of parameters of a new distribution and queuing models with encouraged arrivals and retention of impatient customers. The final part of the book discusses an econometric analysis of banking competition and stability.

It would not have been possible to hold the conference without the support of the faculty and administration of IIM Ahmedabad. I would like to thank Ms. Deepna P. and Ms. Pravida Raja for their help and support in preparation of this book volume. We are specially thankful to Ms. Sagarika Ghosh and Ms. Nupoor Singh of Springer for their constant encouragement and support during the preparation of the conference volume. Finally, I would like to thank all the authors of the chapters of this volume for their valuable contribution.

Ahmedabad, India

Arnab Kumar Laha
Conference Convenor, ICADABAI-2017
Associate Professor
Indian Institute of Management Ahmedabad

Contents

About the Editor

Prof. Arnab Kumar Laha takes keen interest in understanding how analytics, machine learning and artificial intelligence can be leveraged to solve complex problems of business and society. His areas of research and teaching interest include advanced data analytics, quality management and risk modelling. He has published papers in national and international journals of repute in these areas and has served in the editorial board of several journals including *Statistical Analysis and Data Mining: The ASA Data Science Journal*. He was featured amongst India's best business school professors by Business Today in 2006 and Business India in 2012 and was named as one of the "10 Most Prominent Analytics Academicians in India" by Analytics India Magazine in 2014 and 2017. He is the convener of the biennial IIMA series of conferences on Advanced Data Analysis, Business Analytics and Intelligence. He is the author of the popular book on analytics entitled *How to Make the Right Decision* published by Penguin Random House. He has conducted a large number of training programmes and undertaken consultancy work in the fields of business analytics, quality management and risk management.

Part I
Brief Overviews

Machine Learning: An Introduction

Sayan Putatunda

Over the years, with the increase in storage capacity and the ease of vast amount of data collection, smart data analysis has become the order of the day. That is why "machine learning" has become one of the mainstays of the technology field over the past decade or so. This chapter aims to give an overview of the concepts of various supervised and unsupervised machine learning techniques such as support vector machines, k-nearest neighbor, artificial neural networks, random forests, cluster analysis, etc. Also, this chapter will give a brief introduction to deep learning, which is the latest fad in the analytics/data science industry.

1 Introduction

Machine learning (ML) can be described as the science and art by which computers learn things by themselves. We can always program computers to perform tasks and solve problems but for complex problems such as image recognition, speech recognition, etc., it is not possible to manually program all the aspects. So, we need the computer to learn by themselves and recognize different patterns (Domingos 2012). For example, let us consider the classic problem of spam detection. We can program to identify whether an incoming email is a spam or not just by writing conditions if the mail contains certain keywords. So, basically, this is similar to hardcoding or writing rules for all scenarios. But that is not scalable. So, a smarter way to operate would be by making the machines learn to distinguish between a spam and a non-spam using algorithms and feeding it historical email data. Here, "data" acts as a fuel. The machine can be used to then make predictions (i.e., spam/not spam) over unseen examples. This entire operation is achievable using machine learning. In

S. Putatunda (✉)
Indian Institute of Management Ahmedabad, Ahmedabad, India
e-mail: sayanp@iima.ac.in

A. K. Laha (ed.), *Advances in Analytics and Applications*, Springer Proceedings in Business and Economics, https://doi.org/10.1007/978-981-13-1208-3_1

short, machine learning is the phenomenon of enabling machines to learn and make predictions using past experiences (Baştanlar and Özuysal 2014).

The applications of machine learning are ubiquitous today. We can see it being used by companies such as Amazon, Apple, Google, Facebook, Microsoft, Uber, etc., in their products. Amazon Echo/Alexa is one such example. Another very recent example is the face detection feature in Apple iPhone X. Another interesting application of machine learning that companies like Google and Microsoft heavily uses is in Computational Advertising (Yang et al. 2017), which is basically targeted marketing where relevant advertisements are shown for different users when they input their queries in search engines such as Google or Bing. Some other examples of applications of machine learning include product recommendation, web search rankings, fraud detection, vehicle destination and travel time prediction, image processing, language translation, speech recognition and many more.

Machine learning is now being adapted across various industries, viz., banking, high tech, agriculture, manufacturing, Retail, etc. The different machine learning methods/algorithms can be broadly grouped under supervised and unsupervised learning methods. We will discuss these in the later sections. The "Data" can be either structured (i.e., tables, etc.) or unstructured (i.e., text, images, etc.). As far as tools for data science are concerned, SAS was the tool of choice for a longtime but ever since the open source revolution, R and Python are the tools of choice in most of the organizations and in academia.

2 Supervised Learning

In supervised learning, we have a target or response variable (i.e., y) for each independent or predictor variables (i.e., x_1, x_2,..., x_n). The goal is to build a model that can relate the y to x_i (where $i = 1, 2,..., n$) with the objective of understanding the relationship between the target and the independent variables and also to make predictions of y for future observations of x_i (Gareth et al. 2013).

The various types of supervised learning problems can be divided into regression and classification problems. In case of regression, the target or response variable is continuous whereas it is categorical in case of classification. For example, prediction of house prices in a city given the features, viz., house area and household income is a regression problem whereas to predict whether an email is a spam or not given some predictor variables, is a classification problem. A common approach generally used for solving these kinds of problems is the linear models such as linear/logistic regression. However, in this section, we will focus on some of the more advanced methods such as k-nearest neighbor, artificial neural networks, support vector machines, and random forests. See Bishop (2006) and Hastie et al. (2009) for more details on linear models.

2.1 K-Nearest Neighbor (KNN)

The k-nearest neighbor algorithm can work in both regression and classification setting. It is basically an extension of the nearest neighbor rule (Cover and Hart 1967) that can be defined as a simple nonparametric classification technique. Here for a new observation say "M", its distance from other observations is calculated using a distance metric T, and then, M is classified based on the class/category of its nearest neighbor. Let us define this more formally. Suppose there are q training pairs $(x_1, c_1), \ldots, (x_q, c_q)$ where x_i is of dimension p and c_j indicates the class category of x_i. Let us consider M* be a test pattern with unknown class/category. If

$$T(M^*, x_u) = \min\{(T(M^*, x_i))\}$$

where i= 1,..., q, then $M*$ is assigned class c_u (Cover and Hart 1967; Murty and Devi 2011).

The above nearest neighbor rule is extended to KNN when the value of K is greater than 1. In such case, the modus operandi remains the same but instead of one nearest neighbor for a test observation M, we determine its k-nearest neighbors and the M is assigned the final class by taking a majority voting of the classes of all its k-nearest neighbors (Putatunda 2017). The same thing works in regression setup as well where instead of majority voting, we take an average of the target variable value of all the k-nearest neighbors. This method is known as KNN Regression (Navot et al. 2005).

The choice of the distance metric is a very important parameter since it can greatly impact the performance of the KNN model. Some well-known distance metrics used in the literature are Euclidean distance, Chebyshev distance, Manhattan distance, Mahalanobis distance, etc., (Weinberger and Saul 2009). Also, the KNN method is computationally expensive for large datasets. The time complexity for KNN algorithm on a dataset with n observations and p variables is O(np) (Kusner et al. 2014).

2.2 Artificial Neural Networks (ANN)

If we go through the history of evolution of neural networks, we can find that the origin of neural networks can be traced back to the 60s and the 80s when there were a lot of studies to understand and represent mathematically the various information processing in biological systems (Rosenblatt 1962; Rumelhart et al. 1986; Widrow and Hoff 1988). ANN is used for both classification and regression problems. In this chapter, we will focus on a type of ANN known as the "multilayer perceptrons"/"feedforward neural networks" (Bishop 2006).

A Feedforward neural network basically consists of a single or multiple hidden layers between an input layer and an output layer. And each of these layers comprises one or more neurons or nodes. Let us try to understand the structure of a basic neural network with two layers, i.e., a hidden layer and one output layer.

Suppose the input layer comprises T nodes corresponding to the input variables $x_1, \ldots, x_T$. Then

$$q_i^{(1)} = \sum_{j=1}^{T} w_{ij}^{(1)} x_i + w_{i0}^{(1)}$$

where i = 1,…, P and superscript (1) indicates that the mentioned parameters are in the first layer (i.e., the hidden layer). The above equation represents the P linear combinations of the input variables $x_1,\ldots,x_T$, where P is the number of hidden nodes that the hidden layer possesses. $w_{ij}^{(1)}$ is the weight, $w_{i0}^{(1)}$ is the bias, and finally, $q_i^{(1)}$ is known as the activations. Then, a nonlinear differentiable function f is used to transform these activations as shown below:

$$c_i = f\left(q_i^{(1)}\right)$$

where c_i is the output of the hidden nodes. The c_i's are further combined to obtain the output unit activations, i.e., $q_k^{(2)}$ as given in the following equation:

$$q_k^{(2)} = \sum_{s=1}^{Y} w_{ks}^{(2)} c_i + w_{k0}^{(2)}$$

where the superscript (2) denotes that the parameters belong to the second layer of the network (i.e., the output layer) and k = 1; N denotes the total number of outputs. Also, another activation function f^2 is used to transform $q_k^{(2)}$ to generate the network output, i.e., $y_k = f^2\left(q_k^{(2)}\right)$ (Bishop 2006; Laha and Putatunda 2017). Some choices for activation functions as given in the literature are Relu, tanh, etc., and the most common choice for the output activation function is the sigmoidal function.

In the literature, one can find that the approximation properties of the feedforward neural networks are well established. According to Hornik et al. (1989), multilayer feedforward neural networks can be described as "universal approximators".

2.3 Support Vector Machines (SVM)

The support vector machine was first proposed by Boser et al. (1992) as an optimal classification method. Vapnik (1995) later extended this for solving regression problems and thus proposed the method support vector regression (SVR). The basic idea of SVM is that often input data is not linearly separable in the normal feature space and SVM maps the input vectors x into a higher dimensional feature space H using the "kernel trick" (Cortes and Vapnik 1995). In the higher dimensional space H, the data becomes linearly separable (Hastie et al. 2009).

The support vector classifier can be represented as

$$g(\chi) = \alpha_0 + \sum_{i=1}^{n} \beta_i \langle \chi, \chi^* \rangle$$

where β_i and α_0 are the parameters. To estimate the parameters β_i and α_0, we need the $\langle \chi, \chi^* \rangle$, which represents the inner products among all the pairs of training observations as shown in the above equation (Gareth et al. 2013; Hastie et al. 2009). We then replace the inner product part of the above equation of a support vector classifier using a generalization, i.e., a function of the form $K(\chi_i, \chi_i^*)$, which we will refer to as "Kernel".

A Kernel is basically a function that is used to quantify the similarity between a pair of observations (Gareth et al. 2013). For example, a linear kernel as defined below computes the similarity between a pair of observations using Pearson correlation (Gareth et al. 2013).

$$K(\chi_i, \chi_i^*) = \sum_{j=1}^{N} \chi_{ij} \chi_{ij}^*$$

Another example of a very widely used kernel is the radial basis function (RBF) kernel (Gareth et al. 2013). The RBF kernel is defined as

$$K = (\chi_i, \chi_i^*) = \exp\left(-\zeta \sum_{j=1}^{N} (\chi_{ij} - \chi_{ij}^*)^2\right)$$

where $\zeta > 0$ is a parameter.

As mentioned above, the modus operandi for SVM is basically combining kernels with support vector classifier. The space and time complexity of SVM are $O(n^2)$ and $O(n^3)$, respectively (Tsang et al. 2005). Thus, for large datasets, the computation time of SVM is quite high.

2.4 Random Forest

Before getting into the nitty-gritties of random forest, let us first try to understand decision trees and bagging. Decision trees are applicable to both regression and classification problems, i.e., when the target variables are continuous and categorical, respectively. The trees are generally upside down where the predictor space is split into different segments known as the *internal nodes*. The different nodes are connected through *branches*. Moreover, this is a greedy algorithm (Gareth et al. 2013). The major advantage of decision trees is that they are easy to explain but they are quite unstable in terms of predictive accuracy compared to that of other methods such as ANN, SVM, etc. Bagging is an ensemble learning method that was proposed by Breiman (1996). This is used to reduce the variance of a machine learning model.

Here, we take X bootstrapped samples from the training dataset and build our model M on each of the X samples. For final prediction, we either take an average of all the prediction (in case of regression) or take a majority voting of all the predictions (in case of classification).

As mentioned earlier, even though decision trees are easy to interpret, their performance in terms of predictive accuracy is not quite stable compared to that of other standard classification/regression methods. However, the predictive accuracy of trees can be increased by aggregating them, and random forest is one such approach. The random forest was first proposed by Breiman (2001). It is similar to bagged trees where decision trees are built on a number of bootstrapped samples of training dataset and the predictions are averaged/majority voting is taken to arrive at the final predictions. But the only difference here is that not all the predictors in the dataset, i.e., N are used for building trees on each bootstrapped samples. In fact, a random sample of P predictors is taken, and generally, the norm is $P = \sqrt{N}$ (Gareth et al. 2013).

3 Unsupervised Learning

In contrast to supervised learning, here we do not have a target or response variable for each independent variables (i.e., x_1, x_2,..., x_n). For example, a customer segmentation problem (i.e., suppose a marketing manager wants to understand who are his customers who are likely to purchase his new product based on features such as demographic characteristics, purchase behavior, etc.) would be in this category. A prominent method for unsupervised learning is the cluster analysis, which we will discuss in Sect. 3.1. A few other techniques that fall under the unsupervised learning category are the principal component analysis and the market basket analysis. One can refer Hastie et al. (2009) for more details.

3.1 Clustering

The basic idea of clustering is to partition the raw data or observations into different groups such that the observations in a group are similar to each other, i.e., homogeneous but each group has distinct characteristics (Gareth et al. 2013). So, for a customer segmentation problem as discussed earlier, a marketing manager can understand a customer better for a new product by grouping the customers into different clusters and profile them based on various characteristics. This will help the marketing manager to target the relevant customers through campaigns or advertisements.

The different clustering methods can be divided into (a) distance-based (e.g., hierarchical, K-means, etc.) and (b) density-based (e.g., DBSCAN, OPTICS, etc.). In this chapter, we will focus on two of the most widely used clustering methods, i.e., hierarchical and K-means clustering.

Hierarchical Clustering: In this approach, a hierarchy of clusters is created. The two approaches for building a hierarchical clustering model are (a) agglomerative—where observations merge to form clusters on their way up the hierarchy ("bottom-up approach"), and (b) divisive—where it begins with one cluster at the top and then it is broken down further as one goes down the hierarchy ("top-down approach") (Kaufman and Rousseeuw 1990). The final results are shown using a tree-like structure called "dendrogram". Moreover, a measure of dissimilarity is used between sets of observations to determine in which cluster they should belong (Kaufman and Rousseeuw 1990). A commonly used metric for dissimilarity is the "Euclidean distance".
K-Means Clustering: Here, we focus on partitioning the data into K nonoverlapping and distinct clusters. But the prerequisite here is that we need to specify the value of K. The goal of K-means clustering is to have a within-cluster variation as low as possible (Kaufman and Rousseeuw 1990).

The modus operandi is basically that we start by randomly assigning a number to observations from 1 to K and thus create initial clusters. We then iterate by computing cluster centroid for each K cluster and assign each observation to the nearest cluster. The nearness of an observation from the centroid of a cluster is determined using a distance metric such as the Euclidean distance (Kaufman and Rousseeuw 1990). The iteration stops when the cluster assignments do not change further.

4 Deep Learning

Deep learning is basically a form of "representation learning" where it is able to extract high-level features from raw data. In other words, the machine is able to develop complex concepts out of simpler ones (Goodfellow et al. 2016). We have explained the concept of feedforward neural network earlier in Sect. 2.2. A prominent example of deep learning would be a feedforward deep neural network (DNN) where we have many hidden layers comprising quite a few hidden nodes each. Each hidden layer is performing a mathematical transformation of the input and this helps uncover a new representation of the input data (Goodfellow et al. 2016).

Deep learning as an idea is not very new (since most of the algorithms have been there in existence since the 80s) but it is only after 2006 that it has shown great progress mainly due to the rise in computing power. However, the term "deep learning" is quite new. It is well established now that deep learning performs better with more data, and for some problems, it performs even better than some other sophisticated methods such as KNN, SVM, etc. It is especially in fields such as image processing and speech recognition where deep learning has got tremendous success.

Deep neural networks (DNNs) with greater depth can execute more instructions in sequence, and each layer encodes the necessary variation that explains the input data (Goodfellow et al. 2016). Apart from DNNs, there are other powerful deep learning algorithms such as autoencoders, recurrent neural networks (RNN), convolutional neural networks (CNN), etc., that we have not discussed in detail here. An autoencoder is generally used to understand the representation of the data, and thus, it is a form of unsupervised learning. An interesting application of autoencoders is in dimensionality reduction. RNNs and CNNs have been very successful in sequence modeling and image processing, respectively. See Goodfellow et al. (2016) to explore the theoretical concepts of various deep learning algorithms in depth.

5 Conclusion

In this chapter, we have discussed the basic idea of machine learning and supervised/unsupervised learning. We give an overview of some of the well-known machine learning algorithms that fall into the category of supervised and unsupervised learning. We also give a brief introduction to deep learning. Moreover, most of the ML methods come under the category of supervised and unsupervised learning. However, there is one more category, i.e., reinforcement learning which we have not discussed in this chapter. Please refer to Sutton and Barto (1998) for an in-depth understanding of reinforcement learning.

We are now living in the age of Big Data (Manyika et al. 2011). There are some who believe that more data will lead to better model performance but it is always the choice of algorithms and its optimization that leads to models with better prediction accuracy and less execution time. However, while building models we need to ensure that the model should neither *overfit* (i.e., performs well in the training dataset but performs poorly on the unseen data that has not been used for model building) nor *underfit* (i.e., when the model performs poorly or gives bad predictions in the training dataset). This is also known as the "Bias–Variance Trade-off" (Hastie et al. 2009).

Machine learning is a vast and a very dynamic field. Almost every other day, we hear of innovations in this field either in terms of new methodological contributions or in terms of new applications of machine learning in different fields. In sum, these are very exciting times to be a data scientist or a machine learning researcher.

References

Baştanlar, Y., & Özuysal, M. (2014). Introduction to machine learning. *Methods in Molecular Biology, 1107*, 105–128.

Bishop, C. M. (2006). *Pattern Recognition and Machine Learning*. New York Inc: Springer.

Boser, B. E., Guyon, I. M., Vapnik, V. N. (1992). A training algorithm for optimal margin classifiers. In *Proceedings of the Fifth Annual Workshop on Computational Learning Theory, (COLT'92)* (pp. 144–152).

Breiman, L. (1996). Bagging predictors. *Machine Learning, 24*(421), 123–140.

Breiman, L. (2001). Random Forests. *Machine Learning, 45*(1), 5–32.

Cortes, C., & Vapnik, V. (1995). Support-vector networks. *Machine Learning, 20*(3), 273–297.

Cover, P., & Hart, T. (1967). Nearest neighbour pattern classification. *IEEE Transactions on Information Theory*, *13*, 21–27.
Domingos, P. (2012). A few useful things to know about machine learning. *Communications of the ACM*, *55*(10), 78.
Gareth, J., Witten, D., Hastie, T., & Tibshirani, R. (2013). *An introduction to statistical learning with applications in R*. Springer Texts in Statistics.
Goodfellow, I., Bengio, Y., & Courville, A. (2016). *Deep learning*. MIT Press.
Hastie, T., Tibsharani, R., & Friedman, J. (2009). *The elements of statistical learning* (2nd ed.). Springer.
Hornik, K., Stinchcombe, M., & White, H. (1989). Multilayer feedforward networks are universal approximators. *Neural Networks, 2*(5), 359–366.
Kaufman, L., & Rousseeuw, P. J. (1990). *Finding groups in data: An introduction to cluster analysis* (1st ed.). New York: Wiley.
Kusner, M., Tyree, S., Weinberger, K. Q., & Agrawal, K. (2014). Stochastic neighbor compression. In *Proceedings of the 31st International Conference on Machine Learning (ICML-14)* (pp. 622–630).
Laha, A. K., Putatunda, S. (2017). Travel Time Prediction for GPS Taxi Data Streams. W.P.No. 2017-03-03, March 2017, Indian Institute of Management, Ahmedabad.
Manyika, J., Chui, M., & Brown, B. (2011). *Big data: The next frontier for innovation, competition, and productivity*. McKinsey Global Institute, June, p. 156.
Murty, M. N., & Devi, V. S. (2011). *Pattern recognition: an algorithmic approach*. Springer.
Navot, A., Shpigelman, L., Tishby, N., & Vaadia, E. (2006). Nearest neighbor based feature selection for regression and its application to neural activity. In *Advances in Neural Information Processing Systems 18 (Neural Information Processing Systems, NIPS 2005)* (pp. 996–1002).
Putatunda, S. (2017). *Streaming data: New models and methods with applications in the transportation industry*. Ph.D thesis, Indian Institute of Management Ahmedabad.
Rosenblatt, F. (1962). *Principles of neurodynamics: Perceptrons and the theory of brain mechanisms*. Washington: Spartan Books.
Rumelhart, D. E., Hinton, G. E., & Williams, R. J. (1986). Learning internal representations by error propagation. In D. E. Rumelhart, J. L. McClelland, & P. Research-Group (Eds.), *Parallel distributed processing: explorations in the microstructure of cognition* (Vol. 1, pp. 318–362). Cambridge, MA, USA: MIT Press.
Sutton, R.,& Barto, A. (1998). *Reinforcement learning: An introduction*. MIT Press.
Tsang, I. W., Kwok, J. T., & Cheung, P.-M. (2005). Core vector machines: Fast svm training on very large data sets. *Journal of Machine Learning Research, 6,* 363–392.
Vapnik, V. N. (1995). *The Nature of Statistical Learning Theory*. New York, NY, USA: Springer, New York Inc.
Weinberger, K. Q., & Saul, L. K. (2009). Distance metric learning for large margin nearest neighbor classication. *Journal of Machine Learning Research, 10,* 207–244.
Widrow, B., & Hoff, M. E. (1988). Adaptive switching circuits. In J. A. Anderson & E. Rosenfeld (Eds.), *Neurocomputing: Foundations of Research* (pp. 123–134). Cambridge, MA, USA: MIT Press.
Yang, Y., Yang, Y., Jansen, B. J., & Lalmas, M. (2017). Computational advertising: A paradigm shift for advertising and marketing? *IEEE Intelligent Systems, 32*(3), 3–6.

Linear Regression for Predictive Analytics

Arnab Kumar Laha

1 Introduction

Linear regression model is one of the most widely used statistical techniques having large scope of application in business and industry. While this technique was primarily built for understanding how the response variable depends on the predictor variables it is now widely used to predict the value of the response based on known values of the predictor variables. A linear relation between the response variable and the predictor variables is postulated and the unknown constants are estimated based on the given data. In this chapter, we discuss the linear regression method keeping the prediction task in focus. The excellent book by Montgomery et al. (2012) give a detailed account of linear regression analysis and the reader may consult the same for further details and proofs.

The chapter is structured as follows: In Sect. 2, we briefly discuss the linear regression model and two popular approaches to parameter estimation; in Sect. 3, we discuss both point and interval prediction using the linear regression model; in Sect. 4, we discuss hidden extrapolation, which is an important point of concern when using linear regression for prediction purpose; in Sect. 5, we discuss measures of prediction accuracy; in Sect. 6, we discuss the usefulness of dividing the data into training, validation and test datasets and discuss some possible approaches to correction of prediction bias; and in Sect. 7, we suggest how to use Shewhart control chart to monitor the predictive performance.

A. K. Laha (✉)
Indian Institute of Management Ahmedabad, Ahmedabad, India
e-mail: arnab@iima.ac.in

A. K. Laha (ed.), *Advances in Analytics and Applications*, Springer Proceedings in Business and Economics, https://doi.org/10.1007/978-981-13-1208-3_2

2 Linear Regression Model

The main idea behind linear regression modelling is to connect the response variable Y to a set of predictor variables $X_1, \ldots, X_k$ using a linear function. The proposed model is

$$Y_i = \beta_0 + \beta_1 X_{1i} + \cdots + \beta_k X_{ki} + \in_i,$$

where $(Y_i, X_{1i}, \ldots, X_{ki})$ is the ith observation, $i = 1, \ldots, n$. The random variables $\in_i$ are uncorrelated with $E(\in_i) = 0$ and $Var(\in_i) = \sigma^2$. Thus, we have

$$E(Y_i | X_1 = x_{1i}, \ldots, X_k = x_{ki}) = \beta_0 + \beta_1 x_{1i} + \cdots + \beta_k x_{ki}$$

and

$$Var(Y_i | X_1 = x_{1i}, \ldots, X_k = x_{ki}) = \sigma^2.$$

Usually, the random variables $\in_i$ are assumed to follow a normal distribution. This implies $\in_i$'s are independent, and the conditional distribution of Y_i given $X_1 = x_{1i}, \ldots, X_k = x_{ki}$ is normal. In applications, the unknown parameters $\beta_0, \beta_1, \ldots, \beta_k, \sigma$ need to be estimated from the data.

Ordinary least squares (OLS) is a popular approach for estimation of these parameters. In this approach, the estimates of the parameters $\beta_0, \beta_1, \ldots, \beta_k$ are obtained by minimising the sum of squared deviations between Y_i and $\beta_0 + \beta_1 x_{1i} + \cdots + \beta_k x_{ki}$, i.e. we solve the problem

$$\min_{\beta_0, \ldots, \beta_k} \sum_{i=1}^{n} (y_i - \beta_0 - \beta_1 x_{1i} - \cdots - \beta_k x_{ki})^2.$$

The resulting estimates $\hat{\beta}_0, \hat{\beta}_1, \ldots, \hat{\beta}_k$ are unbiased, i.e. $E\left(\hat{\beta}_i\right) = \beta_i$ for all $i = 1, 2, \ldots, k$.

An unbiased estimate of σ^2 is $\sigma^2_{UE} = \frac{1}{n-k-1} \sum_{i=1}^{n} \left(y_i - \hat{y}_i\right)^2$ where

$$\hat{y}_i = \hat{\beta}_0 + \hat{\beta}_1 x_{1i} + \cdots + \hat{\beta}_k x_{ki}, \; i = 1, \ldots, n$$

are the fitted values.

An alternative is to use the maximum likelihood (ML) approach. Assuming that $(x_{1i}, \ldots, x_{ki})$ are non-random for all $i = 1, \ldots, n$, we get the likelihood as

$$L\left(\beta_0, \beta_1, \ldots, \beta_k, \sigma^2\right) = \prod_{i=1}^{n} \frac{1}{\sqrt{2\pi}\sigma} e^{-\frac{1}{2\sigma^2}(y_i - \beta_0 - \beta_1 x_{1i} - \cdots - \beta_k x_{ki})^2}.$$

The MLEs of β_0, β_1, ..., β_k, σ are obtained by maximising this likelihood. Simple calculations show that the MLEs of β_0, β_1, ..., β_k are same as that obtained using the OLS approach. However, the MLE of σ^2 is $\hat{\sigma}^2_{ML} = \frac{1}{n}\sum_{i=1}^{n}(y_i - \hat{y}_i)^2$, which is different from the unbiased estimator $\hat{\sigma}^2_{UE}$ given above.

3 Prediction Using Linear Regression Model

Suppose we are required to predict the value of the response for a new case for which the values of the predictors are known. Let x_1^{new} ,..., x_k^{new} be the known values of the predictors. The predicted value of the response is $\hat{y}_0 = \hat{\beta}_0 + \hat{\beta}_1 x_1^{new} + \cdots + \hat{\beta}_k x_k^{new}$. However, to state the formula for obtaining $100(1-\alpha)\%$ prediction interval we need the matrix notation.

In the matrix notation, the linear regression problem is written as

$$Y_{n\times 1} = X_{n\times(k+1)}\beta_{(k+1)\times 1} + \epsilon_{n\times 1} \text{ where } \epsilon_{n\times 1} \sim N_n\left(0_{n\times 1}, \sigma^2 I_{n\times n}\right).$$

where $N_n\left(0_{n\times 1}, \sigma^2 I_{n\times n}\right)$ is the n-dimensional multivariate normal distribution with mean $0_{n\times 1}$ and variance-covariance matrix $\sigma^2 I_{n\times n}$

The least squares estimate (which is also the MLE) is

$$\hat{\beta}_{(k+1)\times 1} = \left(X^T_{(k+1)\times n} X_{n\times(k+1)}\right)^{-1} X^T_{(k+1)\times n} Y_{n\times 1}.$$

For a new observation having predictor values $X_1 = x_1^{new}$, ..., $X_k = x_k^{new}$ we predict the value of Y^{new} as $\hat{Y}^{new} = \hat{\beta}_0 + \hat{\beta}_1 x_1^{new} + \cdots + \hat{\beta}_k x_k^{new}$. A $100(1-\alpha)\%$ prediction interval for Y^{new} is

$$\left(\hat{Y}^{new} - t_{\frac{\alpha}{2},n-k-1}\sqrt{\hat{\sigma}^2_{UE}(1+f)},\ \hat{Y}^{new} + t_{\frac{\alpha}{2},n-k-1}\sqrt{\hat{\sigma}^2_{UE}(1+f)}\right)$$

where $f = x^T_{0,1\times(k+1)}(X^T_{(k+1)\times n} X_{n\times(k+1)})^{-1} x_{0,(k+1)\times 1}$ and $x^T_{0,1\times(k+1)} = \left(1, x_1^{new}, \ldots, x_k^{new}\right)$ and $t_{\frac{\alpha}{2},n-k-1}$ is the $100\left(1-\frac{\alpha}{2}\right)$ percentile of the t-distribution with $n-k-1$ degrees of freedom.

4 Hidden Extrapolation

While predicting a new response one should be careful about extrapolation which can lead to large prediction errors. In some situations, it may happen that the values of the predictors fall outside the region determined by the data points based on which the regression coefficients have been estimated. This leads to extrapolation which at times may not be apparent to the user giving rise to the term 'hidden extrapolation'.

The smallest convex set containing all the n data points $(x_{1i}, \ldots, x_{ki})$, $i = 1, \ldots, n$ is called the regressor variable hull (RVH). Hidden extrapolation happens when $\left(x_1^{new}, \ldots, x_k^{new}\right)$ lies outside the RVH. This can be detected by checking if $f > h_{max}$, where h_{max} is the largest diagonal element of $H_{n\times n} = X_{n\times(k+1)}(X^T_{(k+1)\times n}X_{n\times(k+1)})^{-1}X^T_{(k+1)\times n}$, and f is defined in Sect. 3. It may be mentioned here that $f \leq h_{max}$ does not imply that the predictors of the new observation is inside the RVH but it assures that it is close to the RVH so that the extrapolation (if it happens) is minor.

5 Prediction Accuracy

Prediction accuracy is of utmost concern when a linear regression model is used for prediction. A useful measure for understanding the prediction accuracy of a regression model is the PRESS statistic (where PRESS is an acronym for prediction error sum of squares) where $PRESS = \sum_{i=1}^{n}\left(y_i - \hat{y}_{(i)}\right)^2$ in which $\hat{y}_{(i)}$ is the predicted value of the response of the ith observation using a model which is estimated based on the (n − 1) data points excluding the ith data point. A low value of the PRESS statistic indicates that the linear regression model is appropriate for the given data and can be used for prediction. The $R^2_{prediction}$ statistic is an R^2-like statistic which is based on the PRESS statistic. It is defined as $R^2_{prediction} = 1 - \frac{PRESS}{SST}$ where $SST = \sum_{i=1}^{n}(y_i - \overline{y})^2$. A value of $R^2_{prediction}$ close to 1 indicates the suitability of the linear regression model for the prediction task.

6 Use of Validation and Test Data

While $R^2_{prediction}$ gives us an idea about the overall predictive ability of the linear regression model, it does not allow us to make any comment about the nature of the prediction errors. An alternative approach is to divide the available data randomly into three parts 'training', 'validation' and 'test'. The 'training' data is used for building the linear regression model, the 'validation' data is used to evaluate the model's predictive performance and do possible bias correction, if felt necessary, and finally, the 'test' data is used to evaluate the predictive performance of the final regression model and obtain the statistical characteristics of the prediction error which may be used to track the model performance over time.

Let D denote the Training data. Then note that

$$E\left(Y^{new} - \hat{Y}^{new}|D\right)$$
$$= E((\beta_0 - \hat{\beta}_0) + (\beta_1 - \hat{\beta}_1)x_1^{new} + \cdots + (\beta_k - \hat{\beta}_k)x_k^{new} + \epsilon^{new}\ |D)$$
$$= E((\beta_0 - \hat{\beta}_0) + (\beta_1 - \hat{\beta}_1)x_1^{new} + \cdots + (\beta_k - \hat{\beta}_k)x_k^{new}|D) + E(\epsilon^{new}\ |D)$$

$$= (\beta_0 - \hat{\beta}_0) + (\beta_1 - \hat{\beta}_1)x_1^{new} + \cdots + (\beta_k - \hat{\beta}_k)x_k^{new}$$

since given D the first term is a constant and since $\in^{new}$ is independent of D the second term is 0.

$$= Bias\left(x_1^{new}, \ldots, x_k^{new}\right).$$

(The above result should not be confused with the fact that the unconditional expectation $E\left(Y^{new} - \hat{Y}^{new}\right) = 0$.)

Again,

$$E((Y^{new} - \hat{Y}^{new})^2|D)$$
$$= E(((\beta_0 - \hat{\beta}_0) + (\beta_1 - \hat{\beta}_1)x_1^{new} + \cdots + (\beta_k - \hat{\beta}_k)x_k^{new} + \in^{new})^2|D)$$
$$= E(((\beta_0 - \hat{\beta}_0) + (\beta_1 - \hat{\beta}_1)x_1^{new} + \cdots + (\beta_k - \hat{\beta}_k)x_k^{new})^2|D) + E((\in^{new})^2|D)$$
$$= ((\beta_0 - \hat{\beta}_0) + (\beta_1 - \hat{\beta}_1)x_1^{new} + \cdots + (\beta_k - \hat{\beta}_k)x_k^{new})^2 + \sigma^2$$
$$= (Bias(x_1^{new}, \ldots, x_k^{new}))^2 + \sigma^2.$$

Thus, the average of the residuals ($\bar{r}$) obtained when the estimated regression equation is applied on the observations in the validation data set is an estimate of the 'mean bias' (MB) and the average of the squared residuals $\left(\overline{r^2}\right)$ estimate σ^2 plus the 'mean squared bias' (MSB). An estimate of MSB can be obtained by subtracting $\hat{\sigma}_{UE}^2$ from $\left(\overline{r^2}\right)$. Moreover, an estimate of the variance of the bias (VB) over the validation data set can be obtained as $\left(\overline{r^2}\right) - \hat{\sigma}_{UE}^2 - \bar{r}^2$. MB and VB together give an indication about the performance of the estimated regression model when used for prediction purpose. A large MB or a large VB indicates that the linear regression model may not perform well when used for prediction purpose.

If MB is large, a simple approach to reduce prediction error is to apply a 'bias correction' such as using $\tilde{Y}^{new} = \hat{Y}^{new} + MB$ for estimating Y^{new}. Another approach to bias correction could be to update the coefficients of the regression equation based on the errors observed in the validation data set. To see how this can be done, let us suppose that there are m observations in the validation data set. We randomly sample (with replacement) t observations from validation data set and compute the average error (err_1), average value of X_1 (m_{11}), average value of X_2 (m_{21}), …, and average value of X_k (m_{k1}). Note that

$$E(err_1) = (\beta_0 - \hat{\beta}_0) + (\beta_1 - \hat{\beta}_1)m_{11} + \cdots + (\beta_k - \hat{\beta}_k)m_{k1}.$$

Writing $\beta_j - \hat{\beta}_j = c_j$, we get $E(err_1) = c_0 + c_1 m_{11} + \cdots + c_k m_{k1}$. Repeating this process k times more, we get a system of (k + 1) equations in (k + 1) unknowns c_0, …, c_k as given below

$$E(err_q) = c_0 + c_1 m_{1q} + \cdots + c_k m_{kq},\ q = 1, \ldots, k + 1.$$

Now using err_q as an estimate of $E(err_q)$, we get (in matrix notation)

$$err_{(k+1)\times 1} = M_{(k+1)\times(k+1)} c_{(k+1)\times 1},$$

where $err^T_{1\times(k+1)} = (err_1, \ldots, err_{(k+1)})$, $c^T_{1\times(k+1)} = (c_0, \ldots, c_k)$ and the qth row of the matrix M is $(1, m_{1q}, \ldots, m_{kq})$. Solving the system of equations, we get

$$\hat{c}_{(k+1)\times 1} = M^{-1}_{(k+1)\times(k+1)}\, err_{(k+1)\times 1}.$$

We can then update the regression coefficients $\hat{\beta}_j$ with $\hat{\beta}_j = \hat{\beta}_j + \hat{c}_j$ and use the same in the regression equation for prediction purpose.

Let r_i^{val} denote the residuals obtained after applying the regression equation to the validation data set. A third approach to bias correction could be to regress the r_i^{val} on the predictor variables to obtain the regression coefficients $\dot{c}_j$ which can then be used to update the training data regression coefficients $\hat{\beta}_j$ with $\dot{\beta}_j = \hat{\beta}_j + \dot{c}_j$.

Among the four prediction approaches discussed above it is found through limited simulation experiments that the first two approaches, i.e. (a) using the linear predictor with coefficients estimated using the training data and (b) adding MB to the prediction obtained in (a) are performing better than the other two when applied to test data. However, among these two approaches, no clear winner could be identified. It may be mentioned here that the simulation experiments were done when all the linear regression model assumptions were met. The other two approaches may turn out to be useful in situations where there is a violation of the regression model assumptions or there is overfitting.

7 Tracking Model Performance

As mentioned earlier the characteristics of prediction errors obtained in the test data set can be used for tracking the model performance when the regression model is deployed operationally. A simple approach is to use a Shewhart mean control chart for individuals. Montgomery (2008) gives a detailed account of various control charts and their application.

For monitoring the predictive performance we can construct a Shewhart mean control chart for individuals with the central line (CL) equal to the average of the prediction errors (APE) obtained in the test data and the LCL and UCL are set at APE − 3 SDPE and APE + 3 SDPE, respectively, where SDPE denotes the standard deviation of the prediction errors in the test data. In many situations (such as in sales forecasting) the true value of the response becomes known after some time and the prediction error can be computed. These prediction errors can be plotted on the control chart in chronological order. When the model is performing well, it is expected that the prediction errors will lie within the LCL and UCL. If at some time point it is seen that the prediction error either falls above the UCL or below the LCL, it indicates a need to check the model thoroughly and if needed update the model with more recent data.

References

Montgomery, D. C. (2008) *Introduction to statistical quality control* (6th ed.). Wiley.
Montgomery, D. C., Peck, E. A. & Vining, G. G. (2012). *Introduction to linear regression analysis* (5th ed.). Wiley.

Directional Data Analysis

K. C. Mahesh

1 Introduction

Directional (angular) data arise as an outcome of random experiments in various ways either by direct measurements such as wind directions used to predict ozone concentrations, departure directions of birds from the point of release or indirectly from the measurements of times which are converted into angles such as arrival times at an intensive care unit. For example, patient arrival times (hh:mm) at an emergency room of a hospital are periodic with period of $24 \times 60 = 1440$ (with 0 being at 00:00). This can be viewed as circular data by transforming the arrival time into a degree using the transformation $\theta = \{(60 \times \text{hh} + \text{mm})/1440\} \times 360$. That is, 4 am is equivalent to 60°. Directional data in two and three dimensions arise in many areas of science and social sciences.

The measurement of the direction of flight of a bird or the orientation of an animal (see Schmidt-Koenig 1965; Batschelet 1981) may be an interesting data to a biologist. A medical researcher may be interested in data on control characteristics like sleep–wake cycles, hormonal pulsatility, body temperature, mental alertness, reproductive cycles, etc. which are periodic in nature to study the circadian rhythms (see Proschan and Follmann 1997). Circular data can also be used to analyse topics such as chronobiology, chronotherapy and the study of the biological clock (see Morgan 1990; Hrushesky 1994). Recently, Gavin et al. (2003) discuss that circular data can be used to analyse cervical orthoses in flexion and extension. Circular data can be used to study geological data such as the study of paleocurrents to infer the direction of flow of rivers in the past (see Sengupta and Rao 1966). In ecological and behavioural studies, angular data is used to study animal orientation and habitat selection (Ginsberg 1986 and Wallin 1986). Other types of directional data arising in meteorology include the time of day at which thunderstorms occur and the times of

K. C. Mahesh (✉)
Institute of Management, Nirma University, Ahmedabad, India
e-mail: maheshkc@nirmauni.ac.in

A. K. Laha (ed.), *Advances in Analytics and Applications*, Springer Proceedings in Business and Economics, https://doi.org/10.1007/978-981-13-1208-3_3

year at which heavy rains occur (see Mardia and Jupp 2000). Directional data found applications in political science in which a circular regression model is used to predict the timing of terrorist incidents with the day of occurrence of political violence in a year treated as a circular variable and entities like attacking groups, target groups, etc. as an important factor (Gill and Hangrtner 2010). In the field of demography, a researcher may be interested in studies like geographic marital patterns (Coleman and Haskey 1986), occupational relocation in the same city (Clark and Burt 1980), and settlement trends (Upton 1986). Studies involving paleomagnetic data, astronomical objects, image analysis, signal processing, etc. can be observed as spherical data. More examples of applications of circular and spherical data analysis can be found in Fisher (1993), Fisher et al. (1987), Jammalamadaka and SenGupta (2001) and Mardia and Jupp (2000).

Since direction has no magnitude, it can be represented as points on the circumference of a unit circle centred at origin with respect to some suitably chosen starting point called 'zero direction' and measured in clockwise or anticlockwise direction called 'sense of rotation'. Due to this representation, observations on such two-dimensional directions are called circular data and are commonly summarised as locations on a unit circle or as angles over a 360° or 2π radians range. Since the numerical representation of these data depends heavily on the above choices of zero direction and sense of rotation, it is important to make sure that our conclusions are also not depending on these arbitrary choices. The above nature of these observations makes directional data analysis substantially different from the standard 'linear' statistical analysis of univariate or multivariate data.

2 Descriptive Statistics on the Circle

In circular data analysis, the observations are represented as angles and generally plotted on a circle having unit radius. Since the angular values depend on the choice of zero direction and the sense of rotation, the usual arithmetic mean as well as standard deviation is not useful. This leads to the fact that in circular data one has to look at measures which are invariant under these choices. A meaningful measure of the mean direction for a set of directions represented as angles say, $\theta_1, \theta_2, \ldots, \theta_n$ is given by $\bar{\theta}_0 = \arctan^*\left(\frac{S}{C}\right)$, where $C = \sum_{i=1}^{n} \cos\theta_i$, $S = \sum_{i=1}^{n} \sin\theta_i$ and $\arctan^*$ is the quadrant-specific inverse of the tangent function which is defined as

$$\arctan*\left(\frac{S}{C}\right) = \begin{cases} \arctan\left(\frac{S}{C}\right) & \text{if } C > 0\ S \geq 0 \\ \pi/2 & \text{if } C = 0\ S > 0 \\ \arctan\left(\frac{S}{C}\right) + \pi & \text{if } C < 0 \\ \arctan\left(\frac{S}{C}\right) + 2\pi & \text{if } C \geq 0\ S < 0 \end{cases} \tag{2.1}$$

(see Jammalamadaka and SenGupta 2001). When both C=0 and S=0, a circular mean cannot be defined which indicates that the data is spread evenly or uniformly over the circle, with no concentration towards any direction. It should also be noted that circular mean direction is rotationally invariant. The length of the resultant vector $R = \sqrt{C^2 + S^2}$ is a useful measure for unimodal data of how concentrated the data is towards the circular mean direction. Another measure of dispersion on the circle is the sample circular dispersion $D_v = n - R$ (see Jammalamadaka and SenGupta 2001), which measures the dispersion of the sample relative to the centre through the sample mean direction. The data is highly dispersed from the centre when the value of R is close to 0 and the observations have more concentrated towards the centre when the values of R is close to n. Apart from the sample circular dispersion, appropriate distance measures on the circle like $D(\theta, \mu) = \pi - |\pi - |\theta - \mu||$ where θ and μ are any two points on the circle will also provide measures of dispersion. The median direction in the circular data is the direction μ which minimises the expected distance measure $E[D(\theta, \mu)]$.

A measure of skewness and kurtosis based on the sample trigonometric moments is given by $\hat{s} = \frac{\hat{\rho}_2 \sin(\hat{\mu}_2 - 2\hat{\mu})}{\left(\sqrt[3]{1-\bar{R}}\right)}$ and $\hat{k} = \frac{\hat{\rho}_2 \cos(\hat{\mu}_2' - 2\hat{\mu}) - \bar{R}^4}{(1-\bar{R})^2}$ (see Fisher 1993, p. 34).

3 Probability Models on the Circle

Let T be a unit circle and Ω a suitable probability space, then a circular random variable Θ can be defined as a mapping from $\Omega \to T$. If the arc $[\gamma_1, \gamma_2]$ where $\gamma_1, \gamma_2 \in T$, then the distribution of Θ is the probability that Θ takes values in this subset of the circle T (see Mahesh 2012). The probability density function (p.d.f) $f(\theta)$ of a circular random variable Θ has the following basic properties: (1)$f(\theta) \geq 0$, (2) $\int_0^{2\pi} f(\theta)\, d\theta = 1$ and (3)$f(\theta) = f(\theta + 2\pi k)$ for any integer k implying that $f(\theta)$ is periodic. The commonly used circular distributions for modelling purpose include Circular Normal (CN), Wrapped Normal (WN), Wrapped Cauchy (WC), Circular Uniform (CU), Cardioid, etc. Recently, some new probability models on the circle have been suggested in the literature (Kato and Shimizu 2004; Pewsey and Jones 2005; Pewsey et al. 2007). One such distribution is the Kato-Jones family of distribution proposed by Kato and Jones (2010). A variety of circular distributions like circular normal, wrapped Cauchy and circular uniform distributions can be derived from the Kato-Jones family.

(a) The Circular Normal Distribution

A symmetric and unimodal distribution on the circle which is widely used for applied work is the circular normal (or von Mises) distribution. A circular random variable Θ is said to have a von Mises or CN distribution with mean direction μ and concentration κ if it has the p.d.f:

$$f(\theta; \mu, \kappa) = \frac{1}{2\pi I_0(\kappa)} e^{\kappa \cos(\theta - \mu)}, 0 \leq \theta < 2\pi \text{ where } 0 \leq \mu < 2\pi \text{ and } \kappa > 0. \quad (3.1)$$

where $I_0(\kappa)$ is the modified Bessel function of the first kind and order zero and is given by $I_0(\kappa) = \frac{1}{2\pi}\int_0^{2\pi} e^{\kappa\cos\theta} d\theta = \sum_{r=0}^{\infty}\left(\frac{\kappa}{2}\right)^{2r}\left(\frac{1}{r!}\right)^2$. The distribution is symbolically represented as $CN(\mu, \kappa)$. If $\kappa = 0$ then the CN distribution can be approximated by the circular uniform distribution which has no preferred direction and when $\kappa \geq 2$, the same can be approximated by the WN distribution, which is a symmetric unimodal distribution obtained by wrapping a normal distribution around the circle. Note that for sufficiently large κ, the CN distribution can be approximated by a linear normal distribution. The trigonometric moments of the circular normal distribution can be obtained by the relation $\varphi_p = A_p(\kappa)e^{ip\mu}$, where p is an integer and $A_p(\kappa) = I_p(\kappa)\, I_0^{-1}(\kappa)$.

The length ρ of the first trigonometric moment is given by $A(\kappa) = I_1(\kappa)\, I_0^{-1}(\kappa)$. By virtue of symmetry of the CN density, the central trigonometric moments are $\alpha_p^* = A_p(\kappa)\cos p\mu$. The function $A(\kappa)$ has many interesting properties as follows: (1) $0 \leq A(\kappa) \leq 1$ (2) $A(\kappa) \to 0$ as $\kappa \to 0$ and $A(\kappa) \to 1$ as $\kappa \to \infty$ and (3)$A'(\kappa) \geq 0$, i.e. $A(\kappa)$ is a strictly increasing function of κ. The Maximum Likelihood Estimate (m.l.e) of the parameter μ is given by $\hat{\mu} = \arctan^*\left(\frac{S}{C}\right)$ and m.l.e of κ say $\hat{\kappa}$ can be obtained as the solution of $A(\hat{\kappa}) = I_1(\hat{\kappa})\, I_0^{-1}(\hat{\kappa}) = \frac{R}{n}$. The m.l.e of μ remains the same whether or not κ is known. On the other hand, the m.l.e of κ is different when μ is known and is given by $\hat{\kappa} = A^{-1}\left(\frac{V}{n}\right)$, V > 0 where $V = \sum_{i=1}^{n}\cos(\theta_i - \mu)$ (see Jammalamadaka and SenGupta 2001). Clearly, $\hat{\kappa} = 0$ for V ≤ 0.

(b) **The Wrapped Normal Distribution**

Another well-known symmetric and unimodal probability distribution used for the modelling purpose on the circle is the wrapped normal distribution obtained by wrapping the normal distribution with parameters μ and σ^2 on to the unit circle. The corresponding p.d.f is given by

$$f(\theta;\, \mu,\, \rho) = \frac{1}{2\pi}\left\{1 + 2\sum_{p=1}^{\infty}(\rho)^{p^2}\cos p(\theta - \mu)\right\},\ 0 \leq \theta < 2\pi,\ 0 \leq \mu < 2\pi,\ 0 < \rho < 1 \tag{3.2}$$

Symbolically, the distribution is represented by $WN(\mu,\, \rho)$ where the parameters μ and ρ are, respectively, the mean direction and the concentration parameter. These parameters arise naturally when wrapping $N(\mu, \sigma^2)$ onto the circle where $\rho = \exp\left(-\sigma^2/2\right)$ (Mardia and Jupp 2000). This distribution is a member of the wrapped stable family of distributions (see Mardia and Jupp 2000).

(c) **The Wrapped Cauchy Distribution**

A wrapped Cauchy distribution is obtained by wrapping the Cauchy distribution on line on to the circle. The p.d.f of WC distribution with parameter ρ is given by

$$f(\theta;\, \rho) = \frac{1}{2\pi}\frac{1 - \rho^2}{1 + \rho^2 - 2\rho\cos\theta},\ 0 \leq \theta < 2\pi,\ 0 < \rho < 1 \tag{3.3}$$

where $\rho = e^{-\lambda}$ and λ is the parameter associated with the Cauchy distribution on the line. WC distribution is unimodal and symmetric and as $\rho \to 0$, it tends to the uniform distribution and as $\rho \to 1$, it is concentrated at the point $\theta = 0$.

(d) **The Circular Uniform Distribution**

This distribution on the circle has the p.d.f given by

$$f(\theta) = \frac{1}{2\pi}, \; 0 \le \theta < 2\pi \tag{3.4}$$

This distribution does not have any preferred mean direction and concentration. The distribution is taken as a null model against which various alternative models can be tested.

e) **Kato-Jones Family of Distributions**

Kato and Jones (2010) proposed a four-parameter family of distributions on the circle by transforming the CN distribution through Mobius transformation. The CN distribution and the WC distribution can be obtained as a special case of this new distribution. The distribution has the p.d.f given by

$$f(\theta; \mu, \nu, r, \kappa) = \frac{1 - r^2}{2\pi I_0(\kappa)} \exp\left\{\frac{\kappa(\xi \cos(\theta - \eta) - 2r \cos \nu)}{1 + r^2 - 2r \cos(\theta - \gamma)}\right\} \times \frac{1}{1 + r^2 - 2r \cos(\theta - \gamma)} \tag{3.5}$$

where $0 \le \theta < 2\pi$, $\gamma = \mu + \nu$, $\xi = \sqrt{r^4 + 2r^2 \cos 2\nu + 1}$ and $\eta = \mu + \arg\{r^2 \cos(2\nu) + 1 + ir^2 \sin(2\nu)\}$ such that $0 \le \theta, \mu < 2\pi$, $\kappa > 0$ $0 \le r < 1$ and $i = \sqrt{-1}$.

A three-parameter family of Kato-Jones distributions can be derived as a special case of the above when $\nu = 0$ or $\nu = \pi$. In this case, the above four parameter distribution reduces to

$$f(\theta; \mu, \kappa, r) = \frac{1 - r^2}{2\pi I_0(\kappa)} \exp\left\{\frac{\kappa\{(1 + r^2) \cos(\theta - \mu) - 2r\}}{1 + r^2 - 2r \cos(\theta - \mu)}\right\} \times \frac{1}{1 + r^2 - 2r \cos(\theta - \mu)} \tag{3.6}$$

where $0 \le \theta, \mu < 2\pi$, $\kappa > 0$ and $0 \le r < 1$. The distribution is symmetric about $\theta = \mu$ and $\mu + \pi$ and is unimodal when $0 \le r < 1$.

Excellent surveys on the probability models can be found in Mardia (1972), Mardia and Jupp (2000), Fisher et al. (1987) and Jammalamadaka and SenGupta (2001).

4 Inference on the Circle

(a) Sampling Distributions

Sampling distributions play a very important role in statistics. The characteristic function is used to derive the sampling distributions of different statistics like sample mean direction, sample resultant length, etc. on the circle (see Mardia and Jupp 2000). It can be shown that if the parent population is von Mises, the joint distribution of the statistics $(\bar{C}, \bar{S})$ is asymptotically normal and the distributions of $\bar{\theta}_0$ and $\bar{R}$ is distributed as χ_1^2 when $\rho > 0$ and χ_2^2 when $\rho = 0$ (see Mardia and Jupp 2000).

(b) **Estimation**

Circular data analysis often involves estimation of circular parameters which assumes values on the unit circle. The special nature of circular data restricts one to define desirable properties of an estimator like unbiasedness on the circle just as in the case of linear data. One has to be very careful in defining expectation on the circle. But since a point θ on the circle can be expressed as the unit vector $X = (\cos\theta,\ \sin\theta)^T$, enables us to define expectation and hence unbiasedness. Let η be a circular parameter of some circular distribution and t be a statistic taking values on the circle. Then, t is an unbiased estimator of η if the mean direction of t is η, i.e. $\|E(\cos t,\ \sin t)\|^{-1}[E(\cos t,\ \sin t) = (\cos\eta,\ \sin\eta)$. An $100(1-\alpha)\%$ confidence interval for the population mean direction μ is given by $\bar{\theta}_0 \pm \cos^{-1}\left(\frac{1-\chi^2_{1,\alpha}}{2\hat{\kappa}R}\right)$.

(c) **Hypothesis Testing**

One of the basic tests on the circle is the test for uniformity of the sample measurements. It is possible that a uniformly distributed data may have significant concentration towards a single direction. This makes the choice of uniformity test very important. Graphically, a uniform probability plot can be used to assess this. Deviations from the 45° line indicate a departure from uniform model. Kuiper (1960) proposed an invariant Non-Parametric (NP) test of Kolmogorov type to test the uniformity of measurements on the circle based on the test statistic $V_n = D_n^+ + D_n^-$, where $D_n^+ = \sup_\theta\{S_n(\theta) - F(\theta)\}$, $D_n^- = \sup_\theta\{F(\theta) - S_n(\theta)\}$, S_n is the empirical distribution function and F is the cumulative distribution function of the uniform distribution. The null hypothesis of uniformity is rejected for large values of V_n. Another commonly used NP test for goodness-of-fit is the Watson's U^2 test which is an analogue for circular data of the Cramer-von Mises test on the real line. This test is based on the mean square deviation rather than the discrepancy between the empirical and cumulative distribution functions. The test statistic in this case is given by $U^2 = n\int_0^{2\pi}\left\{S_n(\theta) - \frac{\theta}{2\pi} - \mu\right\}\frac{d\theta}{2\pi}$, where $\mu = \int_0^{2\pi}\left\{S_n(\theta) - \frac{\theta}{2\pi}\right\}\frac{d\theta}{2\pi} = \frac{1}{2} - \frac{1}{n}\sum_{i=1}^n \frac{\theta_{(i)}}{2\pi}$ and $\theta_{(i)}$ are ordered observations. The null hypothesis of uniformity is rejected for large values of U^2.

Parametric test analogous to those based on standard normal distribution on the circle such as test for mean direction is developed mostly for samples coming from the von Mises population. A Likelihood Ratio Test (LRT) for testing $H_0 : \mu = \mu_0$ against the alternative $H_1 : \mu \neq \mu_0$ where the samples are drawn from a $CN(\mu, \kappa)$ distribution when κ is known and κ is unknown has been developed. The test statistic when κ is known is given by $W = 2n\kappa(\bar{R} - \bar{C})$ and when κ is unknown, the test statistic is $W^* = 2n\hat{\kappa}(\bar{R} - \bar{C})$ where $\hat{\kappa}$ is the m.l.e of κ. The null distribution of W and W^* can be approximated to a chi-square distribution for large n. The LRT rejects the null hypothesis for large values of W and W^*. Most of the parametric tests on the circle are proposed by Watson and Williams (1956). An excellent survey on different tests on the circle can be found on Mardia (1972), Mardia and Jupp (2000), Fisher et al. (1987) and Fisher (1993).

5 Robustness with Circular Data

The theory of robustness was first introduced by Huber (1964, 1965, 1968). Huber (1981) suggest that any statistical procedure should be robust in the sense that small deviations from the model assumptions should affect the performance only slightly. The concept of Influence Curve (IC) or Influence Function (IF) was introduced by Hampel (1974) to show how an estimator responds to the introduction of a new observation which assesses the relative influence of individual observations towards the value of an estimate or test statistics. Excellent surveys on robust inference can be found in Huber (1981), Huber and Ronchetti (2009), Hampel et al. (1986), Marona et al. (2006).

The statistical analysis of circular data is significantly different from those for linear data (see Mardia and Jupp 2000; Fisher 1993). The outlier problem in the directional data is completely different from that in the linear case. One might expect fewer outlier problems on the circle as there is restricted room for an observation to out lie. According to Jammalamadaka and SenGupta (2001), how far an observation is from the mean in directional setup should be judged by using appropriate 'circular distance'. Due to bounded support of angular data, outliers can be detected only when the observations are sufficiently concentrated around a particular point. The angular deviation given by $arc(\theta_i, \alpha) = \pi - |\pi - |\theta_i - \alpha||$ between a data point and the population sample mean or median direction can be used to identify whether the observation is outlying or not.

In classical robustness theory a commonly used tool to check the robustness of an estimator say T is the influence function approach introduced by Hampel (1974) which measures the rate of change in the value of an estimator (or statistical functional) for a small amount of contamination. If the influence function of T evaluated at the underlying distribution F is bounded, then T is said to be B-robust (Rousseeuw 1981). Another useful tool to check the robustness of T is the gross error sensitivity which intuitively measures the largest influence that a small amount of contamination of fixed size can have on the value of the estimator. If the gross error sensitivity of T at F is finite, then T is said to be Bias-robust (B-robust). But on the circle, the underlying distribution has bounded parameter space, and hence, the gross error sensitivity will always be finite. To overcome this problem Ko and Guttorp (1988) introduced the concept of standardised influence function and standardised gross error sensitivity which measures the relative influence of the contaminated observation in units of a functional (see Serfling 2002). In general, this function is assumed to be the dispersion measure. If the standardised gross error sensitivity of T evaluated at the family of distribution $\Im$ is finite, then T is said to be Standardised Bias-robust (SB-robust) at $\Im$.

Ko and Guttorp (1988) also introduced the concept of dispersion measure for directional data and showed that the directional mean and the concentration parameter are not SB-robust. Laha and Mahesh (2011) introduced the idea of trimmed mean on the circle and showed that for the family of $CN(\mu, \kappa)$ distribution, γ-circular trimmed mean where γ is a suitable trimming proportion is SB-robust. Laha and Mahesh (2012) also gave an SB-robust estimator for the concentration parameter of $CN(\mu, \kappa)$ distribution. Recently, Laha et al. (2016) proved that the γ-circular trimmed mean is also SB-robust for a larger family of circular distribution introduced by Kato and Jones (2010) and provide some guidelines for choosing the trimming proportion γ.

6 Conclusion

In this chapter, we give a brief overview of circular data and its various applications in the other areas. We discuss some of the important distributions on the circle which are highly useful for modelling real-life data which are circular in nature. This chapter also briefly covers the inference problem on the circle. Robustness of estimators when the data contains outliers is an important area in the field of linear statistics. Detailed literatures on how to check the robustness of an estimator in circular statistics using influence function approach and some of the recent works on robustness of estimators are also included in this chapter.

References

Batschelet, E. (1981). *Circular statistics in biology*. London: Academic Press.

Clark, W. A., & Burt, J. E. (1980). The impact of workplace on residential relocation. *Annals of the Association of American Geographers, 70,* 59–67.

Coleman, D. A., & Haskey, J. C. (1986). Marital distance and its geographical orientation in England and Wales 1979. *Transactions of the Institute of British Geographers, New Series, 11,* 337–355.

Fisher, N. I. (1993). *Statistical analysis of circular data*. Cambridge: Cambridge University Press.

Fisher, N. I., Lewis, T., & Embleton, B. J. J. (1987). *Statistical analysis of spherical data*. Cambridge: Cambridge University Press.

Gavin, T. M., et al. (2003). Biochemical analysis of cervical orthoses in flexion and extension: A comparison of cervical collars and cervical thoracic orthoses. *Journal of Rehabilitation Research and Development, 40*(6), 527–538.

Gill, J., & Hangrtner, D. (2010). Circular data in political science and how to handle it. *Political Analysis, 18*(3), 316–336.

Ginsberg, H. (1986). Honeybee orientation behaviour and the influence of flower distribution on foraging movements. *Ecological Entomology, 11,* 173–179.

Hampel, F. R. (1974). The influence curve and its role in robust estimation. *Journal of the American Statistical Association, 69,* 383–393.

Hampel, F. R., Ronchetti, E. M., Rousseeuw, P. J., & Stahel, W. A. (1986). *Robust statistics: The approach based on influence functions*. New York: Wiley.

Huber, P. J. (1964). Robust estimation of a location parameter. *Annals of Mathematical Statistics, 35,* 73–101.
Huber, P. J. (1965). A robust version of probability ratio test. *Annals of Mathematical Statistics, 36,* 1753–1758.
Huber, P. J. (1968). Robust Confidence Limits. *Z.Wahrsch. verw. Geb., 10,* 269–278.
Huber, P. J. (1981). *Robust statistics*. New York: Wiley.
Huber, P. J., & Ronchetti, E. M. (2009). *Robust statistics* (2nd ed.). New York: Wiley.
Hurshesky, W. J. M. (Ed.). (1994). *Circadian cancer therapy*. Boca Raton: CRC Press.
Jammalamadaka, S. R., & SenGupta, A. (2001). *Topics in circular statistics*. Singapore: World Scientific.
Kato, S., & Jones, M. C. (2010). A family of distributions on the circle with links to and applications arising from, mobius transformation. *Journal of American Statistical Association, Theory and Methods, 105*(489), 249–262.
Kato, S., & Shimizu, K. (2004). *A further study of t-distributions on spheres*. Technical report, School of fundamental science and technology, Keio University, Yokohama.
Ko, D., & Guttorp, P. (1988). Robustness of estimators for directional data. *The Annals of Statistics, 16,* 609–618.
Kuiper, N. H. (1960). Tests concerning random points on a circle. *Nederlandse Akademiya Westenschappen Proceedings Series, A63,* 38–47.
Laha, A. K., & Mahesh, K. C. (2011). SB-robustness of directional mean for circular distributions. *Journal of Statistical Planning and Inference, 141,* 1269–1276.
Laha, A. K., & Mahesh, K. C. (2012). SB-robustness of concentration parameter for circular distributions. *Statistical Papers, 53,* 457–467.
Laha, A. K, Raja Pravida, A. C., & Mahesh, K. C. (2016). SB-robust estimation of mean direction for some new circular distributions. Statistical Papers. https://doi.org/10.1007/s00362-016-0853-9.
Mahesh, K. C. (2012). *Robustness of estimators and tests with directional data*. Ph.D thesis, Saurashtra University, Rajkot, India.
Mardia, K. V., & Jupp, P. E. (2000). *Directional statistics*. Chichester: Wiley.
Mardia, K. V. (1972). *Statistics of directional data*. New York: Academic Press.
Marona, R. A., Martin, R. D., & Yohai, V. J. (2006), *Robust statistics: Theory and methods*, Wiley.
Morgan, E. (Ed.). (1990). *Chronobiology and chronomedicine*. Frankfurt: Peter Lang.
Pewsey, A., & Jones, M. C. (2005). A family of symmetric distributions on the circle. *Journal of American Statistical Association, 100*(472), 1422–1428.
Pewsey, A., Lewis, T., & Jones, M. C. (2007). The Wrapped t-family of circular distributions. *Australian and New Zealand Journal of Statistics, 49*(1), 79–91.
Proschan, M. A., & Follmann, D. A. (1997). A restricted test for circadian rhythm. *Journal of the American Statistical Association, Theory and Methods, 92*(438).
Rousseeuw, P. J. (1981). A new infinitesimal approach to robust estimation. *Zeitschrift fuer Wahrscheinlichkeit und Verwandte Gebiete, 56,* 127–132.
Schmidt-Koenig, K. (1965). Current problems in bird orientation. In D. Lehman, et al. (Eds.), *Advances in the study of behaviour* (pp. 217–278). New York: Academic Press.
SenGupta, S. & Rao, J. S. (1966). Statistical Analysis of Crossbedding Azimuths from the Kamthi Formation around Bheemarsm, Pranhita-Godavari Valley. *Sankhya Series B., 28,* 165–174.
Serfling, R. J. (2002). *Approximation theorems of mathematical statistics*, Wiley.
Upton, G. J. G. (1986). Distance and directional analyses of settlement patterns. *Economic Geography, 62,* 167–179.

Wallin, H. (1986). Habitat choice of some field-inhabiting carabid beetles (Cleopatra: Carabidae) studied by the recapture of marked individuals. *Ecological Entomology, 11,* 457–466.
Watson, G. S., & Williams, E. J. (1956). On the construction of significance tests on the circle and sphere. *Biometrika, 43*, 344–352.

Branching Processes

Sumit Kumar Yadav

1 Introduction and History

Branching processes is an area of mathematics that attempts explaining situations when a particle or entity can produce one or more entities of similar or different types. The basic idea is that a parent produces offsprings. Then, these offsprings further take the role of parents and produce offsprings, and the process continues either perpetually or until extinction.

Branching process has been largely used for the study of the growth of populations in the past century, after it was introduced by Irenee-Jules Bienayme in 1845. In the past few decades, people have also attempted to apply the theories of branching processes in other domains such as Marketing, Finance, Biomedical Sciences, and Operations Research among others.

The theory of branching process originated from a demographic question: What is the probability that a family name becomes extinct? Social scientists have been interested in the problem of extinction of family lines. To understand why this problem could be of some relevance, we can take the example of the town of Berne, Switzerland. Malthus (1798) mentions that in Berne, 379 out of 487 bourgeois families became extinct in a period of just two hundred years, from 1583 to 1783.

The earlier explanation of the problem was sought in biological or social terms, which attempted at looking at degeneration or effects of wars. However, in 1845, at a meeting of the Societe Philomatique de Paris, famous mathematician Irénée-Jules Bienaymé presented his work on the time taken for a family name to become extinct. He attempted a mathematical approach to the problem and tried to answer the question: Given that a man has 0, 1, 2, … sons, what is the probability that his family line will become extinct? Bienayme managed to relate the probability of extinction

S. K. Yadav (✉)
Indian Institute of Management Ahmedabad, Ahmedabad, India
e-mail: sumitky@iima.ac.in

A. K. Laha (ed.), *Advances in Analytics and Applications*, Springer Proceedings in Business and Economics, https://doi.org/10.1007/978-981-13-1208-3_4

of family name with the average number of male children replacing the number of males of preceding generation (i.e. mean number of offsprings).

Apart from Bienayme, Francis Galton and Henry William Watson also come up as first few names while discussing branching processes. Their contribution came almost 30 years after Bienayme's. Galton posed the extinction question as a problem in the Educational Times of 1873:

> Educational Times (April): 17 PROBLEM 4001: A large nation, of whom we will only concern ourselves with adult males, N in number, and who each bear separate surnames colonise a district. Their law of population is such that, in each generation, a_0 per cent of the adult males have no male children who reach adult life; a_1 have one such male child; a_2 have two; and so on up to a_5 who have five. Find (1) what proportion of their surnames will have become extinct after r generations; and (2) how many instances there will be of the surname being held by m persons.

Upon receiving just one incorrect solution, Galton persuaded his acquaintance Watson, who also was a mathematician to work on the problem. Watson, using the theory of generating functions and functional iteration, attempted to solve the problem. He came to the conclusion that all family lines must die out, which was later proved wrong. Their results were published as Watson and Galton (1874).

Thus, in several branching process books and papers, the process is referred as Bienayme–Galton–Watson Process or BGW Process.

2 Simple Branching Process

We shall start by explaining the basic setup of a simple branching process in the discrete time case.

Refer to the picture shown in Fig. 1. The number in the circle represents the generation number. Thus, the process represented in the figure starts from one particle in the zeroth generation and tells the story till the fourth generation. The only particle in the zeroth generation produces two particles that belong to the first generation before dying. The two particles in the first generation further produce two and three particles, respectively. Thus, we now have five particles that belong to the second generation. The five particles in the second generation produce 2, 0, 2, 1, 1 particles, respectively, that belong to the third generation. The process will thus, continue either ad infinitum or till the chain becomes extinct.

We now take a slight diversion and discuss classical branching processes, of which simple branching process is a special case. In classical branching processes, it is assumed that particles can live for an arbitrary amount of time. They give birth at the time of their death. The lifetime of a particle is denoted by the random variable τ. The distribution of τ plays an important role in the behaviour of the process. We make 3 cases. If τ is exponentially distributed, the process is called Markov Branching Process. If τ is an arbitrary non-negative random variable, the process is called Bellman- or age-dependent branching process. If $\tau = 1$, the process is called simple branching process, or Galton–Watson process. We shall discuss only the third

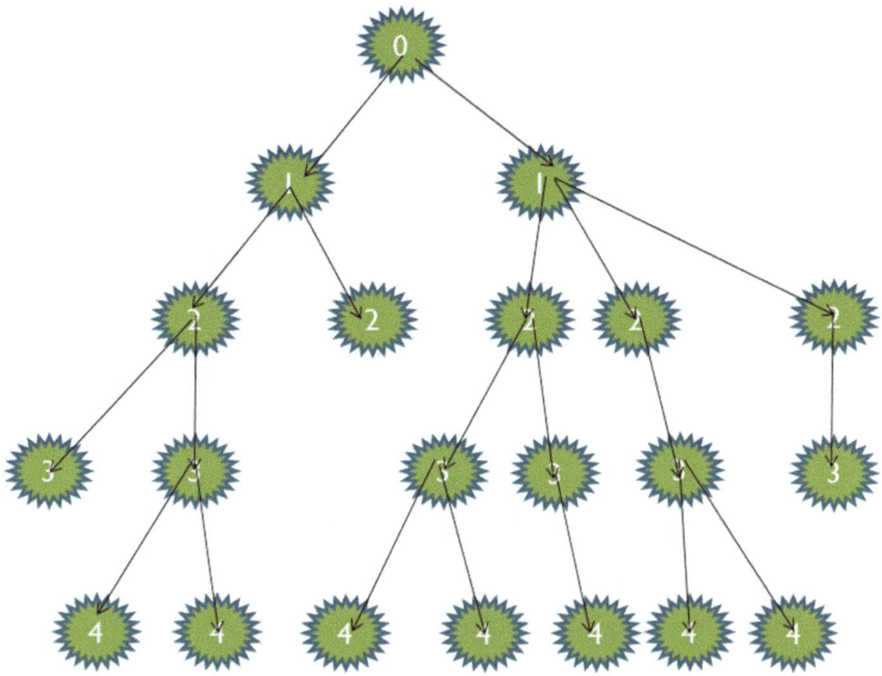

Fig. 1 A simple branching process

case in this chapter as the objective is to give a brief introduction to this important stochastic process. Readers who are interested in the other two cases and even more general types of the branching processes are directed to the references at the end of the chapter (For example, see Harris (1963)).

Typically, the questions that are posed in simple branching process are the number of generations after which the branching process will become extinct, the probability of extinction of the process, the expected number of individuals in a given generation, etc.

Now that we have built an idea about a simple branching process, it would be good to describe it mathematically as follows:

1. The number of particles in nth generation is denoted by Z_n. The process begins with Z_0 individuals in the beginning.
2. A particle survives for one generation and gives birth only at the time of their death
3. Particles can generate offsprings similar to themselves
4. A particle of nth generation produces offsprings which belong to n + 1th generation
5. A probability law is assigned for reproduction of offsprings
6. All individuals follow this law independently of each other

Thus, if a numbering is put on the particles in the nth generation arbitrarily from $i = 1$ to $i = Z_n$, and we denote the number of particles produced by the ith particle as Y_i, the total population of particles in the $n + 1$th generation is given by

$$Z_{n+1} = \sum_{i=1}^{Z_n} Y_i$$

where Y_i are assumed to be independent and identically distributed (i.i.d.) random variables.

Usually, it also helps to consider the probability generating function (PGF) for the branching process to calculate transition probabilities and expected values of some desired quantities. We give a small introduction about PGFs below.

Let a univariate random variable X be such that it takes values 0, 1, 2, 3, …. Then, the PGF of X is $G(s) = E(s^X) = \sum_{k=0}^{\infty} P(X = k) \cdot s^k$, where $|s| \le 1$.

Similarly for the multivariate case, if $\mathbf{X} = (X_1, X_2, \ldots X_d)$ is a vector of discrete random variables taking values in W^d, where W represents the set of whole numbers, then the PGF of $\mathbf{X}$ is $\sum_{k_1, k_2, \ldots, k_d=0}^{\infty} P(k_1, k_2, \ldots k_d) \cdot s_1^{k_1} \cdot s_2^{k_2} \cdot \cdots \cdot s_d^{k_d}$ $= E\left(s_1^{X_1} \cdot s_2^{X_2} \cdot \cdots \cdot s_d^{X_d}\right)$ where $\max(|s_i|) \le 1$.

For a simple branching process, assuming that the process started with one individual, i.e. $Z_0 = 1$, let the probability generating function be represented by $f(s)$. Next, represent the generating function of the nth generation by $f_n(s)$, it can be shown that $f_{n+1}(s) = f_n(f(s))$.

To see the proof, one must note that given $Z_n = k$, the probability distribution function of Z_{n+1} (i.e. the population in the $n + 1$th generation) will be a convolution of the generating function of one individual k times, i.e. $f^k(s)$. This is a standard result in the theory of PGFs. For the proof, one can see Chap. 11 of Feller (1968). Thus, $P(Z_{n+1} = r | Z_n = k) =$ coefficient of s^r in $f^k(s)$.

Also, one could write $P(Z_{n+1} = r) = \sum_{i=0}^{\infty} P(Z_{n+1} = r | Z_n = i) \cdot P(Z_n = i)$. Here, the term $P(Z_{n+1} = r | Z_n = i)$ is given by the coefficient of s^r in the expansion of $f^i(s)$.

Now,

$$\begin{aligned}
f_{n+1}(s) &= \sum_{r=0}^{\infty} P(Z_{n+1} = r) \cdot s^r \\
&= \sum_{r=0}^{\infty} \left(\sum_{i=0}^{\infty} P(Z_{n+1} = r | Z_n = i) \cdot P(Z_n = i)\right) \cdot s^r \\
&= \sum_{i=0}^{\infty} \left(\sum_{r=0}^{\infty} P(Z_{n+1} = r | Z_n = i) \cdot s^r\right) \cdot P(Z_n = i) \\
&= \sum_{i=0}^{\infty} f^i(s) \cdot P(Z_n = i) = f_n(f(s)).
\end{aligned}$$

It must, however, be noted that although the expression for generating function seems very elegant, explicitly deriving a $f_n(s)$ for even simple setups typically gets very complicated.

Another useful result in the study of branching process is the probability of extinction. If the expected number of offsprings of an individual is less than or equal to one, extinction is certain and has probability equal to one. However, if the expected number of individuals is strictly greater than one, the probability of extinction becomes less than one and is given by the smallest non-negative root of the equation $f(s) = s$. For the proof, the reader is referred to Chap. 5 of Haccou et al. (2005).

In the subsequent sections, we shall list variants of the simple branching process and additionally explain in detail some of the variants which are most commonly encountered.

3 Variants of Simple Branching Process

Below we list the variants of simple branching processes that are commonly encountered in literature. We shall go into details of only (a) Bisexual, (b) Varying Environments and (c) Multi-type Branching Process. For a detailed treatment of the rest, one is referred to (Haccou et al. 2005)

i. Overlap of Generations
ii. State Dependence
iii. Dependence on the population itself
iv. Interaction between individuals
v. Branching Processes with migration
vi. Varying Environments
vii. Multi-type
viii. Bisexual.

3.1 Bisexual Branching Processes

When we are considering bisexual branching process, two sexes are involved in the reproduction of offspring, as is the case with the growth of populations in larger organisms including human beings. In some cases, however, this fact can be ignored. For instance, when only females determine the size of the future generation, and males are sufficient in number, modelling using a simple branching process that counts females would also serve the purpose. However, in certain instances, we need to take mating into account explicitly to model the situation accurately.

These kinds of processes were introduced by Daley in the year 1968 as a two-type branching process defined recursively. For a survey of the literature in bisexual

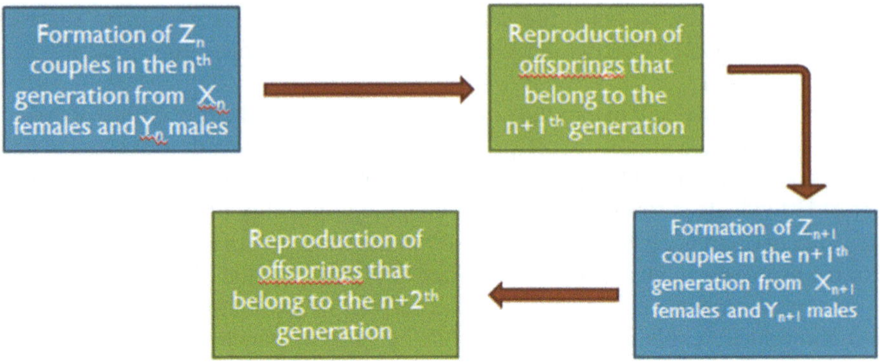

Fig. 2 A bisexual branching process

branching process, the reader is referred to Chap. 20 in González et al. (2010) and Hull (2003).

Usually, a generation is represented by the number of couples (Z_n), thus, the process becomes extinct if $Z_n = 0$. These couples then give birth to the offsprings of the next generation (X_{n+1} females and Y_{n+1} males). These offsprings then form couples of the (n + 1)th generation, which is denoted by Z_{n+1}. The process is also depicted in Fig. 2.

Typically, a mating function is associated with bisexual branching process which determines the number of couples that will be formed given a number of males and females in a generation. The function could be taken to be stochastic or deterministic given the number of males and females in the generation. One example could be of a monogamous setup. Let $R_1(x, y)$ represent the mating function which will give the number of couples as output given the number of females 'x' and number of males 'y' as input. Thus, $R_1(x, y) = \text{minimum}(x, y)$.

Thus, if the mating function is represented by R(x, y), then the number of couples formed in the *n*th generation would be $Z_n = R(X_n, Y_n)$, where X_n and Y_n are the number of females and males in the *n*th generation, respectively. Bruss (1984) noted that the extinction criteria for bisexual branching processes is given by 'average unit reproduction means' $m_j = \frac{1}{j}(E(Z_n|Z_{n-1} = j)), j \geq 1$. If the limit of m_j is less than or equal to one, then for a large class of mating functions, extinction is certain (Daley et al. 1986).

3.2 *Varying Environments*

The simple branching process does not take into account the real-world phenomena that the offspring distribution does not remain same across generations. It may be affected due to various reasons such as resources to feed future generation, changes in lifestyle, weather conditions, etc. This is modelled using varying environments. There are two possible types of variation that are generally discussed. One is deterministic variation, in which the environment changes in a predictable way, other is random variation. We shall only discuss deterministic variation in this chapter.

Suppose that in the nth generation, the population size is Z_n. The mean of the offspring distribution will now depend upon 'n', hence we take it to be a function of n, to be represented by m(n). In the nth generation, all the Z_n individuals would follow offspring distribution with mean m(n).

Hence,

$$E(Z_{n+1}) = E(E(Z_{n+1}|Z_n) = E(m(n) \cdot Z_n) = m(n) \cdot E(Z_n) = \prod_{i=0}^{n} m(i) \cdot E(Z_0)$$

Some examples of such processes could be an environment that varies in a periodic fashion, constantly improving or deteriorating environment.

For constantly deteriorating environment, the mean of the probability law governing offspring distribution follows the following:

$E(X_{n+1}) <= E(X_n)$, where X_{n+1} = number of offsprings of an individual in the $n + 1$th generation, X_n = number of offsprings of an individual in the nth generation.

3.3 Multi-type Branching Processes

A multi-type branching process allows for more than one type of individuals in the population. The individuals can give birth to some or all of the types that exist in the population. Each type can have different offspring distribution.

For most applications which use multi-type branching process model, it would suffice to take the number of types to be finite. However, one could also look at cases where the number of types is infinite, or even the cases where the number of types is uncountable. (For example, if we take height of a person to be a type, as height is a continuous parameter, the number of types is uncountably many.) For a somewhat detailed treatment of these kinds of problems, one is referred to Harris (1963).

For instance, in the figure shown below (Fig. 3), we have two types in the population, let us call them—Grey and Black. The number in the circle indicates the generation number. The picture represents the process from zeroth generation till fourth generation. We start with one individual of type black in the zeroth generation. It gives birth to one individual of type black and one individual of type grey that belong to first generation. The grey type then gives birth to one black and one grey individual belonging to the second generation. The black type then gives birth to two grey and one black individuals that belong to the second generation. Now, the second generation consists of two individuals of type black and three individuals of type grey. The process thus continues in a similar fashion.

Typically, the questions that are posed in multi-type branching process could be the number of generations after which the branching process will become extinct,

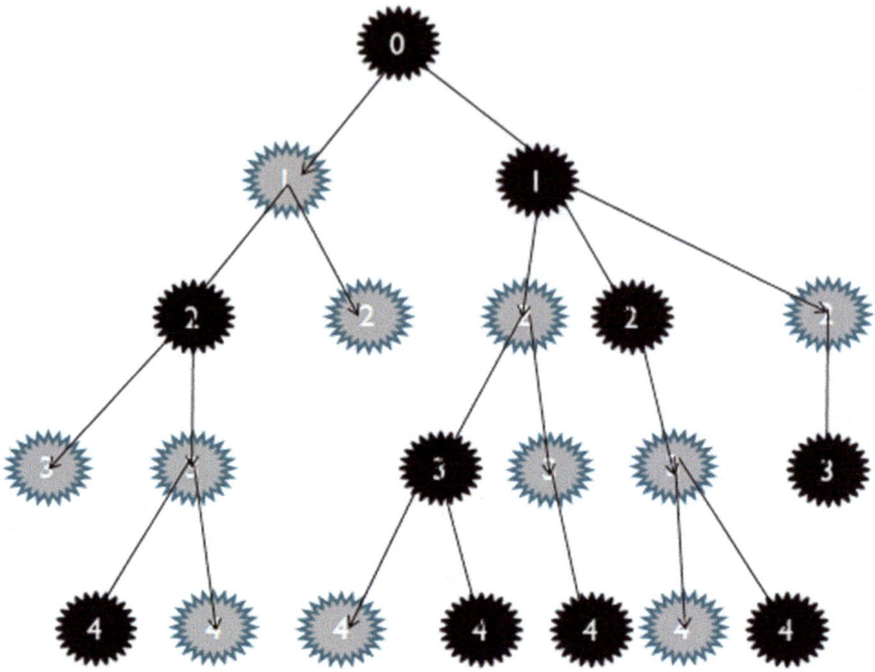

Fig. 3 A multi-type branching process

the probability of extinction of entire branching processes or of some type in the population, the expected number of individuals of each type in a given generation or the ratio of different types in the population for a given generation, etc.

As is the case with simple branching processes, it is also helpful to discuss the generating functions in the context of multi-type branching processes as well.

The multi-type branching process is defined by a set of vector-valued random variables $\mathbf{Z_0}$, $\mathbf{Z_1}$, $\mathbf{Z_2}$, $\mathbf{Z_3}$, **…**. The ith component of $\mathbf{Z_n}$ represents the numbers of individuals of type 'i' in the nth generation. Let there be a total of 'k' possible types of individual in the population.

We represent by $p^i(r_1, r_2, \ldots, r_k)$ the probability that an individual of type 'i' has r_1-offsprings of type 1, r_2-offsprings of type 2, …, r_k-offsprings of type k.

Hence, if we start with just one individual of type 'i' in the zeroth generation, then the probability generating function of $\mathbf{Z_1}$ would be

$$f^i(s_1, s_2, \ldots s_k) = \sum_{r_1, r_2, \ldots, r_k = 0}^{\infty} p^i(r_1, r_2, \ldots, r_k) \cdot s_1^{r_1} \cdot s_2^{r_2} \ldots s_k^{r_k}, \quad |s_j| \leq 1 \text{ for all } j$$

The generating function of $\mathbf{Z_n}$, in this case, will be denoted by $f_n^i(s_1, s_2, \ldots s_k)$ which can also be represented as $f_n^i(\boldsymbol{s})$. An analogous result of the simple branching process in this case is

$$f_{n+1}^{i}(s) = f^{i}\left[f_n^1(s), f_n^2(s), \ldots, f_n^k(s)\right]$$

The proof is an extension and similar to the proof as was done in simple branching process case and is thus omitted. However, one can refer to Chap. 2 in Harris (1963).

To understand the growth of population, it would be convenient to make a matrix of mean offsprings of each type from each type. We shall call it the 'mean matrix' and denote it by M. Let $M = \left(m_{hj}\right)_{h,j=1}^{k}$, where m_{hj} = expected number of offsprings of type 'j' from an individual of type 'h'.

$$\begin{bmatrix} m_{11} & m_{12} & m_{13} & \cdots\cdots & m_{1,k-1} & m_{kk} \\ m_{21} & m_{22} & m_{23} & \cdots\cdots & m_{2,k-1} & m_{2,k} \\ m_{31} & m_{32} & m_{33} & \cdots\cdots & m_{3,k-1} & m_{3,k} \\ \vdots & \vdots & \vdots & \ddots\ddots & \vdots & \vdots \\ \vdots & \vdots & \vdots & \ddots\ddots & \vdots & \vdots \\ m_{k-1,1} & m_{k-1,2} & m_{k-1,3} & \cdots\cdots & m_{k-1,k-1} & m_{k-1,k} \\ m_{k,1} & m_{k,2} & m_{k,3} & \cdots\cdots & m_{k,k-1} & m_{k,k} \end{bmatrix}$$

The expected number of type j individuals in generation n satisfies

$$E\left(Z_{nj}\right) = E\left(E\left(Z_{nj}|\mathbf{Z}_{n-1}\right)\right) = E\Big(\sum_{h=1}^{k} Z_{n-1,h} \cdot m_{hj}\Big) = \sum_{h=1}^{k} E(Z_{n-1,h}) \cdot m_{hj}$$

Let $\boldsymbol{E(Z_n)} = (E(Z_{n1}), E(Z_{n2}) \ldots E(Z_{nk}))^T$ be the vector of expectation of individual types. From here, it is easy to derive $\boldsymbol{E(Z_n)}^T = \boldsymbol{E(Z_0)}^T \cdot M^n$.

In a multi-type branching process, one can also study the extinction properties. The calculations are a bit more complicated than the simple branching process case. We just provide some basic results here. For a further detailed treatment of this using eigenvalues and eigenvectors of M, one is referred to Chap. 2 of Haccou et al. (2005).

Multi-type branching process can be divided into two parts—indecomposable and decomposable. If any initial population composition can lead to any other composition, (i.e. has a non-zero probability) we say that the process is indecomposable. It is somewhat intuitive to now believe that for indecomposable processes, all types will grow at the same rate. One can refer (Mode 1971) for the proof. This rate is given by ρ, which is the largest positive eigenvalue of the 'mean matrix'. If $\rho <= 1$, extinction happens with probability 1. If $\rho > 1$, there is a finite probability of the population never getting extinct. The extinction probability depends on the number and type of individuals present in the initial population. We omit the discussion of decomposable multi-type branching processes.

4 Applications

In this section, we list down the areas we encountered in our literature survey in which branching processes has been applied. In finance, it has been applied to model stock prices, and also for barrier option pricing. For detailed reading, one is referred to Dion and Epps (1999), Epps (1996) and Mitov et al. (2009). Also, the theory has been used in the field of operations research by Chakraborty (1995) for studying properties of genetic algorithms and by Dumas and Robert (2001) to provide bounds while studying throughput of a resource-sharing model. Bruss (1978) provides a framework that can be possibly used in biomedical applications. Also, a large number of books and papers are devoted to applications of branching processes in biology. Pakes (2003) and Kimmel and Alexrod (2002) are among the popular books for biological applications of branching process.

Recently, van der Lans et al. (2010) have successfully applied the theory of branching process in marketing domain. The problem they attempted is to predict the reach of a marketing campaign using branching process model. A firm sends mail to a set of people for marketing purposes. They also incentivise the customers to forward this email to their contacts. Thus, the branching process starts. Now, the idea is to predict how many people would come to know about the campaign. So, we would be interested to know the total particles in the branching process before it becomes extinct. They also tested their model on an actual marketing campaign in which 2,00,000+ people had participated (Fig. 4).

To conclude the chapter, we note that in the past century, branching process has been an active area of research. There still are many interesting and challenging theoretical problems that need to be solved. Apart from this, modelling using branching

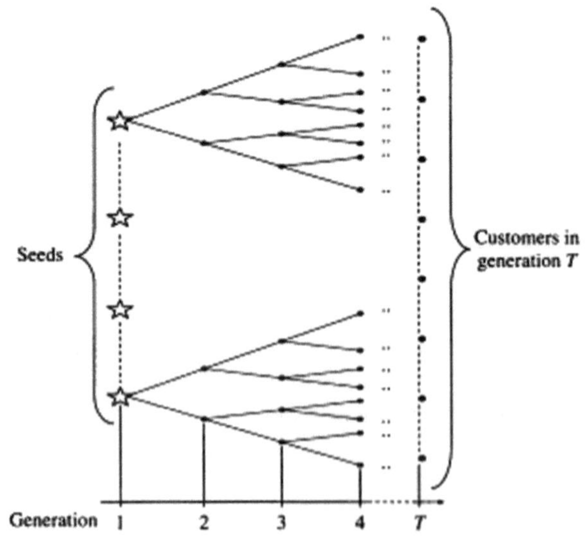

Fig. 4 Spread of message in viral marketing campaign. van der Lans et al. (2010)

processes has proved to be quite useful in several applications. Interested readers can explore modelling some phenomena using a simple branching process model or any of its variant.

References

Bruss, F. T. (1978). Branching processes with random absorbing process. *Journal of Applied Probability, 15*(1), 54–64.
Bruss, F. T. (1984). A note on extinction criteria for bisexual Galton-Watson processes. *Journal of Applied Probability, 21*(4), 915–919.
Chakraborty, U. K. (1995). A branching process model for genetic algorithms. *Information Processing Letters, 56,* 281–292.
Daley, D. J., Hull, D. M., & Taylor, J. M. (1986, September). *Journal of Applied Probability, 23*(3), 585–600.
Dion, J. P., & Epps, T. W. (1999). Stock prices as branching processes in random environments: estimation. *Communications in Statistics—Simulation and Computation, 28*(4), 957–975. https://doi.org/10.1080/03610919908813587.
Dumas, V., & Robert, P. (2001). On the throughput of a resource sharing model. *Mathematics of Operations Research, 26*(1), 163–173.
Epps, T. W. (1996). Stock prices as branching processes. *Communications in Statistics. Stochastic Models, 12*(4), 529–558. https://doi.org/10.1080/15326349608807400.
Feller, W. (1968). *An introduction to probability theory and its applications* (Vol. 1). Wiley.
González, M., Puerto, I. M., Martínez, R., Molina, M., Mota, M., & Ramos, A. (Eds.). (2010). *Workshop on branching processes and their applications* (Vol. 197). Springer Science & Business Media.
Harris, T. E. (1963). *The theory of branching processes.*
Haccou, P., Jagers, P., & Vatutin, V. A. (2005). *Branching processes: Variation, growth, and extinction of populations.*
Hull, D. M. (2003). A survey of the literature associated with the bisexual Galton-Watson branching process. *Extracta Mathematicae, 18*(3), 321–343.
Kimmel, M., & Alexrod, D. E. (2002). *Branching processes in biology. Interdisciplinary Applied Mathematics* (Vol. 19). http://doi.org/10.1016/S0070-2153(06)75012-8.
Malthus, T.R. (1798). An Essay on the Principle of Population, as it affects the future improvement of society, with remarks on the speculations of Mr. Goodwin, M. Condorcet and other writers. London: John Murray.
Mitov, G. K., Fabozzi, F. J., Rachev, S. T., & Kim, Y. S. (2009). Barrier option pricing by branching processes. *International Journal of Theoretical and Applied Finance, 12*(7), 1055–1073.
Mode, C. J. (1971). *Multitype branching processes.* New York: American Elsevier.
Pakes, A. G. (2003). Biological applications of branching processes. *Handbook of Statistics, 21*(1), 693–773. https://doi.org/10.1016/s0169-7161(03)21020-8.
van der Lans, R., Van Bruggen, G., Eliashberg, J., & Wierenga, B. (2010). A viral branching model for predicting the spread of electronic word of mouth. *Marketing Science, 29*(2), 348–365. https://doi.org/10.1287/mksc.1090.0520.
Watson, H. W., Galton, F. (1874). On the probability of the extinction of families. *Journal of the Royal Anthropological Institute, 4,* 138–144. Great Britain Ireland.

Part II
Predictive Analytics Applications

Click-Through Rate Estimation Using CHAID Classification Tree Model

Rajan Gupta and Saibal K. Pal

Abstract Click-Through Rate (CTR) is referred to as the number of clicks on a particular advertisement as compared to the number of impressions on it. It is an important measure to find the effectiveness of any online advertising campaign. The effectiveness of online advertisements through calculations of ROI can be done through the measurement of CTR. There are multiple ways of detecting CTR in past; however, this study focuses on machine learning based classification model. Important parameters are judged on the basis of user behavior toward online ads and CHAID tree model is used to classify the pattern for successful and unsuccessful clicks. The model is implemented using SPSS version 21.0. The dataset used for the testing has been taken from Kaggle website as the data is from anonymous company's ad campaign given to Kaggle for research purpose. A total of 83.8% accuracy is reported for the classification model used. This implies that CHAID can be used for less critical problems where very high stakes are not involved. This study is useful for online marketers and analytics professionals for assessing the CHAID model's performance in online advertising world.

Keywords Click-through rate · Online advertisements · Classification tree Click estimation · Mobile ads

1 Introduction

The Click-Through Rate (CTR) is specified as the proportion of number of clicks out of the number of impressions for an online advertisement. CTR is an important

---

R. Gupta (✉)
Department of Computer Science, Faculty of Mathematical Science,
University of Delhi, North Campus, New Delhi 110007, India
e-mail: guptarajan2000@gmail.com

S. K. Pal
Defence R&D Organization, SAG Lab, Metcalfe House, New Delhi 110054, India
e-mail: skptech@yahoo.com

A. K. Laha (ed.), *Advances in Analytics and Applications*, Springer Proceedings in Business and Economics, https://doi.org/10.1007/978-981-13-1208-3_5

factor which helps in determining what advertisement is needed to be displayed and in what sequence. The choice of a correct advertisement along with the appropriate display sequence is significant influencers of the clicks made by user on that particular advertisement (Koonin and Galperin 2003). Impression of an advertisement is how the content of advertisement influences the user ant attracts him for clicking the advertisement. The advertisements which include expert evidence and statistical evidence have high CTR than the advertisements which involve causal evidence only (Haans et al. 2013). CTR outlines the number of users who click the specific website through the advertisement from the other applications (apart from search engines) like e-mail campaign, html link, search engines, and advertising. But with the number of clicks one has to check whether the clicks are generating revenues for the business or not otherwise all the clicks are worthless (Hopewell 2007).

Most search engines aim at maximizing the ad quality (which is measured by clicks) to be able to maximize revenue, and their ordering of advertisements by most search engines takes place based on their expected revenue. Also, the chance that an advertisement will get clicked on drops very prominently with the position of the ad, thus, the accuracy with which we estimate this CTR, is very important as it has a significant on the revenues that can be earned.

1.1 Need for CTR

The prominent question is how the description of an advertisement can influence the user or searcher that may result in high CTR. It helps the business in converting leads and thereby helps in increasing the revenue of the business. It also helps in forecasting digital performance of various campaigns, user traffic on their website, and identification of search keywords targets. A report based on keyword search statistics communicates that majority of the searches on platforms like Google results in listing of paid advertisements to attract clicks from the users (Catalyst and Martineau 2013).

The probability of clicking on advertisement drops significantly as the query of users does not match with the searched results. It could be as high as 90% as per few studies (Richardson et al. 2007). Therefore, the accuracy with which we estimate the CTR of a website or an advertisement can have a significant effect on revenue of the business. When a query is typed by the user, search engine matching occurs for all the available keywords which results in appropriate advertisement display. Further, these ads can be distinguished into three categories, namely, exact match, phrase match, and broad match. Exact match can be defined as exact similarity between the user's query and keywords present in the search engine. Phrase match can be defined as the occurrence of the keyword as the subset of the query. Similarly, broad match can be defined as the close similarity between the keyword and the advertisement that appeared on the screen (Trofimov et al. 2012). On the basis of these evaluations, a relevance judgment will be done that will indicate the relevance of each document (Carterette and Jones 2007).

A ranking with a single perfectly relevant document might have low CTR than one for the same query that lists may somewhat relevant documents. Advertisements need to be ranked with already existed advertisements as ranking can have strong influence on the users and advertisers satisfaction.

1.2 Factors Impacting CTR

There are broadly two factors that influence the results of the queries on the search engine, viz., selection of the advertisements to be displayed and the order of the advertisement to be displayed, i.e., ranking need to take place. Advertisers specify under what circumstance the advertisements will be shown. Ranking of advertisements is done in order to place "best performing advertisement first". If a user is shown an advertisement, will it be attractive enough to grab a click form the user? This is a very important question for any sponsored search advertising. "*We find that click-through rates are higher for advertisements involving expert/statistical evidence than for advertisements involving causal evidence*" (Haans et al. 2013).

Display Position

The number of advertisements that match a query or to say that an advertisement that is going to appear on specific search keywords by the user is far more than the number of advertisements that can actually be displayed (Richardson et al. 2007).

There are slots available which the advertisers look at occupying for the advertisements they are selling to search engines. The number of valuable slots available for eligible advertisements is very less. For instance, majority of the users searching information on engines seldom move ahead of the first page of the engine's search results. If they do not find what they are looking for, they try a new set of keywords. In such a case, the number of advertisements which can be possibly displayed on this page corresponding to this query is limited. CTR is greatly affected by the position of an advertisement on this page. As the visibility goes on decreasing with the lower positioning of the advertisement, CTR also sees a declining pattern corresponding to the position. This is to say that the advertisements placed at lower slots are relatively less impactful.

Search Engine Advertisement

It is the focus of online marketers designing advertisements to design in a way that they are able to persuade the users to click on the posted advertisement and make every effort to turn these users into buyers. Advertisers write persuasive text messages in their advertisements to get them clicked. But the quality of their argument is of critical importance which helps in deciding the outcome of this process of persuasion. The advertisers make claims in their advertisement—presentations. This evidence becomes a part of the claim as the support required for improvement of the quality of the argument presented by the advertiser. Evidence may be referred to as "*data (facts or opinions) presented as proof for an assertion*" (Reynolds and Reynolds 2002, p. 429). There are different types of evidences that are used by the advertisers to improve the effectiveness of advertisement texts. These can be causal, statistical, anecdotal, and expert evidence (Hornikx 2005). General idea about the evidence has been given in the literature and is formulated by Rieke and Sillars (1975) as follows:

a. Causal evidence—explaining the occurrence of an effect;
b. Statistical evidence—a numerical representation of numbers;
c. Anecdotal evidence—usage of case stories, examples, or illustrations;
d. Expert evidence—citing experts in order to enhance the credibility of the advertisement.

Thus, for users querying on the search engines, conduction of an early stage search, with ads getting involved in expert evidence having high statistical evidence or credibility thereby increasing the verification of the information (Lindsey and Yun 2003), may have higher CTRs as compared to causal evidence-based ads, which are not based on simple rejection or acceptance cues. But, on the other hand, central route-based users, i.e., those who browse with the purpose of searching particular information, are in a position to grab underlying information and meaning related to the content of the message in a much better way. For this set of users, the causal argument is considered as a precious information source and thus their chance of conversion of a causal advertisement is more than any other kind of advertisement. Based on the environment and circumstances, the two different types of evidences may perform differently, with one being superior to the other. The most important task for any advertisement is to decide its objective. If the aim is increasing traffic, statistical or expert evidences must be used in an ad as these result in more number of clicks but if the aim is to get conversions, the causal evidence works the best on this metric and outweighs all others (Haans et al. 2013).

Banner Advertisement

Content elements include emotional appeals and use of incentives. Design elements include color, animation, and interactivity. Effective Internet banner advertisement builds brand value and thereby increases the efficiency of advertisements. Advertisers are inclined to modify the information on the advertisements based on the extent of their involvement. If they are highly involved, they use cognitive approach to make evaluations and factors such as color and sound do not make impact. But in case of low involvement, users barely pay attention to the content but just scan through. There is only subconscious level of engagement of the user. The interactive design of a banner advertisement has a substantial effect on its CTR as it facilitates two-way communication. The animation and the color it uses can also elicit some kind of emotional feeling in the users (Lohtia et al. 2003). In either case, it is either the design or the content of the advertisement that affects the CTR directly.

User Interactivity

This is one factor that primarily differentiates the traditional advertisement methods from the new-age online advertising. "*The point and click nature of the online medium makes it easy for frustrated or bored visitors to head off to other websites; therefore, ads displayed on websites need to capture web surfers' and web searchers' attention through an emotional, rational, or mixed (i.e., rational and emotional) appeal*"

(Singh and Dalal 1999). "*Interactive advertising enhances brand awareness and usually results in higher click-through rates than other forms of online advertising*" (Rosenkrans 2010; Lemonnier 2008). A unifying principle behind advertisement based on good media content is its ability to interact with audience/user (Lemonnier 2008). An advertisement with better media-based content can attract and convert more number of consumer attention as compared to the static banners, and can be helpful in enhancing interactivity, along with allowing online users to engage with e-commerce based transactions without leaving the web hosting infrastructure for the advertisement (Briones 1999; Li and Bukovac 1999).

Relevancy of Keywords
Whenever any user puts in a query, the matching happens in the search engine with all the keywords and searched relevant advertisement for the searched query are also browsed. The relevancy of the keyword is decided by the degree of match between the query asked and the result optimized by the search engine. The relevancy of the keyword depends on some factors like word count, body of the word count, title of the advertisement, relevancy of capital letters in the advertisement, length of the query asked, etc. These factors decide the rankings of the advertisement on the basis of the relevance of the keywords used in the content of the advertisement.

Cyclic Fluctuations
CTR estimation can be done from the available historical data with an assumption of notable exceptions: infrequent searches, time-to-time occurrences, and isolated events. Here, cyclic fluctuations are complicated and widely studied. As per a study, approximately 34% of the keywords in the query show some degree of periodic behavior, but approximately 90% show no periodic changes (Regelson and Fain 2006).

Offer
Most people, who are surfing the web, are often trying to look for things they are not able to get offline. They search for better deals, discounts; they search for what the brand has for its online advertisement offer (Lohtia et al. 2003). Offers sent via e-mail advertisements build a brand relationship with its customers and have become a source of engagement. It has become a place to search for new products and talent.

1.3 Issues with CTR

The query issued by the user and advertisements placed on the basis of the query is carried out by web mining and ranking of similar advertisements, with best advertisement among them will be displayed on the top by the search engine. By optimizing the relevance of user, query may or may not lead to clicks. This is done in order to show the correlation between the clicks and the query rewrites. CTR of an advertisement gets highly affected by the popularity of the keywords (Jerath et al. 2014). CTR is predicted by the investigators by collecting historical click information as it provides tangible and intangible examples of user behavior. Sometimes sufficient historical data is available for CTR estimation which gives a reliable estimation, and with the clustering of the searched information the investigators could predict CTR. Even after estimating CTR from historical data, it may vary because of the smaller number of searches, ineffective impression, and thus a smaller sample size of data for estimating CTR (Regelson and Fain 2006).

Nowadays, CTR prediction is also done for news queries that should be displayed when it is highly relevant to the query that has been asked by the user. This growing trend of commercializing specialized content such as news, products, etc. furnished with web search results introduces challenges by mixing news search results with the regular search results panel (König et al. 2009).

Despite the fact that all forms of online marketers are striving hard to incorporate all elements that add efficiency to their advertisement, make it effective enough to not only gain a click but also get the desired lead; there are studies, elaborated in the literature review that the rate at which the advertisement gets a click is a lot more than the advertisement generates conversion. Also, studies show that the CTR for a mobile advertisement is more than the advertisement for desktop browsing (Ackley 2015).

1.4 Research Purpose

The research purpose for the current work is as follows.

a. To study click patterns for individuals on the desktop and mobile ads, and
b. To develop a machine learning based prediction model to estimate the click-through rate.

The study has been divided into five sections. In the initial section, we have already given the background about CTR. The next section will give an overview of the kind of work done in this field. It is followed by Sect. 3, i.e., methodology of the study. Then the results of the experiments are presented in Sect. 4 followed by Sect. 5 that describes the discussion of results obtained and conclusion.

2 Literature Review

Advertisement refers to any paid communication about a product or a service, which is not personal. Advertisement started with print media such as newspaper and magazines, and then it moved on to radio broadcast and television. With the evolution of Internet, the focus shifted toward online advertisement but now with increasing adoption of mobiles and other portable devices, the advertisers are now shifting their interest toward mobile advertisement. Mobile advertisement is being considered as the fastest growing platform for advertisement (Bart et al. 2012; Wang et al. 2011).

After clicking on the advertisement, the user is redirected to the website of the advertiser and then user makes decision of purchasing the product or buying the services provided on the website. Our definition of CTR does not involve purchasing of the product by the user after clicking on the advertisement.

Sponsored search occupies two-fifths of the overall online market of advertising. According to a study done by Catalyst and Martineau (2013) on Google Desktop CTR, 48% of the searches result in organic click on page one and remaining 52% searches result in paid clicks. Thus, there is high possibility for a user clicking on a sponsored ad (Catalyst and Martineau 2013).

The position of an advertisement in the sponsored search list can impact the CTR of the advertisement. CTR reduces with different positions, and conversion rate initially goes up and then comes down for larger keywords (Agarwal et al. 2011). Joachims et al. (2005) also talk about biasness of trust which results in more clicks on advertisement ranked higher on search engine. A study done by Catalyst and Martineau (2013) on Google Desktop CTR also suggests that order in which the advertisement is displayed affects its CTR. Thus, position of an advertisement can be considered as integral influencer which impacts the CTR of an advertisement.

The content and design of an advertisement can impact the CTR of an advertisement banner (Lohtia et al. 2003). The advertisement size and design help in increasing the CTR for ads (Sigel et al. 2008). The authors stated that an advertisement banner of size 160 × 160 style achieved the highest CTR when compared to advertisement banner of size 728 × 90 and 300 × 250. The authors further stated that the advertisement banner of size 160 × 160 performed better than 728 × 90 in the interaction rate and 300 × 250 size advertisement banner achieved the highest interaction rate, and thus it shows that it becomes relevant to study the size of advertisement banner for increasing the efficiency of the advertisement.

There is a close relationship between visual appearance of an advertisement banner and user response (Azimi et al. 2012). The authors conducted different experiments to find out the relationships of visual features for prediction of click-through rate, along with the performance classification and ranking. There are many factors that can lead to user clicking on an advertisement. According to the authors Richardson et al. (2007), factors like reputation, attention capture, relevancy landing page quality, etc. impact the CTR on an advertisement. Further, the authors developed a model for predicting the probability that the advertisement will be clicked by the user on the basis of above factors.

Zorn et al. (2012) talk about the influence of language and animation on banner CTR. The authors conducted experiment across two website types to study the influence of language and animation on CTR. The authors reported that language had no impact on the two websites while search sites portrayed high-level differences among the static and animated sites.

Lohtia et al. (2003) studied the dependence of CTR on design of the advertisement and content of the message. To study the characteristics of design, the authors studied interactivity, color, and animation. The authors further stated that CTR is impacted by interactivity in the advertisement, and use of animation and emotion improved the CTR for B2C advertisement banner but for B2B banner advertisement, CTR decreased and using moderate level of color was found to be more effective than using high or low level of color in banner advertisement. There is a close relationship between visual appearance of an advertisement banner and user response (Azimi et al. 2012).

According to Tucker (2010), the banner advertisements that provide privacy control of personalized information are likely to attract more users toward the advertisement. The study conducted by the author indicated that clicks can be doubled on advertisement that provides privacy control and uses unique private information to personalize their message. Most of the firms based on Internet collate humungous information about the web visitors and use the information to create personalized advertisement (Tucker 2010). Using personalized information about user can lead to negative reaction from consumer which can cause consumer to avoid the advertisement's appeal (White et al. 2008). Cleff (2007) also talks about the mobile advertisement's privacy issues. As per the author, the success of m-advertising is determined by the industrial and legislative initiative's development and execution. The author further added that users should have control in some form on the data in their phone and there should be a mechanism for choice of mobile advertisement in their phones. Trust is an important factor that decides whether a user is going to click on advertisement or not (Young and Wilkinson 1989). A study done by Davis et al. (2011) suggests that user's trust is impacted by factors like reputation of the vendor and structural assurance.

Bart et al. (2012) suggest that product type and product involvement are two important determinants of user's intent of purchasing the advertised products. The authors further stated that presence of these two factors provides more exposure to advertisement. Effectiveness of online advertisement is also impacted by determinants like Internet skills and usage, content of the advertisement, location of ad, and income (Mohammed and Alkubise 2012). The authors' findings suggested that location of advertisement is the most important determinant of the online advertisement's effectiveness.

Mobile advertisement is the process of advertising on wireless devices (Chen, Wu and Li 2014). The authors' findings further suggest that quality and creativity impacts the performance of product marketing, and it positively affects sales performance of the product. A study done by Catalyst and Martineau (2013) suggests that paid advertisements on mobile are found to be more effective than desktop as the screen size is small and results are viewed in limit. Free applications available on the Google Play (earlier know as Android Market) and App Store from Apple follow a revenue model in which the free application includes advertisement which gets inserted in the application itself and it is shown at different positions during the usage (Vallina et al. 2012). 73% of applications available on Google Play are free (Leontiadis et al. 2012), and thus it can be considered that free applications attract larger number of users, and hence larger number of downloads as compared to paid applications (Hamburger 2014). The mobile advertisement model consists of three actors: user who uses the application, the developer who expects benefits from usage of the application, and advertisement network helping the developer with compensation in exchange of advertisements of user's interest (Leontiadis et al. 2012).

The prior literature (Agarwal et al. 2011; Catalyst and Martineau 2013; Lohtia et al. 2003; Azimi et al. 2012) reveals that advertisement position impacts the CTR of an advertisement. The display and contents of advertisements, and use of animation, language, and privacy controls are some factors that can impact the CTR of an advertisement. It also indicates that the advertisers are now more focused toward mobile advertisement because of increasing number of mobile user.

Most studies done on mobile advertisement have been focused upon the estimation of number of clicks on the advertisement but very less research has been done on conversion of those of clicks. Click-through rate only provides information regarding the number of clicks on a particular advertisement; it is not an effective measure to study the number of users who actually visited advertiser's website with intent of purchasing the product or using the services provided on the website. In this chapter, we will be building a CTR model through machine learning which will be able to predict whether an advertisement will be clicked or not by the user.

3 Methodology

The study has been designed to forecast the clicks by user on an online advertisement. A total 0.8 million of transactional data has been used to prepare and test the model. Classification tree model has been used to develop the click estimation using SPSS version 21.0. The data is available from Avazu which has shared its data with Kaggle (https://www.kaggle.com/c/avazu-ctr-prediction/details/timeline). The available data fields are the ad identifier, binary symbol for click or non-click, time format, position of banner, site details (id, domain, and category), app details (id, domain, category), mobile device details (id, ip, model, and type), type of connection, and some anonymized categorical variables.

The sample data was portioned in two groups with eighty percent as training group and twenty percent as test group. The first group was used for model development, while the other group was used for testing the rules generated. Two parameters were used to check the model's significance—Accuracy and precision. The total number of correct predictions out of total elements is defined as accuracy (AC) and can be defined as follows:

$$\frac{TruePositive + TrueNegative}{TruePositive + FalsePositive + TrueNegative + FalseNegative}$$

Classification trees are easy to implement technique and majorly implemented in areas like retail, health care, BFSI, etc. CHAID classification tree algorithm has been used in this study which has segmented the set into different groups. "*These segments, called nodes, are split in such a way that the variation of the response variable (categorical) is minimized within the segments and maximized among the segments. After the initial splitting of the population into two or more nodes (defined by values of an independent or predictor variable), the splitting process is repeated on each of the nodes. Each node is treated like a new sub-population*" (Ramaswami and Bhaskaran 2010). There is hierarchical output generated from the CHAID modeling, as shown in the analysis section. And the tree has been used to forecast the Click-Through Rate (CTR).

4 Results

Each node in Fig. 1 contains the details of node id (ID), number of data objects (N), and the possible outcomes of "CTR and non-click-through rate." The tree begins from the topmost decision node (ID = 0) with (N = 799,999) instances of the dataset, and the whole dataset is divided further on the basis of variable that is likely to affect customer intend of clicking a particular website or not. The topmost node suggests that site category is the utmost important variable that is likely to affect customer's intention to click a particular website. The first branch stemming out of the tree, i.e., (ID = 1 to 9), represents the occurrence of all those events that are likely to be generated. It can be seen that ID = 1 and ID = 5 have the best CTR for the site category, while ID = 3, 4, 7, 8 have the next best set of CTR.

Figure 2 showcases the further bifurcation of (ID = 1) on the basis of site id and subsequently with device connection. This is the second most important variable that advertisers should take into consideration while running ads online. ID = 1 containing 177,984 instances is further split into 18 nodes (ID = 1 to 18), each node representing different site ids with their respective CTR and non-click-through rates. Here, Node 10 is further split into two more nodes (ID-37 and 38), on the basis of predictor

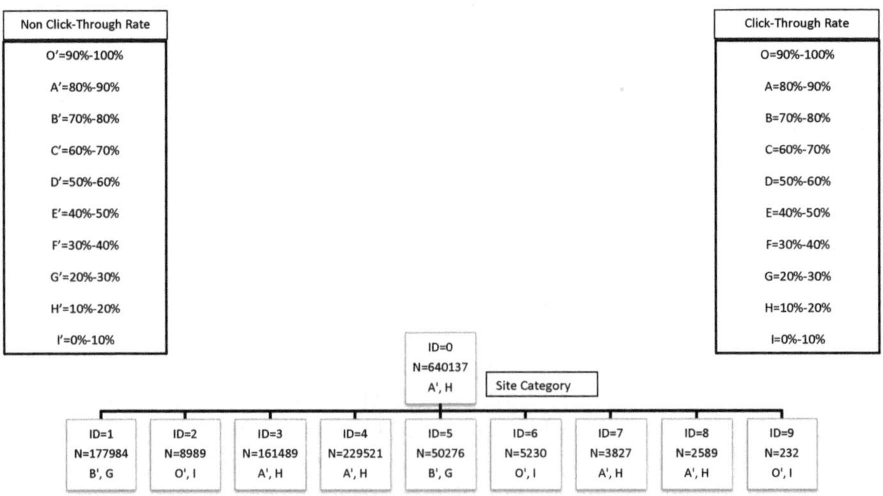

Fig. 1 First-level classification tree for the dataset based on clicks and non-clicks

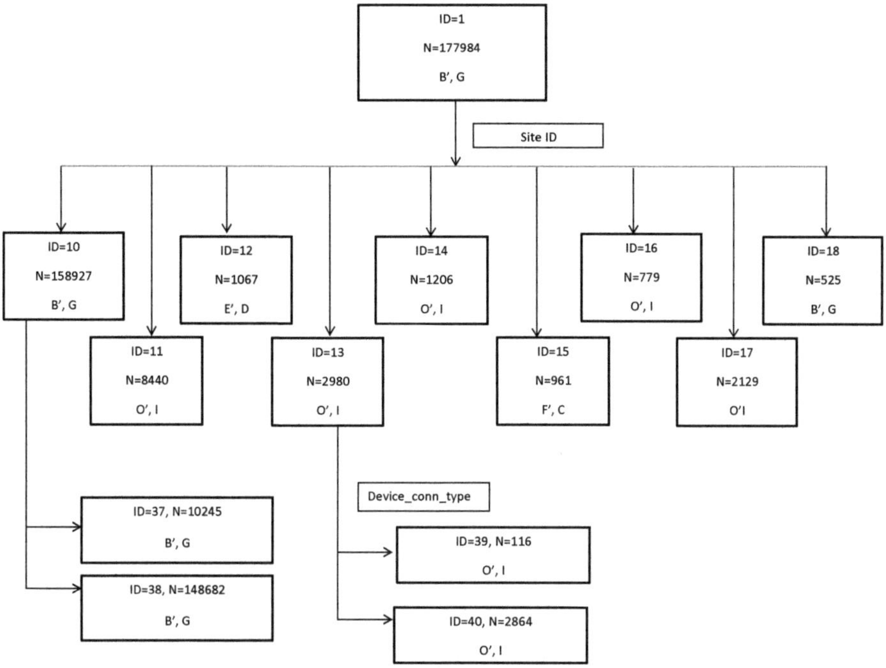

Fig. 2 Classification tree nodes at level 2 based on site ID

variable device connection type and Node 13 is also further split into two different nodes (ID-39 and 40), on the basis of predictor variable device connection type.

It can be concluded that for site category (ID = 1) and site id (ID = 15), the CTR is 63.8% which is higher when compared to others. Similarly, more nodes were tested for the dataset. Model validation was done with the help of split sample method. The sample data was divided into two sub-groups, test and training, with the first being used to test the rules generated out of the second group. The results revealed that the overall model prediction accuracy of 83.83% was achieved from the CHAID technique (accuracy for test set and train set was found to be approximately similar to each other), which suggests that it seems that CHAID is a fairly efficient way of classification model for CTR prediction, and was found to be better than the analysis presented by Ramaswami and Bhaskaran (2010).

5 Conclusion

This study has reviewed the difficulty of estimating CTR for advertisements using CHAID model. CHAID method was useful in visualizing the relation between CTR and other related factors. The decision rules have been formulated to estimate CTR.

Some of the important variables have been presented in the study that affects the probability of an advertisement to be clicked. This study would be useful for online advertisement agencies and marketing managers to take decisions regarding the placement of an online advertisement. Also, it would be useful for managers to obtain better ROI for various ad campaigns.

References

Ackley, M. (2015). Quantifying the mobile revolution: Mobile advertising progress in 2014. Search Engine Land. Retrieved January 15, 2017, from https://searchengineland.com/quantifying-mobile-revolution-2014-benchmark-mobile-advertising-progress-216696.

Agarwal, A., Hosanagar, K., & Smith, M. D. (2011). Location, location, location: An analysis of profitability of position in online advertising markets. *Journal of Marketing Research, 48*(6), 1057–1073.

Azimi, J., Zhang, R., Zhou, Y., Navalpakkam, V., Mao, J., & Fern, X. (2012, April). The impact of visual appearance on user response in online display advertising. In *Proceedings of the 21st International Conference Companion on World Wide Web* (pp. 457–458). ACM.

Bart, Y., Stephen, A. T., & Sarvary, M. (2012). Which products are best suited to mobile advertising? A field study of mobile display advertising effects on consumer attitudes and intentions. *Journal of Marketing Research, 51*(3), 270–285.

Cleff, E. B. (2007). Privacy issues in mobile advertising. *International Review of Law Computers and Technology, 21*(3), 225–236.

Briones, M. (1999). Rich media may be too rich for your blood. *Marketing News, 33*(7), 4–5.

Catalyst & Martineau, A. (2013). Google CTR Study How User Intent Impacts Google Click-Through Rates. Retrieved January 25, 2017, from https://www.catalystdigital.com/wp-content/uploads/GoogleCTRStudy-Catalyst.pdf.

Carterette, B., & Jones, R. (2007). Evaluating search engines by modeling the relationship between relevance and clicks. In *Advances in Neural Information Processing Systems* (pp. 217–224).

Chen, L. Y., Wu, C. C., & Li, M. T. L. (2014). Utilizing the technology acceptance model to explore the effects of mobile advertising on purchase intention. *Displays, 4*(5).

Davis, R., Sajtos, L., & Chaudhri, A. A. (2011). Do consumers trust mobile service advertising? *Contemporary Management Research, 7*(4), 245–270.

Doubleclick (2005). Search before the purchase: Understanding buyer search activity as it builds to online purchase. www.doubleclick.com.

Haans, H., Raassens, N., & van Hout, R. (2013). Search engine advertisements: The impact of advertising statements on click-through and conversion rates. *Marketing Letters, 24*(2), 151–163.

Hamburger, E. (2014). Indie smash hit Flappy Bird racks up 50 K per day in ad revenue. *Päivitetty, 5*(1), 2014.

Hopewell, N. (2007). Click-Through Rates 101. *Copyright American Marketing Association 2007 of the ACM, 42*(8), 91–98.

Hornikx, J. (2005). A review of experimental research on the relative persuasiveness of anecdotal, statistical, causal, and expert evidence. *Studies in Communication Sciences, 5*(1), 205–216.

Interactive Advertising Bureau (IAB) & Pricewaterhouse Coopers International (PwC). (2009). Internet ad revenues at $10.9 billion for first half of '09 from http://www.iab.net/about_the_iab/recent_press_releases/press_release_archive/press_release/pr-100509.

Jerath, K., Ma, L., & Park, Y. H. (2014). Consumer click behavior at a search engine: The role of keyword popularity. *Journal of Marketing Research, 51*(4), 480–486.

Joachims, T., Granka, L., Pan, B., Hembrooke, H., & Gay, G. (2005, August). Accurately interpreting click through data as implicit feedback. In *Proceedings of the 28th Annual International ACM SIGIR Conference on Research and Development in Information Retrieval* (pp. 154–161). ACM.

König, A. C., Gamon, M., & Wu, Q. (2009, July). Click-through prediction for news queries. In *Proceedings of the 32nd International ACM SIGIR Conference on Research and Development in Information Retrieval* (pp. 347–354). ACM.

Koonin, E. V., & Galperin, M. Y. (2003). Principles and methods of sequence analysis. In *Sequence—Evolution—Function* (pp. 111–192). US: Springer.

Lemonnier, J. (2008). Rich media. *Ad Age, 79*(11), 48.

Leontiadis, I., Efstratiou, C., Picone, M., & Mascolo, C. (2012, February). Don't kill my ads!: Balancing privacy in an ad-supported mobile application market. In *Proceedings of the Twelfth Workshop on Mobile Computing Systems and Applications* (p. 2). ACM.

Li, H., & Bukovac, J. L. (1999). Cognitive impact of banner ad characteristics: An experimental study. *Journalism and Mass Communication Quarterly, 76*(2), 341–353.

Lindsey, L. L. M., & Yun, K. A. (2003). Examining the persuasive effect of statistical messages: A test of mediating relationships. *Communication Studies, 54*(3), 306–321.

Lohtia, R., Donthu, N., & Hershberger, E. K. (2003). The impact of content and design elements on banner advertising click-through rates. *Journal of advertising Research, 43*(04), 410–418.

Mohammed, A. B., & Alkubise, M. (2012). How does online advertisements affects consumer purchasing intention: Empirical evidence from a developing country. *European Journal of Business and Management, 4*(7), 208–218.

Nikki Hopewell. Click-Through Rates 101. *Copyright American Marketing Association 2007 of the ACM, 42*(8), 91–98.

Ramaswami, M., & Bhaskaran, R. (2010). A CHAID based performance prediction model in educational data mining. arXiv:1002.1144.

Regelson, M., & Fain, D. (2006, January). Predicting click-through rate using keyword clusters. In *Proceedings of the Second Workshop on Sponsored Search Auctions* (Vol. 9623). Retrieved January 12, 2017, from https://pdfs.semanticscholar.org/d80c/03be2ef28a94229473c9c8484ae98d0cd003.pdf.

Reynolds, R. A., & Reynolds, J. L. (2002). Evidence. In Dillard, J. P. & Pfau, M. (2002). *The persuasion handbook: Developments in theory and practice*. Sage Publications.

Richardson, M., Dominowska, E., & Ragno, R. (2007). Predicting clicks: estimating the click-through rate for new ads. In *Proceedings of the 16th International Conference on World Wide Web (WWW'07)*. ACM, New York, NY, USA, pp. 521–530. https://doi.org/10.1145/1242572.1242643.

Rieke, R. D., & Sillars, M. O. (1975). *Argumentation and the decision making process* (p. 55). New York: Wiley.

Rosenkrans, G. (2010). Maximizing user interactivity through banner ad design. *Journal of Promotion Management, 16*(3), 265–287.

Sigel, A., Braun, G., & Sena, M. (2008). The impact of Banner ad styles on interaction and click-through rates. *Issues in Information Systems, 9*(2), 337–342.

Singh, S. N., & Dalal, N. P. (1999). Web home pages as advertisements. *Communications ACM, 42*(8), 91–98. http://dx.doi.org/10.1145/310930.310978.

Trofimov, I., Kornetova, A., & Topinskiy, V. (2012). Using boosted trees for click-through rate prediction for sponsored search. In *Proceedings of the Sixth International Workshop on Data Mining for Online Advertising and Internet Economy (ADKDD'12)*. ACM, New York, NY, USA, Article 2, 6 pp. http://dx.doi.org/10.1145/2351356.2351358.

Tucker, C. (2010). Social networks, personalized advertising, and privacy controls. *Journal of Market Research, 51*(5), 546–562.

Vallina-Rodriguez, N., Shah, J., Finamore, A., Grunenberger, Y., Papagiannaki, K., Haddadi, H., & Crowcroft, J. (2012, November). Breaking for commercials: characterizing mobile advertising. In *Proceedings of the 2012 ACM Conference on Internet Measurement Conference* (pp. 343–356). ACM. https://doi.org/10.1145/2398776.2398812.

Wang, X., Li, W., Cui, Y., Zhang, R. B., & Mao, J. (2011). Click-through rate estimation for rare events in online advertising. In *Online multimedia advertising: Techniques and technologies* (pp. 1–12). https://doi.org/10.4018/978-1-60960-189-8.ch001.

White, T., Zahay, D., Thorbjornsen, H., & Shavitt, S. (2008). Getting too personal: Reactance to highly personalized email solicitations. *Marketing Letters, 19*(1), 39–50.

Young, L. C., & Wilkinson, I. F. (1989). The role of trust and co-operation in marketing channels: A preliminary study. *European Journal of Marketing, 23*(2), 109–122.

Zorn, S., Olaru, D., Veheim, T., Zhao, S., & Murphy, J. (2012). Impact of animation and language on banner click-through rates. *Journal of Electronic Commerce Research, 13*(2), 173–183.

Predicting Success Probability in Professional Tennis Tournaments Using a Logistic Regression Model

Saurabh Srivastava

Abstract With a global audience of over 1 billion, professional tennis is the most widely followed individual sports in the world. The present study attempts to model the probability of success for a tennis player in a men's singles tournament of a given type (ATP 250, ATP 500, ATP Masters and Grand Slams) so as to enable his management team to take better decisions with respect to his calendar planning. The model in this study tries to arrive at the probability of success in a given category of the tournament by modelling the success of an athlete in that tournament (measured by his ability to reach the quarterfinals), using the logistic regression method. The scorecard that is built uses five variable categories to arrive at the probability of success, which can be used to rank order the tournaments in a given category for a player, and can be subsequently augmented through a linear programming method to help a player arrive at the most optimum selection of tournaments.

Keywords Sports analytics · Logistic regression · Tennis

1 Introduction

With a global audience of over 1 billion, professional tennis is the most widely followed individual sports in the world. The total prize money in 2016 for its biggest events, called the Grand Slams, ranged from 31.1 million USD (Australian Open) to 46.3 million USD (US Open), making it the most lucrative individual sports for athletes. Tennis is also one of the most geographically spread out sports in the world, with its venues present in more nations than any other sports except for football, and its top 100 ranked players coming from over 45 different countries.

Professional Tennis tournaments are usually organized based on gender and number of players. The most common configuration includes men's singles, women's singles, men's and women's doubles (where two players play on each side of the

S. Srivastava (✉)
EXL Analytics, Gurugram, India
e-mail: saurabh.dse@gmail.com

A. K. Laha (ed.), *Advances in Analytics and Applications*, Springer Proceedings in Business and Economics, https://doi.org/10.1007/978-981-13-1208-3_6

net), and the mixed doubles (in which teams are formed of two players of opposite genders). The present study is based on the men's singles professional tournaments, which are managed by the Association of Tennis Professionals (ATP) and are divided into four categories (ATP 250, ATP 500, ATP Masters, and Grand Slams). Within each category, the prize money and the ranking points remain more or less comparable, but from one category to another the prize money and ranking points on offer differ significantly (the lowest being in ATP 250 and the highest being in the Grand Slams).

Therefore, winning the more advanced category tournaments is more rewarding. However, the probability of losing in an early stage is also high in the advanced category tournaments as they are usually intensely competitive. In case of an early exit, an athlete usually faces a triple blow—(a) time loss (player has to wait till the next week before he can participate in another tournament), (b) financial loss (due to cost of travel and fees incurred for participation), and (c) opportunity loss (points and prize money which he/she could have otherwise earned had they played a more suitable tournament where he could have advanced to later rounds).

This chapter begins by looking at the existing studies on the subject, reviewing some of the research on the similar issue of finding success/victory probability in tennis tournaments. It then goes on to discuss the model it proposes—the data sources, variables, and the appropriateness of the underlying modelling methodology. It finally reviews the results of the model and suggests ways on how this model can be further improved.

2 Literature Review

Although professional tennis is one of the most competitive and lucrative sport, limited research in the realm of sports analytics has been done on it. Three research works, however, are important. First is that of Klaassen and Magnus (2003), wherein the authors have attempted to predict the winning athlete during an ongoing tennis match, using a fast and flexible statistical model based on a fuzzy logic algorithm. However, this study, as acknowledged by the authors themselves, had a limitation that it would be able to predict victory in an ongoing match only and could not predict match victory beforehand. The second is the study by Boulier and Stekler (1999) who have tried to check, through Brier scores and using data related to two sports—US collegiate basketball and professional tennis—whether or not the seeding (ranking) of a player is a good predictor of his/her victory. Their study validated the hypothesis that rankings are good victory predictors for these sports, but did not talk about any other factor that might also be responsible for winning the match.

The third important study is again from Magnus and Klaassen (1999) in which they test the hypothesis that the person serving first in the match has a higher probability to win it. Their study, using a simple Bayesian error rate to study misclassification, found that this hypothesis could only be accepted for the first set of the match while in the subsequent sets, the player serving first had a higher probability of losing that set instead of winning it. This study again was based on one variable, which is also decided during a match on the basis of a toss, and thus could not be used for prior planning.

It was found that while the existing studies in the area are novel in their approaches, they use a limited number of variables and are not able to point to a comprehensive range of determinants of match victory. Moreover, the variables they use cannot be used for prior planning. The present study attempts to overcome these limitations.

3 Model, Data and Results

The present model is the first layer of a two-stage model, which attempts to provide an optimum solution to an athlete or his/her management team in choosing his events. The first stage (subject of the present paper) attempts to predict the probability of success of a tennis athlete in the event, while the second stage would be performing the final selection based on four factors—cost of travel, time taken to reach the tournament city, participation fees, and benefits from tournaments (prize money, ranking points, etc.). The second stage of the model would be based on a linear programming-based optimization algorithm.

The present study attempts to model the probability of success for a tennis player in a men's singles tournament so as to enable his management team to take better decisions with respect to his calendar planning. If a player chooses a tournament of a type higher than his calibre, he risks having an early exit from the tournament, leading to a significant loss of points and prize money. On the other hand, if he chooses a tournament of a type lower than his calibre, he may not be able to move ahead fast in the rankings and lose on both the prize money in the current tournament and more favourable draws in future ones. Therefore, an accurate knowledge of his chance of succeeding in a tournament of a particular category can be very helpful for an athlete in obtaining the best possible results. This study attempts to develop a model to aid decision-making precisely in this area.

In this study, we try to arrive at the probability of success by modelling the success of an athlete in a tournament (measured by his ability to reach the quarterfinals). This variable x takes a value of 1 when a player reaches the quarterfinals and 0 when he does not. The choice of quarterfinals as the cut off stage for determining a player's success in the tournament is arbitrary, though it is guided by the logic that quarterfinals usually arrive in the latter half of the tournament and are widely considered as a mark of accomplishment. The model uses publicly available ATP tournament data for training and validating the model, as detailed (Table 1).

Sanity checks to attest Quality and Integrity of Data on the following parameters: (1) Completeness of Information/Missing Value, (2) Outlier/Extreme Value Study, (3) Duplicate Record and (4) Distribution were done before proceeding with model development. Match-level statistics were pooled and adjusted to provide tournament-level information. Variables used only for indexing purposes were dropped, along with other match-related variables which clearly seemed to have no direct or indirect

Table 1 Data usage for training and validation

Purpose	Vintage	% of data used
Training	2015	70
Out of sample (OOS) validation	2015	30
Out of time (OOT) validation	2016	100

impact on match outcomes. Data, thus, obtained was used for model development using principles of logistic regression.

Event was defined as 'Qualification to the Quarter Finals of a Tournament'. The hypothesis to be tested was whether factors such as Type of Tournament, Points, Physiological Characteristics (such as Age, Height and Weight), Player Rank, Surface of Play and Roof Characteristics (Indoor/Outdoor) impact the probability of an athlete in reaching the Quarter Final stage of a Tournament. Signs in Table 2 represent the hypothesized relation of the variable with the outcome (reaching Quarterfinals of the

Table 2 Variable description

#	Variable	Description	Category	Expected sign	Validated	Significant
1	Ranking	ATP ranking of the player	Numeric	Negative	Yes	Yes
2	Points	ATP points of the player	Numeric	Positive	Yes	Yes
3	Surface	Grass, clay, hard, or carpet	Categorical	NA-categorical		Yes
4	Tournament type	ATP250, ATP500, ATP masters, or Grand Slam	Categorical	NA-categorical		Yes
5	Roof characteristics	Indoor or outdoor	Categorical	NA-categorical		Yes
6	Age	Age of the player	Numeric	Negative	No	No
7	Height	Height of the player	Numeric	Positive	No	No
8	Weight	Weight of the player	Numeric	Negative	No	No
9	Draw size	No of opponents	Numeric	Negative	Yes	No

Tournament). Variables with p-value of less than 0.05 were considered significant in the model. Only the variables that were significant were allowed to stay in the model.

The modelling methodology is based on logistic regression. Logistic regression measures the statistical relationship between the dependent variable and one or more independent variables by estimating probabilities through a logistic function. Logistic regression is generally thought of as a more suitable method for developing models, where there is a binary response variable and the predicted values are probabilities and are restricted to (0, 1).

Assuming $p(\overline{x})$ as the probability of win, the model in this study constructs the following equation:

$$\log\left(\frac{p(\bar{x})}{1-p(\bar{x})}\right) = \beta_0 + \beta_1 x_1 + \cdots + \beta_p x_p + \varepsilon$$

Here, β_0 represents the intercept, β_p represent the coefficients of the various variables (x_p), and ε the error term. The probability of success in reaching quarterfinals can thereafter be obtained using the following equation:

$$p(\bar{x}) = \frac{\exp(\beta_0 + \beta_1 x_1 + \cdots + \beta_p x_p)}{1 + \exp(\beta_0 + \beta_1 x_1 + \cdots + \beta_p x_p)}$$

After training our model, the estimates for the various parameters are calculated to arrive at the final equation for $p(\overline{x})$ (please refer to Table 3). The values thus obtained for the various observations of the model are rank ordered based on their $p(\overline{x})$ values and grouped into deciles for evaluating the model parameters (Table 4).

The Variance Inflation Factor (VIF) represents the severity of multi-collinearity in a regression analysis. It is an index which measures how much the variance (the square of the estimate's standard deviation) of an estimated regression coefficient has increased because of collinearity. Cut Off of VIF in this study was kept at 10 to avoid multi-collinearity. However, all variables which entered the model (based on significance test) had a very desirable VIF value of less than 2 (Table 4).

The C-statistic (or the 'concordance' statistic or C-index) is a statistic that is used to measure the goodness of fit for binary outcomes in a logistic regression model. The model had a very desirable c statistic of 0.886, indicating good model strength (Table 5).

In order to validate the results, the model's lift was studied across the Training, Out of Sample (OOS), and Out of Time (OOT) datasets (refer to Table 6). The model not only performed well on these parameters for all three datasets, but also showed stability in lift results.

Table 3 Model estimates and significance

#	Variable	Description	Estimate	Prob Chi Sq	Wald Chi Sq
1		Intercept the expected mean value of Y, given X = 0	−3.5453	<0.0001	468.2977
2	ATP250	Indicator—when tournament type is ATP250	2.6564	<0.0001	437.1616
3	ATP500	Indicator—when tournament type is ATP500	1.5221	<0.0001	117.1053
4	Grand Slam	Indicator—when tournament type is Grand Slam	−0.4433	0.0205	5.3667
5	Points	ATP points of the athlete	0.0006	<0.0001	374.5926
6	Rank	ATP rank of the athlete	−0.0017	<0.0001	165.2119
7	Indi_Clay	Indicator—when playing surface is clay	0.4108	<0.0001	25.1634
8	Indi_Indoor	Indicator—when the match is played indoor	0.2224	0.0157	5.841

Table 4 Variable correlation, contribution, and variance inflation factor

#	Variable	Correlation	Variable contribution	Variance inflation factor (VIF)
1	Intercept	–	–	0
2	ATP250	0.14956	0.23799	1.87703
3	ATP500	0.03406	0.09896	1.48064
4	Grand Slam	−0.14901	0.03164	1.55394
5	Points	0.47848	0.22901	1.15711
6	Rank	−0.21658	0.35254	1.14118
7	Indi_Clay	0.0378	0.03447	1.12884
8	Indi_Indoor	0.09764	0.01539	1.22021

Table 5 Model performance statistics

#	Concordance	Value	Other stats	Performance
1	Percent concordant	88.5	Somers' D	0.772
2	Percent discordant	11.3	Gamma	0.774
3	Percent tied	0.2	Tau-a	0.222
4	Pairs	10,362,632	c	0.886

Table 6 Model lift

Sample	Lift at decile 1	Lift at decile 2
Train	0.448	0.666
Out of sample validation	0.449	0.659
Out of time validation	0.462	0.69

4 Conclusion

This research provides the most insightful variables deciding the success probability of a player in the match, and attempts to throw light on their relative importance. While the model confirmed the previous studies that players' competency indicators (such as Ranking and Points) continue to be the leading factors in deciding their performance, it showed that other factors such as playing condition and tournament category play an important role as well. Physiological and Demographic characteristics of athletes, such as Age, Height, and Weight were found to be statistically less significant in determining their ability to reach quarterfinals of the event, while player performance attributes (rank and points) were found to be most significant, followed by tournament characteristics (indoor or outdoor, surface, and tournament category). In order to develop a robust calendar management tool for an athlete, the current model can be complemented by other factors such as cost of travel, time taken to reach the tournament city, participation fees, and benefits from tournaments (prize money, ranking points, etc.).

References

Boulier, B. L., & Stekler, H. O. (1999). Are sports seedings good predictors? An evaluation. *International Journal of Forecasting, 15,* 83–91.

Hosmer, D. W., Lemeshow, S. (2000). *Applied logistic regression* (2nd ed.). Wiley.

Klaassen, F., & Magnus, J. (2003). Forecasting the winner of a tennis match. *European Journal of Operational Research, 148,* 257–267.

Magnus, J. R., Klaassen, F. J. G. M. (1999). On the advantage of serving first in a tennis set: Four years at Wimbledon. *Journal of the Royal Statistical Society, Series D (The Statistician), 48,* 247–256.

Hausdorff Path Clustering and Hidden Markov Model Applied to Person Movement Prediction in Retail Spaces

Francisco Romaldo Mendes

1 Introduction

Current advances in technology allow for the efficient capturing and storage of high-resolution and high-frequency person movement data. The advent of Wi-Fi position triangulation has allowed us to capture human movement with a great deal of accuracy inside a closed urban structure, e.g., a university or a shopping mall. While there have been significant advances in our ability to capture this data, advances in robust modeling techniques have been largely absent. Inferences are drawn mainly by visual techniques (heat maps, path plotting, etc.) in technology-driven applications. In this paper, we aim to present our theoretical insights based on person movement data collected from Deloitte University, West Lake Texas. We outline two independent approaches in this paper. The first aims to model person movement and develop an appropriate prediction mechanism, while the second aims to classify people based on their movement history.

2 Data Description

We collect the movement data of 1,425 users, who have logged into Deloitte University's Wi-Fi. The exact x and y coordinates are calculated by triangulating the signals from two Wi-Fi routers every 3 s. For a given user, we would have data of the following form (MAC Address, Latitude, Longitude, Time), where MAC is an identifier that uniquely identifies a user's mobile device. Every third second, a new row of data is added to the aforementioned form. We have 435,290 such rows of data. For a given user, a "chain" of observations is a sequence of time-ordered

F. R. Mendes (✉)
Deloitte Consulting LLP, New York, USA
e-mail: frmendes@deloitte.com

A. K. Laha (ed.), *Advances in Analytics and Applications*, Springer Proceedings in Business and Economics, https://doi.org/10.1007/978-981-13-1208-3_7

geo-location 3-dimensional vectors of the following form: {(*Latitude*$_1$, *Longitude*$_1$, *Time*$_1$),...,(*Latitude*$_T$, *Longitude*$_T$, *Time*$_T$))}. Where T refers to the last observed time point in the dataset.

3 Preliminaries

We define a few mathematical preliminaries to aid our discussion. For a given individual "k", we define the following set: $p_k = \{(x_1, y_1, t_1), \ldots, (x_T, y_T, t_T))\}$ such that $t_i < t_{i+1}$. For a given real space define the set of labels: L = {"A", "B", "C", "D",...}. Each label corresponds to a specific room in the retail space. For example, the label set could be written as L = {"Cinema Hall", "Coffee Shop", "Park Spot 453",...}. In order to maintain conciseness, we choose label names "A", "B", and so on rather than more illuminating label names.

4 Part 1: Movement Prediction

The movement prediction approach takes place in using two independent methods applied successively as follows:

Part 1. Room assignment using k-means clustering. The objective is to find out the rooms inside Deloitte University. The basic idea is that the rooms inside D.U. will be densely populated with co-ordinate vectors. We use the k-means algorithm to discover such rooms.
Part 2. Movement prediction using the Hidden Markov Model. After we have assigned every point in the coordinate space of a room, we translate every given users movement data to room data. The details of which are enclosed in the following sections. Once, we have translated every user's movement data from a sequence of coordinates to a sequence of rooms, we can then use the Hidden Markov Model trained on such discrete symbol sequences to predict newly observed sequences.

4.1 Room Assignment

One of the major challenges with person movement is that due to the continuous nature of the data and the complex layout of most urban structures, it is not easy to assign rooms to a specific collection of (x, y) coordinates. For example, let us say we have two pairs of points (x_l, y_l) and (x_h, y_h) such that $x_l 6 = x_h$ and $y_l 6 = y_h$. Both these points may correspond to the same physical room. The simple way to go about this would be to use some set of inequalities on the real plane to assign rooms to the points. This approach suffers from two drawbacks, this may be prohibitively complex for large spaces with a large number of small rooms, the second more subtle problem with this approach is that it would eliminate any inferences we could draw from the spatial data on how people actually use the room. As an example, consider how people use a supermarket or a grocery store, while from a construction point of view the grocery store is a single large "room", people may not be visiting all corners of that large "room". In order for us to actually predict or draw inferences from movement data, we must not only gather their movement data but also assign it a label that makes sense from a business point of view. In this context it would be the specific section that the coordinate refers to, e.g., "Fresh Vegetables" or "Cereal" (Fig. 1).

The approach we propose actually allows us to infer what the rooms are based on the observed data, not only this is computationally more efficient but also allows retail space owners to understand how people actually use their retail space. We choose a distance metric (in most practical applications, the Euclidean metric should be

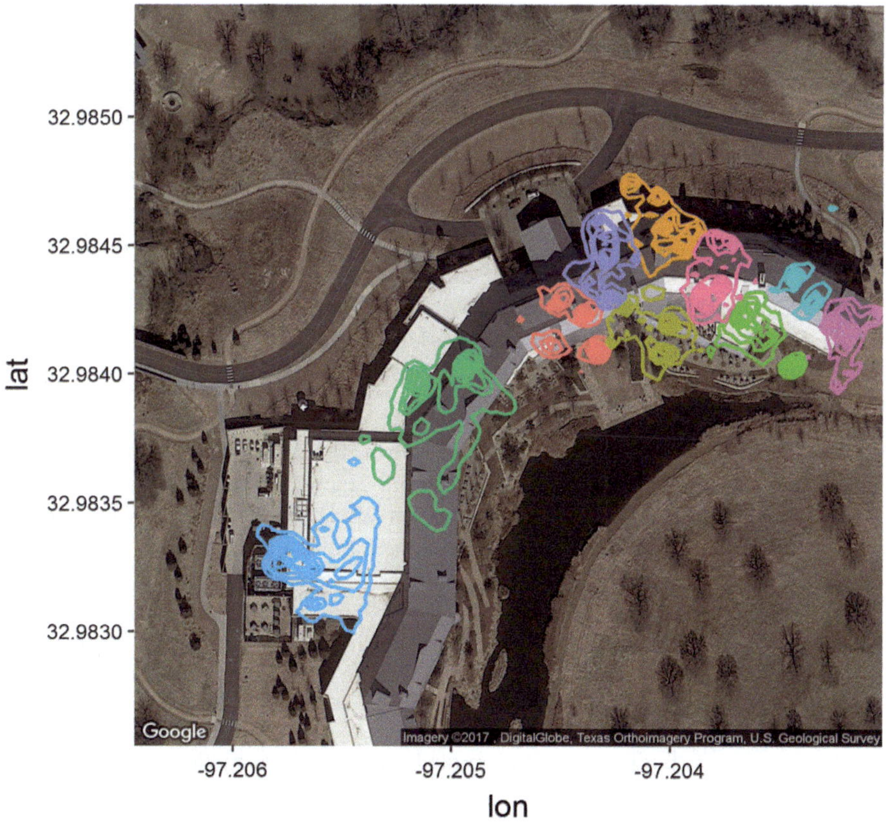

Fig. 1 Colors represent rooms (here we have 10 centers) and topography represents density of observations (Color figure online)

sufficient) and cluster the data based on the distance metric starting with an arbitrary number of cluster centers (usually, we set the number of cluster centers as the number of large rooms we see in the data. Choice of cluster centers depends critically on what level of granularity we want in our predictions. We will deal with this issue later in the text. The process of room assignment essentially defines a one to one mapping from the real space to the space of labels for each of the rooms in the data, i.e., $f\colon R \times R \rightarrow$ L.

4.2 Movement Modeling Using Hidden Markov Models

Recall that the path is defined as follows: $p_k = \{(x_1, y_1, t_1), \ldots, (x_T, y_T, t_T))\}$, where x, y were GPS coordinates and t was the time stamp. Now, every point in that path has been assigned to a room, hence the sequence is now discrete and given by a sequence of labels, e.g., P = {A, A, A, B, B, C, D, A ... A, C, C, D, ... D}, where every label represents the room to which the corresponding co-ordinate was allocated using the k-means clustering algorithm. Note that "P" represents a path in terms of a sequence of room labels and "p" represents the path in terms of the coordinate space. In our example paths, it is necessarily true that $(x_1, y_1) \in A$ and $(x_T, y_T) \in D$. Also note that, the sequence denoted by P is ordered in the time-sense, i.e., first observation occurred before the second, second before third, and so on. These path sequences will form long chains of observations for any given individual. This naturally suggests the use of the discrete version of the Hidden Markov Model. For a given individual, we train a Hidden Markov Model of order 1 (we can conceptually extend this to higher order Markov models as well) having two states on the movement data of the whole dataset. Before we fit a Hidden Markov Model, we must completely specify the following parameters (for a further discussion on Hidden Markov Models and their specification see Rabiner):

1. N = 2 (Hidden States play an important role in Hidden Markov Theory, for a more complete discussion refer to Rabiner, in our case it can be intuitively thought of the time of day, i.e., the sequence of room labels are very likely to be different depending on the tie of the day. We assume here that there are two major states, which emit two very different kinds of Markov sequences depending on the time of day).
2. M = L (Number of rooms, i.e., number of symbols we observe, as standard for a Markov chain the observed labels must come from a predefined set of limited labels).
3. $A = \begin{bmatrix} S_{11} & S_{12} \\ S_{21} & S_{22} \end{bmatrix}$—The transition probabilities between hidden states S_i and S_j, where $a_{ij} = P[q_{t+1} = S_j | q_t = S_i]$, $0 < i, j < N - 1$, where i, j denote the positions in the A matrix., where q_t denotes the state which occurs at time "t". Not that even though the variable q_t can assume values S_i and S_j at any time "t"

we cannot observe this, it is purely theoretical (for a more complete discussion refer to Rabiner).

4. $B = \{b_j(k)\}$—The probabilities of the observable labels, i.e., rooms L_k in state S_j.
5. $\Pi = \{\pi_i\}$—The initial hidden state probabilities, where $pi_i = P[q_i = S_i | t = 0]$ $0 < i < N - 1$.

5 Part 2: Path Clustering

Customer paths offer a huge amount of high-dimensional data that can be used to draw insights into their behavior. Simply visualizing these patterns of movement can give several powerful insights into customer behavior. However, in this section, we seek to formalize some techniques so that we can employ these techniques over large data sets fairly easily. Clustering essentially defines $p_i\ p_j \forall\ p \in c_k$ where p_i and p_j are paths and c_k is a cluster. In this context, our algorithm works in several stages

Step 1: Path Simplification using the Ramer–Douglas–Peucker (RDP) algorithm.
Step 2: Defining a distance metric (we use Hausdorff distance) and using this distance metric to cluster the data.
Step 3: Choosing an appropriate clustering algorithm to cluster various data points.

6 Experimental Results

In this section, we discuss our experimental results on the Deloitte University Westlake Texas Campus. We assign 6 symbols to the 6 main areas of the Deloitte University Campus. We present the experimental results for a USER ID "USCRESTRONMVSC". We have an observation sequence for this user which is 12,000 observations long, we train the HMM on 9600 observations and then test on the remaining 2400 observations. We report the following:

For N = 2, we chose two hidden states because this gave the best results and they correspond roughly to the two states we see in our data, "Workhours" and "After-Work Hours".

M = {"A", "B", "C", "D", "E", "F"} Here, we label the major areas of Deloitte.

University Westlake, Texas using the integer labels 1 through 6 (shown in Fig. 2). These labels are shown in Fig. 2. The major areas are as follows:

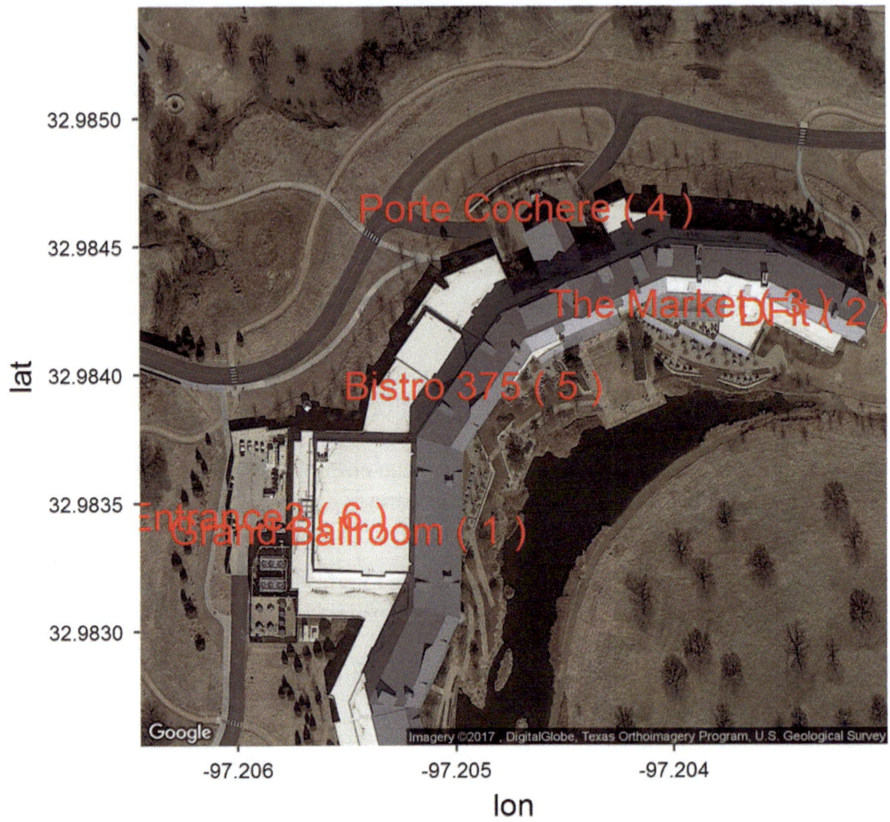

Fig. 2 Major locations in Deloitte University

A. Entrance 2, an alternate entry/exit point.
B. Grand Ballroom, largest conference hall in the campus.
C. DFit which is the gymnasium.
D. Bistro 375.
E. The Market which is a large open cafeteria.
F. Porte Cochere which is the main entrance.

$$A = \begin{bmatrix} 0.9897389 & 0.0102611 \\ 0.13262620 & 0.8673740 \end{bmatrix}$$

In keeping with our theme, that the two states in our model represent "Work-Hours" (State 1) and "After-Work Hours" (State 2), the state transition matrix matches our intuition, as the Diagonal Transitions are of highest probability and the off-diagonals are of fairly low probability.

$$B = \begin{bmatrix} 0.009762765 & 0.32275700 & 0.31065118 & 0.009762765 & 0.009762765 & 0.33730352 \\ 0.086580087 & 0.08658009 & 0.086580087 & 0.086580087 & 0.5670099567 & 0.08658009 \end{bmatrix}$$

The emission probability matrix also matches our intuition as 5 represents The Market (large cafeteria), where people go in the evening. This confirms that the model is picking up on aspects of human behavior as we would expect. As the emission probability of The Market is highest during State 2 and the emission probabilities of the other five locations are fairly high during State 1, i.e., "Work-Hours".

$$P = \begin{bmatrix} 0.7 & 0.3 \end{bmatrix}$$

The initial state probabilities are also skewed towards "Work-Hours" because the Wi-Fi triangulation devices are switched off shortly after 9 pm. We use the Hidden Markov Model described in the parameters above to predict the next in a sequence of observations using the algorithm described below:

Step 1: Train Hidden Markov Model up to observation sequence of length T.
Step 2: Simulate T+1, 10^5 times call this sequence $T = \{T_1', T_2', T_3', \ldots, T_m'\}$
Step 3: Take a vote over the set τ and choose the prediction that occurs the most frequent.
Step 4: Call this T′+1, evaluate against T+1 (the actual value and give it a score of 1 if correct else 0).
Step 5: Repeat step 1 for the observation chain up to T+1.

We report an accuracy of 65% for this user. Such a prediction algorithm would allow us to predict a user's next location based on their current location. In this context, accuracy is not always the best measure of usefulness, it is merely a guide when designing algorithms. As any targeted advertising based on this algorithm would also behave as a nudge to the user, users may just boost the usefulness of the algorithm by using any byproduct of the algorithm as a behavioral nudge rather than rejecting it entirely. We could also boost the accuracy by increasing the order of the Markov chain used to make the prediction.

7 Part 2: Path Clustering (Experimental Results)

Path Clustering is an extremely complex process, as one can see by the sheer number of calculations needed between all permutations and combinations between the sets of each path. Path Cleaning provides a simplification over the initial raw path that is an accurate representation of the raw path. The RDP algorithm works by calculating a new path consisting of a straight segment from given start and end points and either checking if all the points in between are not too distant or including the most distant point as a necessary endpoint, cutting the proposed segment into two and repeating recursively on the two smaller segments. In our case, we use an unconstrained path simplification but a constrained path simplification is the best practice as it will take into account all obstacles, which may contain important person movement information.

Clustering algorithms necessitate the development of a distance metric that judges the similarity of two objects in the dataset. Any two objects in our dataset are paths and the similarity between any two objects is defined by the Hausdorff distance. Hausdorff distance which has been discussed above is a basic measure of similarity between sets.

Choice of the clustering algorithm, we use the DBScan algorithm primarily because it does not necessitate the knowledge of the number of clusters beforehand.

Possible extensions to the method described above include extending the dimensions of the path vector, in order to integrate higher dimension variables, such extensions would be simple extensions from a theoretical standpoint but may be computationally expensive in practice. For example, a user path vector could include {x, y, t, W, D...}, where W denotes total money spent at similar {x, y}, co-ordinate on the previous trip in the same supermarket, D denotes total distance traveled since Wi-Fi first triangulated position. Such higher dimension path vectors would allow us to get a greater insight into user behavior.

8 Experimental Results of Path Clustering Using Hausdorff Distance

Here, we present a few examples of path clustering using our algorithm using the Hausdorff distance. Figure 3 shows three representative paths that could provide the motivation for the need for path clustering. The green path shows the movement of a user who has entered Deloitte University and used The Market, Bistro 375, and the Fit. The pink path shows a user who has used Bistro 375 and The Market. The blue path denotes the most usual Deloitte University user, who uses the Grand Ballroom during business hours perhaps to attend a conference and then moves to The Market for a lunch break and perhaps moves outside near Porte Cochere. Our algorithm as described in the preceding sections will first simplify the paths and then cluster them based on some similarity metric.

Figure 4 shows a larger number of representative paths from 3 similar user groups after simplification, the algorithm is able to segregate the users into 3 classes.

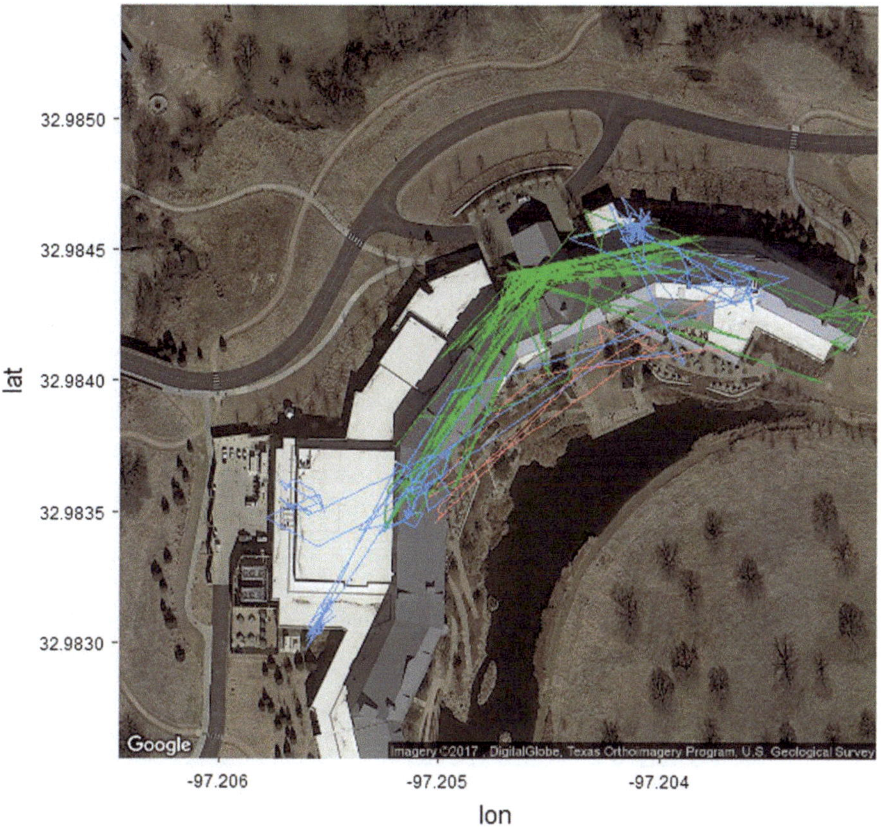

Fig. 3 Path clustering without simplification

9 Conclusions and Further Work

This paper presented a novel technique for modeling movement data of a large population as well as techniques to draw inferences about individuals using their path traces. We report an accuracy of 65% for a Hidden Markov Model of order 1. We expect that optimized models may yield higher prediction accuracy. Our results for a small number of paths show that our algorithm efficiently segregates paths into various path types. We recommend using HMMs of higher orders as these accurately model human behavior better as human movement usually has memory ≥ 1.

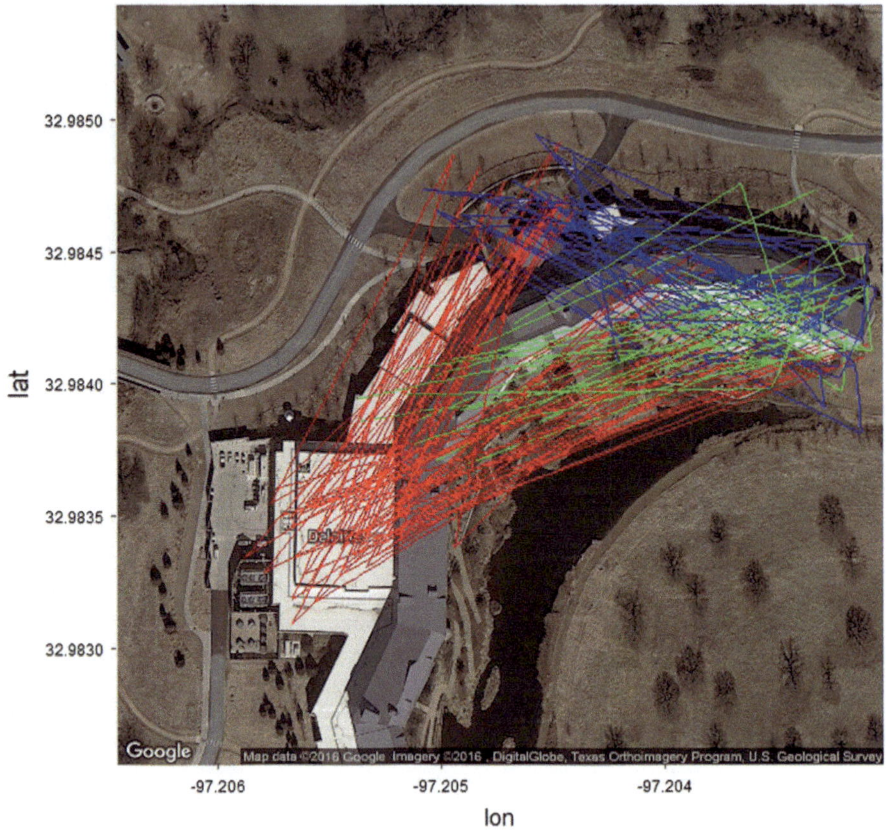

Fig. 4 Path clustering example showing 3 clusters after simplification

References

Algorithms for the reduction of the number of pints required to represent a digitized line or its caricature. *Cartographica: The International Journal for Geographic Information and Geovisualization, 10*, 112–122 (1973).

Bauckhage, C., Sifa, R., Drachen, A., Thurau, C., & Hadiji, F. (Aug 2014) Beyond heatmaps: Spatio-temporal clustering using behavior based partitioning of game levels. In *CIG'14: IEEE Conference on Computational Intelligence and Games pages* (pp. 1–8).

Birney, E. (2001) Hidden Markov models in biological sequence analysis. *IBM Journal of Research and Development, 45*(3/4).

Campbell, J., Tremblay, J., & Verbrugge, C. (2014) Clustering player paths. In *Proceedings of the 9th International Conference on Foundations of Digital Games*.

Gellert, A., & Vintan, L. (2006) Person movement prediction using hidden Markov models. *Studies in Informatics and Control, 15*(1).

Rabiner, L.R. (Feb 1989) A tutorial on hidden Markov models and selected applications in speech recognition. *Proceedings of the IEEE, 77*(2).

Campbell, J., Tremblay, J. Unity 3d—Clustering Open Source Implementation.

Improving Email Marketing Campaign Success Rate Using Personalization

Gyanendra Singh, Himanshu Singh and Sonika Shriwastav

Abstract Email marketing provides one of the best methods for direct communication with consumers. However, the success rate of an e-mail marketing campaign is often low because of its generic content and inadequate segmentation of customers. This paper aims to showcase the application of a two-step personalization process to improve effective open and click rates for email marketing campaigns. Consumer behaviour is monitored over a period of time in terms of email opens and click pattern. This behaviour is stream-lined into keywords sorted in order of user preference. The keywords are updated at regular intervals to account for behavioural changes in user preference. While sending the email, keywords relevant to the campaign are picked individually for each user. These keywords are used to form attractive subject lines using probabilistic language models such as noisy channel model (Mark and Charniak, Proceeedings of the 42nd annual meeting on association for computational linguistics, 2004) and hidden Markov model (Lawrence and Juang, Fundamentals of speech recognition, 1993).

Keywords E-mail marketing · Open rate · Natural language processing
Language modelling

1 Introduction

As of 2016, the Internet penetration in India has reached 462 million i.e. almost 34.6% of the overall population (http://www.internetlivestats.com/internet-users/india/). Sixty eight percent of the total internet users in India are connected through e-mail (http://www.huffingtonpost.com/2012/03/27/email-connects-the-w

---

G. Singh · H. Singh (✉) · S. Shriwastav
Experian Credit Information Company of India Private Limited, Mumbai, Maharashtra 400070, India
e-mail: himanshu.singh@experian.com

G. Singh
e-mail: gyanendra.singh@experian.com

A. K. Laha (ed.), *Advances in Analytics and Applications*, Springer Proceedings in Business and Economics, https://doi.org/10.1007/978-981-13-1208-3_8

orld_n_1381854.html). The largest e-retailers like Flipkart have reported a user base of 100 million in September 2016 which are targeted through email for marketing. This user base has grown at a whopping rate of 33% within the past 6 months (http://www.business-standard.com/article/companies/flipkart-grows-user-base-to-100-million-116092100216_1.html). This presents an opportunity to reach out directly to the interested customers. However, such a high level of growth doesn't come without challenges. In the past years, rate of unsubscription has also shown an unprecedented growth. Inbox report by Octane research states that 'Emails not being relevant' was stated as the top reason for unsubscription by nearly 46% of consumers during the years 2012–2014 (Octane Marketing 2015). Hence, there is a need to understand the difference between e-mail advertising and conventional form of advertising such as TV commercials and newspaper advertisements. The former needs to be tailored to suit the need of the consumers. Personalization has already been proven as a game changer for a variety of online stores such as Netflix and Amazon. Netflix identifies its personalization algorithms and segmentation strategies (metadata tagging) to be a key differentiator among its competitors (https://blog.kissmetrics.com/how-netflix-uses-analytics/). It has been reporting savings of approximately 1 billion dollars per year through combined effect of personalization and recommendation (Gomez-Uribe and Hunt 2016).

In Digital Marketer Survey 2016, Experian (2016) states having single customer view as today's biggest challenge being faced by marketing organisations. The report also states that emails with personalized subject lines show a 32% lift in open rates as compared to industry benchmarks. The lift was measured by taking a count of unique customers which had opened the email using personalized subject lines against generic subject lines used in the marketing industry. The personalization techniques currently in use involve including the individual's name in subject line or changing offer type with a generic subject body based on consumer behaviour. These methods take care of individual's needs but still a lot needs to be done to enhance the overall experience. The present work aims to set-up an automated process for delivering unique subject lines taking a single customer view.

A typical email used for marketing consists of a subject line and a URL on the inside along with some info-graphics or written information. Few of the useful parameters used in this study are as follows:

- Subject Lines of Mails: The subject with which promotional mailers have been sent out to customers.
- Number of Mails Sent: The number of emails sent to a user with a specific subject line.
- Number of Mails Opened: The number of emails with a specific subject line opened by the user.
- Number of Mails Clicked: The number of emails with a specific subject line clicked by the user.
- Click URL: Tagged URL of the landing page.

As depicted in Table 1, the dataset used in this study is derived from a user base of 2.2 million with a total of 46 million emails.

Table 1 Statistical overview of email marketing data

Field	Statistics
Total no. of mails sent	46 million
No. of distinct mail ids	2.2 million
No. of mails opened	4.1 million
No. of mails clicked	0.9 million

The frequency of mails sent, opened and clicked help highlight the success rate of a mailer and the subject lines and click URLs bring consumer interests to the fore.

2 Language Modelling

A sentence primarily consists of words and phrases. Each word in a sentence is associated with the words coming before it. A sentence can be represented as a combination of words:

$$\text{Sentence (S)} = w_1 w_2 w_3 \ldots w_n \quad (1)$$

Hence, probability of occurrence of a sentence in a corpus can be described using chain rule as follows:

$$P(S) = P(w_1) \times P(w_2/w_1) \times P(w_3/w_2 w_1) \times \cdots \times P(w_n/w_{n-1} w_{n-2} \ldots w_1) \quad (2)$$

$P(w_2/w_1)$ is the probability of occurrence of second word after the first word.

$$P(w_2/w_1) = n(w_1 w_2)/n(w_1) \quad (3)$$

$n(w_1 w_2)$ represents the total number of appearances of the second word after the first word and $n(w_1)$ determines the frequency of the first word in the corpus. Similarly, likelihood of other terms in Eq. 2 can also be computed. But, it might not be possible to calculate the likelihood of all terms in a sentence even with huge corpora such as web. It is because number of valid combinations of words to form phrases can be practically infinite. Markov showed that the likelihood of a sequence can be estimated with significant accuracy without going too far into the past (Markov 1913). Hence, we can say that probability of word in a sentence depends only on the previous word. Such a model represents a bi-gram model. Similarly, trigram and n-gram models can be formed by using last two words or last $n - 1$ words respectively. Using bigram model, probability of a sentence can be estimated as:

$$P(S) = P(w_1) \times P(w_2/w_1) \times P(w_3/w_2) \times \cdots \times P(w_n/w_{n-1}) \quad (4)$$

While computing probability for practical purposes, we need to add start and stop symbols to a sentence so that we can consider the probability of starting/ending the sentence with a given word.

$$\begin{aligned} P(S) = {} & P(w_1/\text{<start>}) \times P(w_2/w_1) \times P(w_3/w_2) \\ & \times \cdots \times P(w_n/w_{n-1}) \times P(w_1/\text{<start>}/w_n) \end{aligned} \tag{5}$$

$$\begin{aligned} P(S) = {} & P(\text{<start>}w_1) \times P(w_1 w_2) \times P(w_2 w_3) \ldots P(w_n\text{<stop>})/P(w_1) \\ & \times P(w_2) \times \cdots \times P(w_n)) \end{aligned} \tag{6}$$

Hence, chain rule of probability and the Markov assumption can be used to estimate the probability of formation for any sentence. The probability of occurrence of a sentence can be modelled as a product of bigram probabilities over unigram probabilities.

In this study, our objective is to generate an automated subject line from user specific keywords. The first step is to create a corpus of all available subject lines. Then add start and stop symbols to each line in corpus. The likelihood of appearance of any n-gram is estimated as the log of fraction for count of the given n-gram over total count of n-grams in corpus. As a part of convention, language model probabilities are always displayed in log format.

$$P(\text{n-gram}) = \log_2(\text{No. of occurrences of n-gram/Total no. of n-gram in corpus})$$

Subject lines can be generated from a keyword in two parts: Let the keyword be w_k and the start and stop keywords are represented as <start> and <stop> keywords. In the first part, we start at the keyword and compute the next most likely word associated with it until stop phrase is found:

$$S_1 = w_k \ldots \text{<stop>}$$

In the second part, we start from the keyword to establish the most likely word which should appear before it and thus continuing till the start phrase is found:

$$S_2 = \text{<start>} \ldots w_k$$

Thus, we arrive at our subject line by combining the two parts together:

$$S_{2\,U}S_1 = \text{<start>} \ldots w_k \ldots \text{<stop>}$$

A Markov chain represents a group of interconnected states with a set of possible transitions and probabilities associated with each transition (Baker 1990). In an observed Markov chain, the states represent the observed instances as in case of words in a sentence. Hence, n-gram models are a representation of observed Markov models. In hidden Markov models, the states represent the underlying instances such as Parts of Speech (POS) tags associated with the words. Thus, it provides a method to attach the probability of an observed instance with the causal state.

Markov models have been used extensively for the tasks of text generation (Jelinek 1985), speech recognition (Bahl et al. 1983) and machine translations (Brants et al. 2007) but their potential for the developing personalized subject lines is yet to be explored.

3 Two-Step Personalization Process

3.1 Step 1

The analysis was carried out to assess the behavioural pattern of more than 2 million email users. Promotional e-mails with generic subject lines were sent to all users. It is assumed that email opening by user depends on the subject line composition. The data was aggregated at the subject level and the number of opened mails was counted against the total number of mails which were sent. Each subject line was broken into sets of unigrams, bigrams and trigrams which can be referred to as n-gram sets. Each n-gram in the subject line was assigned the same count for opens as the subject line itself. Then the data was aggregated at the level of n-gram to calculate the total mails opened against the total number of mails sent for each user.

Based on the count, separate lists of most relevant sets of n-gram were prepared for every user. A similar set of n-grams was prepared based on relevance for the marketing campaign. While running the marketing campaign, the n-gram sets from the campaign are matched with the relevant n-grams associated to the user. The matching n-gram would be used to generate an automated subject line based on the corpus created using subject lines.

3.2 Step 2

The next level of personalization is done to optimize the content of mail. For the mails, which reported a click on the mentioned URL, landing-page URL was stored. The URL was already tagged with the location (city) and contained specific words related to the offer. Based on the tagged information, user preference was identified in terms of categories and subcategories e.g. the subcategories within food would comprise sweets, barbeque, drinks, fast-food, fine-dine and restaurant.

The arrangement of categories is done inside the email in the form of set of blocks. The arrangement of the blocks is done to make it as intuitive as possible. Each block will display the sub-categories based on the user preference. The increase in effective open and click rates is used as a metric to evaluate the effectiveness of the proposed technique.

Table 2 Typical bi-grams (modified) and their respective open rate

Bi-grams	Open percentage (%)
Buckle for	18.1101
Off Sale	15.9984
80% off	15.0343
Be brave	14.6860
Kicks off	14.6174
ISL starts	14.5958
Kids time	14.4451

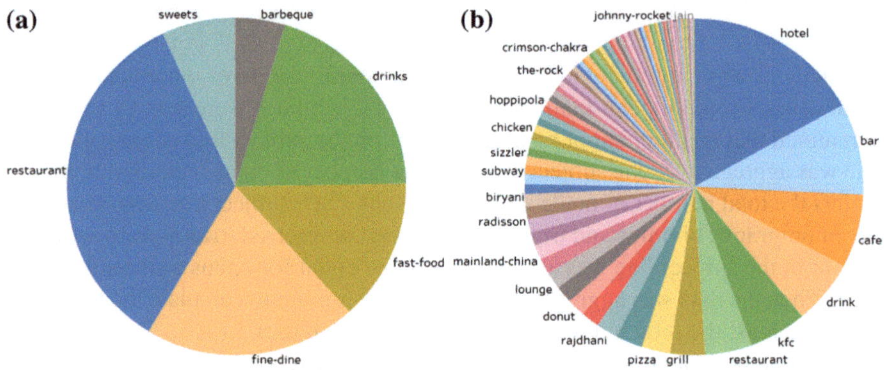

Fig. 1 Classification of food (as a type) into: **a** sub-types **b** keywords

4 Results and Future Scope

Table 2 shows a list of few bigrams with highest open percentages. Slight modifications have been made to the list to ensure confidentiality. It can be seen that certain bigrams show an open rate of 18% which is much higher as compared to overall open-rate of 8.9%. It also suggests that overall open-rate of any campaign can be increased by using selective bigrams which show a high open-rate and omit the use of bigrams with low open-rate.

The next part of the analysis resulted in segmentation of keywords into categories and sub-categories. The data consisted of 3.2 million URLs for 0.3 million distinct email-ids. After cleaning of URLs, the data was reduced to a set of 2 million URLs for 0.25 million distinct email-ids. The landing page URL was parsed using regular expressions to identify the cities and specific keywords. The data was aggregated based on the keywords and sorted based on frequency. These keywords were further classified into subtypes which were classified into individual types. Figure 1 shows a layout for the classification of one of the types.

Table 3 shows a list of desirable keywords associated with email-id profiles:

The performance of any n-gram model is dependent on n-gram's length which is used for modelling (Taylor and Black 1998). In general, trigram models perform better than a bigram model. The performance of the model is also dependent on the

Table 3 List of keywords and subject lines generated from bi-gram model

Email	Keywords	Suggested subject-line from bi-gram model
Email-1	Fly	Wednesday means savings @ rs. 199 and fly higher
Email-2	Dhamaka	Zor ka dhamaka deals
Email-3	Travel	Travel bestsellers buy one + up to rs. 30,000 off!

size of the corpus. An increase in size of corpus will also require more computational power. Another drawback for such a kind of model is that it cannot be used on new consumers as we don't have their preference for keywords. For new customers, we need to first use generic subject lines to learn about their preference. The model can be improved by using the combination of keywords as compared to a single keyword for forming the sentence. The on-going further development is focussed on application of hidden Markov models to improve grammatical performance and limit the size of subject lines.

References

Bahl, L. R., Jelinek, F., & Mercer, R. L. (1983). A maximum likelihood approach to continuous speech recognition. *IEEE Transactions on Pattern Analysis and Machine Intelligence, 5*(2), 179–190.

Baker, J. K. (1990). Stochastic modeling for automatic speech understanding. In A. Waibel & K. F. Lee (Eds.), Readings in Speech Recognition, pp. 297–307.

Brants, T., Popat, A. C., Xu, P., Och, F. J., & Dean, J. (2007). Large language models in machine translation. In *EMNLP/CoNLL*.

https://blog.kissmetrics.com/how-netflix-uses-analytics/, 20 Feb 2017.

http://www.internetlivestats.com/internet-users/india/, 20 Feb 2017.

Experian Marketing Services (2016). The 2016 Digital Marketer Survey.

Gomez-Uribe, C. A., & Hunt, N. (2016). The Netflix recommender system: Algorithms, business value, and innovation. *ACM Transactions on Management Information Systems (TMIS) 6*(4), 13.

http://www.business-standard.com/article/companies/flipkart-grows-user-base-to-100-million-116092100216_1.html, 20 Feb 2017.

http://www.huffingtonpost.com/2012/03/27/email-connects-the-world_n_1381854.html, 20 Feb 2017.

Jelinek, F. (1985). Markov source modeling of text generation. In *The Impact of Processing Techniques on Communications*. Springer, Netherlands, pp. 569–591.

Lawrence, L. R., & Juang, B. H. (1993). Fundamentals of speech recognition.

Mark, J., & Charniak, E. (2004). A TAG-based noisy channel model of speech repairs. In *Proceedings of the 42nd Annual Meeting on Association for Computational Linguistics*. Association for Computational Linguistics.

Markov, A. A. (1913). Essai d'une recherche statistique sur le texte du roman "Eugene Onegin" illustrant la liaison des epreuve en chain ('Example of a statistical investigation of the text of "Eugene Onegin" illustrating the dependence between samples in chain'). Izvistia Imperatorskoi Akademii Nauk (Bulletin de l'Acad´emie Imp´eriale des Sciences de St.-P´etersbourg), 7, 153–162.

Octane Marketing (2015). Inbox, Annual State of Marketing in India, Inbox-Report 2015.

Taylor, P., & Black, A. W. (1998). Assigning phrase breaks from part-of-speech sequences. *Computer Speech & Language, 12,* 99–117.

Predicting Customer Churn for DTH: Building Churn Score Card for DTH

Ankit Goenka, Chandan Chintu and Gyanendra Singh

Abstract Customer churn is when the customers either switch from their incumbent service provider to its competitor or stop using the service altogether. Per existing research, it costs five to six times more to acquire customers than to retain existing customers, and this emphasizes the importance of managing churn by organization. In this paper, we build Churn prediction model for one of India's largest Direct to Home (DTH) operator, for its customer base. We use data provided by the DTH operator to build the model. Given the varied base of customers, the data was segmented in smaller homogenous chunks, with similar profile and behaviour. To achieve this, we conducted segmentwise analysis to determine the factors that affect churn differently in different segments. This indicated the need to have different models. Analysis showed that age on the network was the key segmentation driver and hence separate models were built for each segment. We further applied various data mining techniques such as logistic regression, random forest and gradient boosting method to build the model for each segment. Gradient boosting method outperformed both logistic regression and random forest on measures of AUC and Gini. The proposed model correctly classifies churn customers between 76 and 78% depending on the segment. The primary drivers of churn across all the segments for DTH are the age of the customer, type of package subscribed, longest delinquency, maximum amount recharged and maximum valid days for which customer has recharged their set up box. The paper also shows that customer experience while interacting with operator and quality of device are equally important attributes.

Keywords DTH · Churn · Machine learning

A. Goenka · C. Chintu (✉) · G. Singh
Experian Credit Information Company of India Private Limited,
Mumbai 400070, Maharashtra, India
e-mail: Chandan.Chintu@experian.com

A. Goenka
e-mail: Ankit.Goenka@experian.com

G. Singh
e-mail: Gyanendra.Singh@experian.com

A. K. Laha (ed.), *Advances in Analytics and Applications*, Springer Proceedings in Business and Economics, https://doi.org/10.1007/978-981-13-1208-3_9

1 Introduction and Motivation

In India, DTH industry is one of the most competitive markets, with companies facing fierce competition not only from organized market players but also from local vendors who have a substantial market share in their own regions (Bansal 2015). Given the cost incurred in acquiring new customers and long breakeven cycle, it is imperative for the businesses to engage and retain their customers. Customer churn, defined as a customer who switches from their incumbent service provider to either its competitor or stops using the service altogether, further affects the business due to revenue leak. Moreover, once the customer leaves, the organization's ability to cross-sell other products/services to the customer also reduces, leading to further potential business loss. In this context, churn prediction, i.e. the ability to identify customers who are likely to leave in near future, thus becomes an important business tool. Businesses which accurately predict the customers who are likely to churn, and, thereby offer additional benefits and services in an effort to engage and retain those customers, are likely to be more successful.

In this paper, we develop Churn prediction model for one of India's largest DTH operators using the customer data available with the operator. The factors that tend to impact customer churn are derived from customer level attributes as well as company-specific performance measures such as customer service, quality of instruments, etc. This paper also demonstrates results of various data mining techniques on churn prediction and validates results on a subpopulation.

The rest of the paper is organized as follows: Sect. 2 describes the literature review on Churn followed by Sect. 3 that covers the dataset used in the model. Section 4 describes the methodology of churn prediction model followed by Sect. 5 where we produce results of the analysis. In Sect. 6, validation results are presented followed by conclusion.

2 Literature Survey

Churn identification and prediction are important aspects for any business across industry verticals, and, more importantly for communication industry. In academic literature, significant research work has been conducted in churn prediction. Kamalraj and Malathi (2013) provide a survey of different techniques used in the communication sector to analyse the reason for churn and build churn predictive model. Their survey indicates that most of the papers tend to use churn as a binary problem and use neural network, decision tree and regression analysis and cluster analysis techniques for churn prediction with limited application of other techniques. Tsai and Lu's (2010) paper reviews patents along with 21 related studies published from 2000 to 2009 and compares them in terms of the domain dataset used, data preprocessing and prediction techniques considered, etc. Lazarov and Capota (2007) have provided a comparative analysis of all the techniques that are predominantly used in churn prediction techniques such as regression analysis, Naive Bayes, decision trees, neural networks, etc. Portela and Menezes (2009), in their paper, used duration model to show that time to churn can be factored in churn prediction model, making the business target customers at appropriate time. They also showed that the spend variables are a major factor influencing churn. Usage and subscription condition do not seem to affect churn in fixed telecom market of Portugal. Lu, Junxiang made use of customer survival and hazard function to estimate and gain knowledge about customer churn for telecommunications industry.

In review of papers, we find that most of the work done on Churn is not specific to DTH industry and does not capture the dynamics of customer churn for India. The closest literature that we could find in Indian context was the work done in the prepaid telecom services by Rajeswari and Ravilochanan (2014). Their paper provided insights for customer churn behaviour in India's prepaid telecom industry using survey data and regression techniques. Reasons for churn included issues related to technology, network coverage, Internet/data speed and complaint resolution. Furthermore, work by Sharma and Panigrahi (2011) covers churn prediction in wireless cellular services where they used neural network based approach for predicting churn in cellular wireless services. Their model predicted customers who are likely to churn with 66% accuracy. However, the dataset that they used is not India specific and hence limited in application to our study.

In this paper, we have used data of more than 10 million customers to derive Churn prediction model. We have used various data mining and regression techniques to benchmark the performance of model with each other and derive the best model with highest accuracy.

This paper contributes to existing literature by providing insights into factors that affect churn in Indian market. Further segmentation analysis conducted in the paper demonstrates the need to have different models for subpopulations, as the factors that affect churn vary. The paper also validates the model on multiple aspects to demonstrate the robustness and consistency of the model. The sections below cover the dataset considered for churn model development followed by methodology, results and validation.

3 Dataset

Our dataset consists of 14 months of month-on-month information starting from 1 July 2015 to 31 August, 2016. Brief description of the dataset used for churn model is provided below.

- **Subscriber base Dataset**: The dataset contains information on full customer base and covers information such as demographic details (which includes age, district, circle), date of activation, package subscribed, number of DTH connections, number of point of contact, type of set up box (for example, HD, Normal, Recording HD, etc.) and customer class (Home, individual, commercial, hotel, etc.). Additionally, this data also contained information on other products held by the customer along with the existing DTH connection.

- **Payment Dataset**: The dataset provides payment-related information of customer such as date of recharge, amount of recharge, mode of recharge, expiry date of package, etc.
- **SMS Dataset**: The dataset provides information on customer's SMS interaction with DTH service provider. It contains the date and facility availed (upgrade, downgrade, remove package, etc.). This dataset also provides the response sent by service provider.
- **Customer care Dataset**: The dataset contains information service request calls made by the customer to the call centre of the DTH operator. It contains information on number of times a customer interacted with the call centre, type of call (service related, non-service, etc.), source of call (voice, email, escalation, etc.), call centre details (location of call centre), type of issues raise such as video related, HD related, time taken to resolve an issue, etc.
- **Suspension Dataset**: The file contains information about customers who did not recharge the package by due date. The data file indicates the delinquency behaviour of customer.
- **Churn Dataset**: This dataset contains information about the customers who churned and their date of churning. Churn is defined if the customer has not recharged their package for more than 90 days from their recharge due date.

These datasets are our primary source of information and above datasets were further used to generate factors/variables that were used to predict churn.

4 Methodology

4.1 Sampling Strategy and Base Creation

Dependent variable: Our dependent variable is binary, i.e. it takes the value '1' if the customer has churned and '0' otherwise. Churn is defined if the customer has not recharged for more than 90 days from the recharge due date. Please note given that customer can have multiple accounts, i.e. customer may have multiple DTH accounts, churn is defined at account level and not customer level.

Sample design: Our sample design represents design principles used typically in risk scorecard development. Figure 1 depicts the method of variable generation and performance classification. Our observation window is a 6-month period and performance window is of 3 months. For example, for a customer who has due date in the month of January 2016, information from 1 July 2015 to 31 December 2015 (6 months period) is considered as observation window and the period from recharge due date to 30 April 2016 (3 months period) is defined as performance window. All the predictive characteristics are generated based on the information from the observation window and customers are defined as churn or no-churn based on their

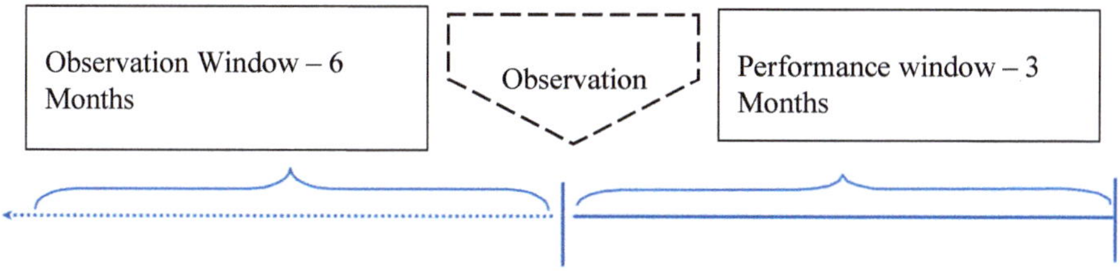

Fig. 1 Character variable generation and sample design process

performance during performance window. A pooled dataset is created by conducting similar analysis data from other months and basis customer recharge due month.

4.2 *Database Creation*

Since the data shared comprised of different files covering customer payment behaviour, demographic information, SMS data record, interaction with customer care, number of past suspension and information regarding other product with par-

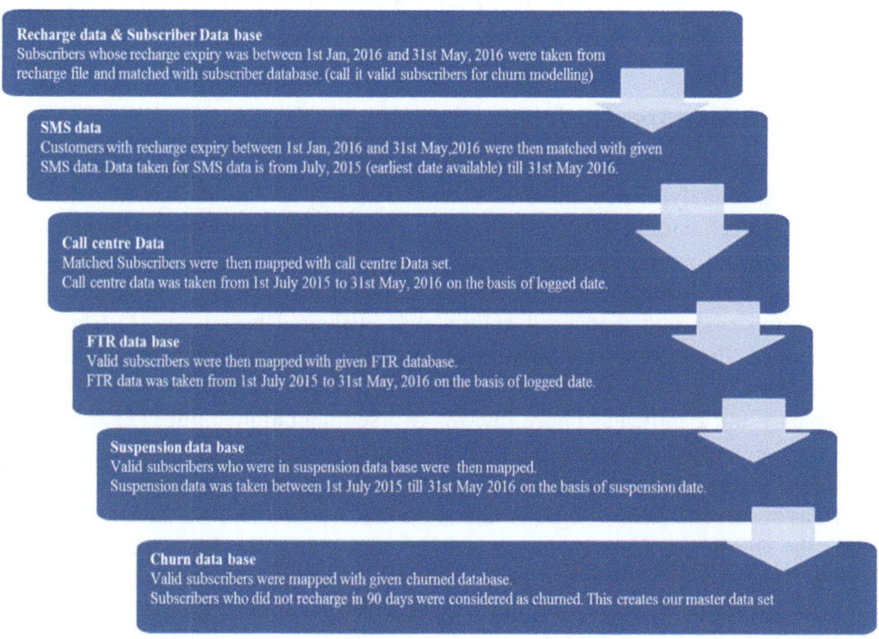

Fig. 2 Data merging step

ent company, master data file has been created for purpose of model development (see Fig. 2).

Churn = f (Demographic details, payment behaviour, SMS record, customer care interaction, number of past suspension and other product information).

4.3 Character Variable Generation

Based on the information available in the dataset, more than 300 predictive factors were created. All factor generation was conducted using the last 6 month's data record of the customer, i.e. the observation period. The broad categories of variables that were created are:

- Payment patterns over a time period
- Demographic profile
- Service requests and resolutions
- Package and set top box information
- Geo profiling
- Mode of recharge (easy recharge, ATM based, IVR, etc.)
- Valid days (Number of days between date of recharge and package expiry date.)
- Preferred recharge (Mode of recharge that customer has used more frequently.)
- Message error (SMS sent by DTH Company in response to customer's SMS.)

4.4 Segmentation Analysis

Segmentation analysis is conducted to identify sub-population/segments that may have distinct behavioural patterns, i.e. either the churn rate is different or the factors that affect churn are different across segments. In this paper, we have used expert segmentation analysis techniques where, first, list of segment splitters was discussed with the business and then each splitter was evaluated based on: (1) Churn rate difference (2) Proportion of population in each segment (3) Difference in factors that impact churn rate.

Tables 1 and 2 indicates churn rate for two-segmentation splitter (1) Age on the network, i.e. number of years spent with current service provider and (2) Nature of subscription plan of customer. Following was the result of our segmentation analysis.

It was observed that the longer the association of a customer with the DTH provider lesser the churn rate. Clearly, we can see that customers who have spent more time with DTH service provider have the lowest churn rate. In terms of duration of package, interestingly customers who have subscribed to annual package have shown a relatively higher churn rate compared to non-annual package subscriber; but the percentage of population in annual subscriber is very thin. Therefore, we have seg-

Table 1 Segmentation on the basis of annual subscription

Annual subscription status		
Attribute bins	% Population (%)	Churn rate (%)
No annual plan	97.81	3.78
Annual plan	2.19	4.36
Total	100.00	3.79

Table 2 Final segmentation on the basis of age on network (AON)

Final segmentation criteria for circle 1			
Age on network (in year)	Attribute bins	% Population (%)	Churn rate (%)
	1 year	11.81	5.78
	1 year–2 year	17.50	3.72
	2 year–4 year	29.16	3.66
	Greater than 4 year	41.53	3.35
	Total	100.00	3.79

mented our dataset into four group based on Age on Network (AON). In the results section, we demonstrate how the factors that affect churn rate across each segment vary indicating the need to perform segmentation on the large customer base.

4.5 Modelling Methodology

4.5.1 Logistic Regression (LR)

LR is one of the most used regression methodologies in binary choice dependent variable problem. The methodology has been used extensively in modelling default risk, marketing propensity models, recovery modelling, etc. The key advantage of the LR is its ease of interpretation and closed form solution of its predicted probabilities.

4.5.2 Random Forest (RF)

Breiman (2001) introduced RF models to overcome the shortcoming of decision tree models. Generally while building a single decision tree for the model, it tends to over-fit the data and hence biased model may be generated. RF overcomes this problem in two ways: first, by sampling data randomly with replacement, and second, by choosing few variables out of all variables for classification at each node of the decision tree.

If the population sample size in 'n', RF creates 'N' sample sets each of size 'n' with replacement: this suggests that many sample points may get repeated in the sample set 1, 2…N. Now RF builds 'N' decision trees with 'N' sample sets. After training and generation of trees, the class of the unknown sample is decided by vote of all trees developed which is essentially the 'Mode' class.

4.5.3 Gradient Boosting Method (GBM)

Boosting algorithm is a machine learning approach that works by combining many weak classifiers to arrive at highly accurate predictor. The weak classifiers are factors whose error rate is only better than random guessing. In GBM, the learning procedure consecutively fits new models to provide a more accurate prediction of the response variable. Once the tree is built, it analyses the accuracy of the prediction. Based on the current accuracy, it finds out areas where the accuracy is not as desired. The newly built tree learns from the mistakes done previously and tries to make the model more robust. GBMs are extremely flexible and have been used in multiple applications. Hastie, Friedman and Tibshirani, in their work, showed that depending upon the application and number of factors, random forest models and GBM model will tend to perform similar. In the situation where there are a large number of variables but the relevant variables are few, random forest model will tend to perform poorly; however GBM model may not get impacted due to it (Hastie et al. 2013).

In this paper, LR, RF and GBM models have been built and evaluated on accuracy measures such as Gini, AUC (see Sect. 6.2) on both training and development dataset. The next section details the results of the model and provides a comparison of different techniques on the churn dataset.

5 Results and Findings

5.1 Comparison of Result at Population Level

Table 3 provides the comparative results of the various techniques on the train and test data. Train and test data has been derived by partitioning the data into two random samples of 80% (Train) and 20% (Test). Gini, which is a measure of accuracy of the model and its ability to discriminate between Churn and non-churn population, indicates that GBM method outperforms both the RF and LR model.

Further, when we evaluate cumulative capture rate (see Fig. 3), which implies how much percentage of good and bad our model can capture given a population level. The graph below represents the percentage of churn capture rate by the total population. For example, the bottom 20% of population (by score) contains almost 60% of the churn. Thus, I may action on that 20% of the population to reduce my churn, as it will be more focused.

Table 3 Model comparison results at population level

Method	Training GINI (%)	Testing GINI (%)
Random forest	49.14	48.82
Gradient boosting method	52.06	51.35
Logistic regression	48.19	49.03

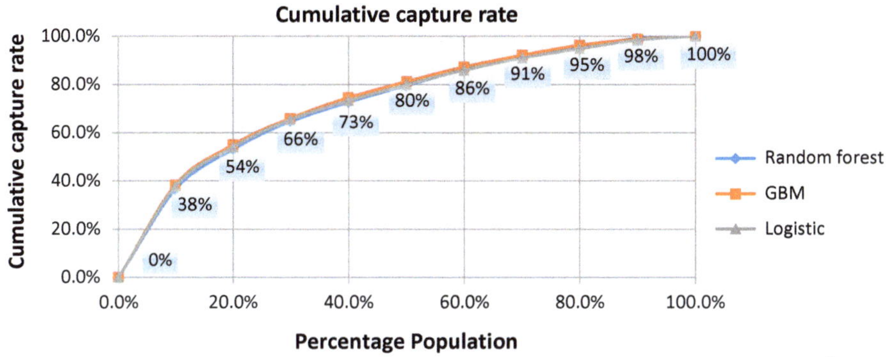

Fig. 3 Cumulative capture rate (ROC) curve at population level

Table 4 Model comparison results at segment level

Results for various prediction techniques for different segment				
Age on network	Parameter	LR (%)	RF (%)	GBM (%)
Less than 1 year	Training GINI	45.03	43.79	*50.37*
	Testing GINI	44.79	43.73	*45.78*
	AUC	72.40	71.87	*74.87*
1–2 year	Training GINI	42.08	44.21	*46.95*
	Testing GINI	44.31	43.59	*48.54*
	AUC	72.16	71.79	*74.27*
2–4 year	Training GINI	45.62	45.44	*47.76*
	Testing GINI	47.22	42.47	*49.44*
	AUC	73.61	71.24	*74.72*
Greater than 4 year	Training GINI	48.70	47.32	*51.03*
	Testing GINI	48.34	49.74	*52.03*
	AUC	74.17	74.87	*76.01*

In Fig. 3, we can see that GBM method marginally outperforms other methods. GBM also shows highest AUC (see Table 4) in comparison to other methods.

5.2 Segment Level Model Result

Population data was divided into subsegment based on the AON of each account. Further, different models were developed using data mining and regression techniques. Segmentwise comparison of results using each of the technique shows that GBM continues to perform better across each segment in terms of accuracy measures. In case of LR and RF, model performance varies across segment. In 'AON less than 1-year' segment, LR models perform better than RF model, however in the 'AON between 1 and 2 year' segment, RF model performs better. In other segments, LR performs marginally better than RF model For the purpose of demonstration of determinants, the rest of the paper results is shown using GBM methodology. The results for other techniques indicate similar results and hence have been excluded from further section.

Please note from the Table 4, performance of each technique across segment varies indicating that inherent drivers across segment vary. In the Table 5, we present top 25 characteristics that were found to have high importance as per GBM model across each segment. As mentioned earlier, not each factor appears in each of the segment, rather some of the variables even though they appeared across segments, their relative important across segment varies.

On further review, we find that five factors that are most predictive in all the four segments are maximum amount recharge, maximum valid days, longest delinquency, type of packages subscribed and age of the customer. In addition to above factors, type of box and manufacturer of box are also important attribute. Manufacturer of box indicates the quality of picture that the customer will be exposed to and if the customer does not find the quality to be good, the customer is likely to quit. Customer interaction at call centre or through SMS is also important predictor, though the impact varies across the segment. Interestingly, customer interaction through SMS mode is far more predictive and consistently predict churn. This is important, as increasingly customers are using non-voice means to interact and if the operator does not provide service to the level expected by customer, the customer is likely to move out of the system.

In addition to above findings, the results also indicate that factors that impact churn are not uniform across segments and hence while managing churn, it is important to understand the underlying subsegment while conducting the analysis. The different factors across segment indicate that the expectation of customer across segment varies and hence organizations will need to have different churn management strategies.

Table 5 List of predictive characteristics across segments

Age on network					
Group	Variables	1 year	1–2 year	2–4 year	GT 4 year
Customer relationship value	Total number of recharges	No	*Yes*	No	No
	Max amount recharged	*Yes*	*Yes*	*Yes*	*Yes*
	Average amount recharged	No	*Yes*	*Yes*	No
Variation in recharge amount	Standard deviation of recharge amount	*Yes*	No	No	No
	Coefficient of variation of recharge amount	No	No	No	*Yes*
Duration of recharge	*Max valid days*	*Yes*	*Yes*	*Yes*	*Yes*
	Minimum valid days	No	No	*Yes*	No
	Average valid days	No	No	No	*Yes*
	Annual package versus others	*Yes*	No	No	No
Variability in recharge duration	*Longest delinquency*	*Yes*	*Yes*	*Yes*	*Yes*
	Coefficient of variation of valid days	No	*Yes*	*Yes*	No
	Percentage delayed recharge	No	No	*Yes*	*Yes*
Mode of recharge	Total number of recharge using ATM	*Yes*	No	No	*Yes*
	Number of activation recharge	*Yes*	No	No	No
	Number of recharge by first preferred mode	*Yes*	No	*Yes*	No
	Total number of recharge using easy recharge	No	*Yes*	No	*Yes*
Cash back offer	Total number of cash back	*Yes*	No	*Yes*	No
	Cash back	No	No	No	*Yes*
SMS error message	Number of error messages send	*Yes*	*Yes*	No	*Yes*
	Number of suspension message received	No	No	*Yes*	No

(continued)

Table 5 (continued)

Age on network					
Group	Variables	1 year	1–2 year	2–4 year	GT 4 year
Suspension related	Total number of suspension	*Yes*	No	No	No
	Number of suspension in last 6 months	No	*Yes*	*Yes*	*Yes*
Manufacturer of STB box	Manufacturer of STB box	*Yes*	No	*Yes*	*Yes*
	MODEL of STB Box	No	*Yes*	*Yes*	*Yes*
Other product of parent company	Company's another product	*Yes*	*Yes*	*Yes*	*Yes*
Type of package subscribed/ viewing behaviour	Base pack type	*Yes*	No	No	No
	Type of package	*Yes*	*Yes*	*Yes*	*Yes*
Demographic details	Customer class (ex. Home, individual, commercial, etc.)	No	*Yes*	No	*Yes*
	Age of customer	*Yes*	*Yes*	*Yes*	*Yes*
Number of DTH connections	Number of DTH connections	*Yes*	No	No	*Yes*
Point of contact	Total number of Point of Contact	*Yes*	No	No	No
Customer care interaction variables	Source of interaction with Call centre	*Yes*	*Yes*	No	No
	Total different source of interaction	No	*Yes*	No	No
	Number of call within 30 days	*Yes*	No	No	No
	Call centre	No	No	*Yes*	*Yes*
	First level issue	No	*Yes*	No	*Yes*
	Second level issue	No	No	*Yes*	No
	Issues related to STB Box	No	No	*Yes*	No

Table 6 Model validation results across circle

Gini value for model validation for different circles				
Age on network (AON)				
Circle	Less than 1 year (%)	1–2 year (%)	2–4 year (%)	Greater than 4 year (%)
Circle 1	55.6	47.7	50.6	52.1
Circle 2	56.6	50.2	53.3	54.1
Circle 3	44.3	50.0	52.7	54.9

6 Validation Result

It has been well established in research and businesses that customer purchasing and spending behaviour vary widely based on different months, geography, etc. For example, during April–May, DTH subscription generally tends to be higher because of IPL matches and churn increases soon after as the event ends. Similarly, during exam months such as February and March, churn rate increases, as the parents do not renew the package owing to examination performance concerns. To address the concern, in addition to validation of model on train and test data, further subsegment validation of the model was conducted. Below section presents results for circle and monthwise validation.

6.1 *Circlewise*

Following are the test result based on GBM for three different circles across four segments (Table 6).

From the Gini table, we can see that model performs consistently across most of the circles, except for the segment 'AON Less than 1 year' for Circle 3 where we find significant drop in Gini (44.3%). We have also presented cumulative capture rate for three circles and for all the four segments on which we have built our model in the Appendix section (Please see appendix section "Validation Table Circle Wise" for more details).

6.2 *Monthwise*

Further to check the consistency of model across months, we have tested our model for different months and results are shown in Table 7.

Except for some large jump in first segment (AON = Less than one year), we can see that Gini has remained consistent for all the month. Cumulative capture rate for different months and AON segment are provided in the Appendix section (Please

Table 7 Model validation results across month

Gini value for model validation on circle				
Age on network (AON)				
Data time period (based on due date)	Less than 1 year (%)	1–2 year (%)	2–4 year (%)	Greater than 4 year
Training Gini (Jan–May 2016)	53.7	48.1	52.5	52.1
Testing Gini (Jan–May 2016)	54.9	49.6	52.4	53.3
Gini for January 2016	69.1	54.2	58.5	56.6
Gini for February 2016	60.5	55.6	58.6	57.5
Gini for March 2016	53.3	53.6	55.3	53.9
Gini for April 2016	52.9	49.3	52.2	52.9
Gini for May 2016	51.7	46.7	49.0	50.7

see appendix section "Cumulative Capture Rate (ROC) Curve at Segment Level for Different Circles" for more details).

7 Conclusion

Churn leads to significant loss of revenue for companies and given the cost of acquiring new customers, it is important to have a proactive customer churn management. Churn prediction model provides an important tool to organization to identify customers who are likely to churn thereby providing opportunity to the organization to engage and retain the customer. The current paper provides framework for developing Churn model for a DTH customer base. The key aspects of the model are providing insights into factors likely to affect churn and performance of various data mining techniques. In addition, the paper demonstrates for large customer base that conducting segmentation analysis allows to understand factors that have differential impact across segment, thereby enabling organization to have different strategies and look at different drivers. The model accurately classifies between 76 and 78% churner across segment. The model framework and results are also consistent across circle and are able to capture seasonality in churn rates.

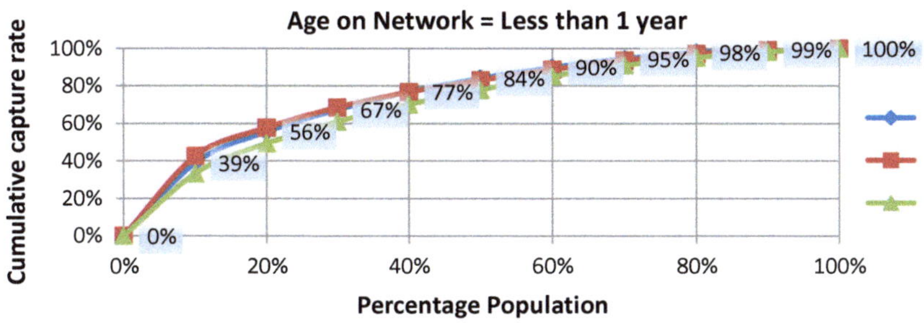

Fig. 4 Cumulative capture rate (ROC) curve at segment level: AON less than 1 year

As a future scope of work, alternative modelling techniques such as neural network, SVM models can be explored. While these techniques have been applied in other papers, but using the same for DTH industry will probably give more insights. Further, building churn model is the first step of the customer churn management, and proactive churn management requires further development of segment-specific business strategies and customer offers that will induce the customer to stay with the player. An extension of the paper will be work done on defining framework for customer retention basis the model proposed. This should cover the aspect of defining personalized offers, best channel of contact, etc.

Appendix

Validation Table Circlewise

See Table 8.

Cumulative Capture Rate (ROC) Curve at Segment Level for Different Circles

See Figs. 4, 5, 6 and 7.

Table 8 Cumulative capture rate versus population across circle

Cumulative capture rate at different percentage population (circlewise)

% Population

AON	Circle	10%	20%	30%	40%	50%	60%	70%	80%	90%	100%
Less than 1 year	Circle 1	39%	56%	67%	77%	84%	90%	95%	98%	99%	100%
	Circle 2	43%	58%	69%	77%	83%	89%	94%	97%	99%	100%
	Circle 3	33%	49%	61%	70%	78%	85%	91%	95%	98%	100%
1 year–2 year	Circle 1	37%	53%	64%	73%	80%	87%	92%	96%	99%	100%
	Circle 2	38%	54%	65%	74%	81%	87%	92%	95%	99%	100%
	Circle 3	36%	54%	66%	75%	82%	88%	93%	96%	99%	100%
2 year–4 year	Circle 1	38%	55%	66%	74%	81%	87%	92%	96%	99%	100%
	Circle 2	41%	57%	68%	76%	83%	88%	92%	96%	99%	100%
	Circle 3	39%	56%	68%	77%	83%	89%	94%	97%	99%	100%
Greater than 4 year	Circle 1	39%	56%	67%	77%	84%	90%	95%	98%	99%	100%
	Circle 2	43%	59%	68%	76%	82%	88%	92%	96%	99%	100%
	Circle 3	43%	59%	70%	77%	84%	89%	93%	97%	99%	100%

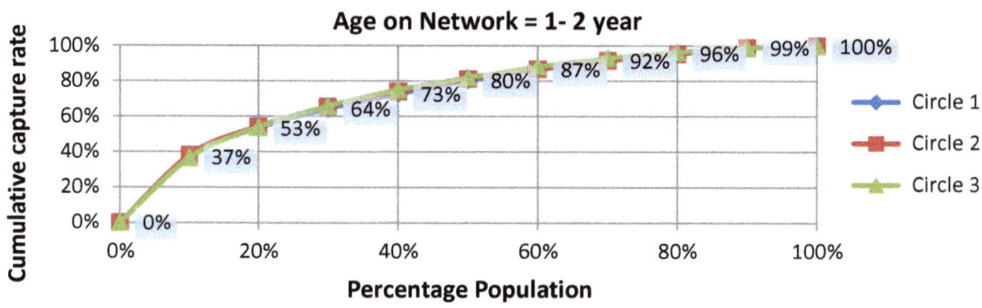

Fig. 5 Cumulative capture rate (ROC) curve at segment level: AON between 1 and 2 year

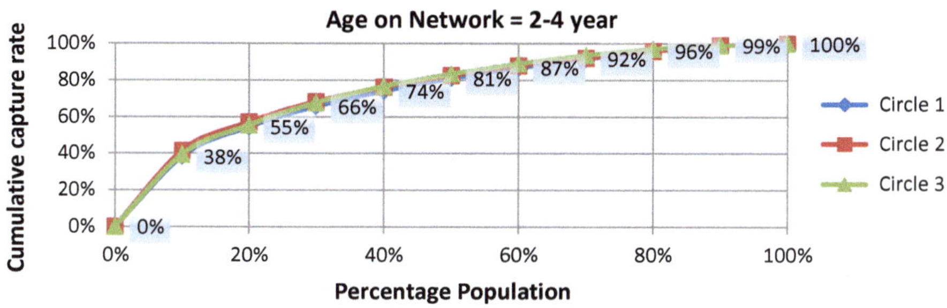

Fig. 6 Cumulative capture rate (ROC) curve at segment level: AON between 2 and 4 year

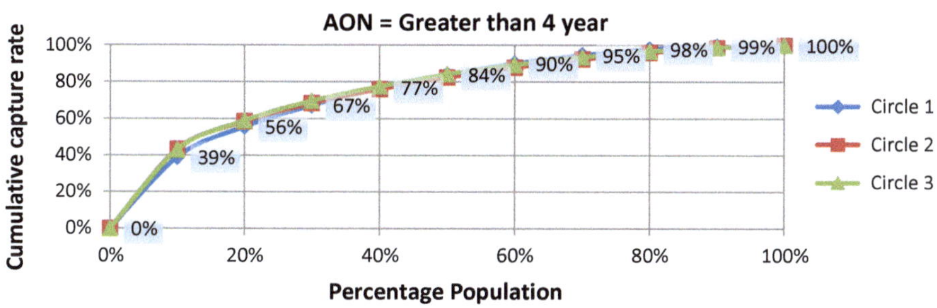

Fig. 7 Cumulative capture rate (ROC) curve at segment level: AON greater than 4 year

Validation Table Monthwise

See Table 9.

Table 9 Cumulative capture rate versus population across month of recharge

Cumulative capture rate at different percentage population (monthwise)											
Percentage population											
AON	Month	10%	20%	30%	40%	50%	60%	70%	80%	90%	100%
Less than 1 year	January 16	56%	69%	78%	85%	91%	95%	97%	99%	100%	100%
	February 16	46%	61%	71%	79%	86%	91%	96%	98%	99%	100%
	March 16	39%	54%	65%	75%	82%	88%	94%	97%	99%	100%
	April 16	36%	53%	65%	75%	82%	88%	93%	97%	99%	100%
	May 16	35%	53%	65%	74%	82%	88%	93%	97%	99%	100%
1 year–2 year	January 16	42%	57%	68%	76%	83%	88%	94%	98%	100%	100%
	February 16	42%	57%	69%	78%	84%	90%	95%	98%	100%	100%
	March 16	42%	57%	67%	76%	83%	88%	92%	97%	99%	100%
	April 16	38%	54%	64%	73%	80%	86%	91%	95%	98%	100%
	May 16	37%	53%	62%	71%	78%	84%	89%	94%	98%	100%
2 year–4 year	January 16	44%	60%	71%	80%	86%	91%	96%	99%	100%	100%
	February 16	44%	60%	71%	80%	86%	91%	96%	98%	100%	100%
	March 16	42%	58%	68%	77%	84%	89%	94%	97%	99%	100%
	April 16	40%	56%	67%	75%	81%	87%	92%	95%	98%	100%
	May 16	38%	54%	65%	73%	80%	85%	90%	95%	99%	100%
Greater than 4 year	January 16	43%	59%	69%	78%	85%	90%	95%	98%	100%	100%
	February 16	44%	59%	70%	78%	85%	91%	95%	98%	100%	100%
	March 16	42%	57%	67%	76%	82%	88%	93%	97%	99%	100%
	April 16	42%	57%	67%	75%	82%	87%	92%	96%	99%	100%
	May 16	41%	55%	65%	73%	80%	87%	91%	96%	98%	100%

Cumulative Capture Rate Curve at Segment Level for Different Months of Recharges

See Figs. 8, 9, 10 and 11.

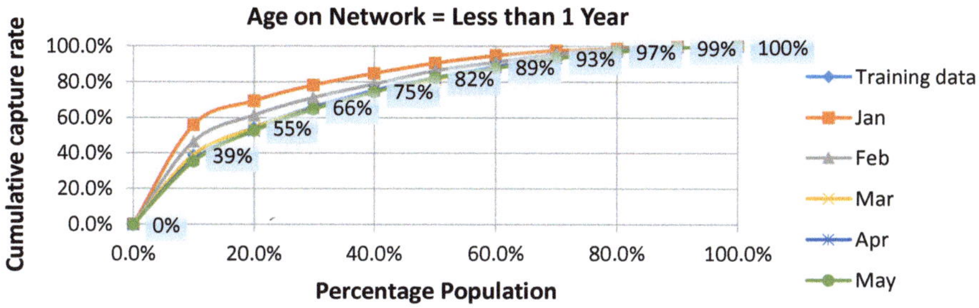

Fig. 8 Cumulative capture rate (ROC) curve at segment level: AON less than 1 year

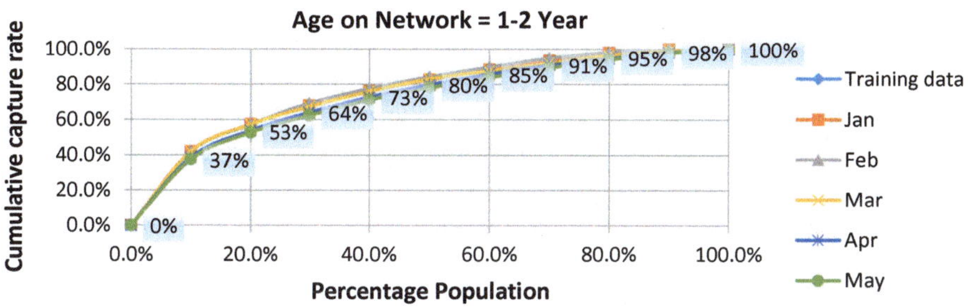

Fig. 9 Cumulative capture rate (ROC) curve at segment level: AON between 1 and 2 year

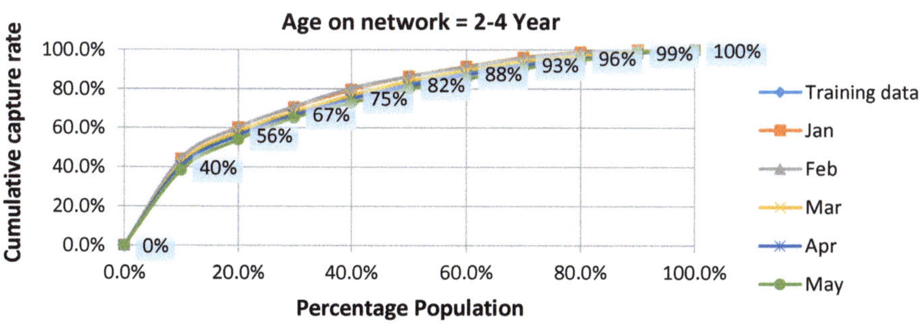

Fig. 10 Cumulative capture rate (ROC) curve at segment level: AON between 2 and 4 year

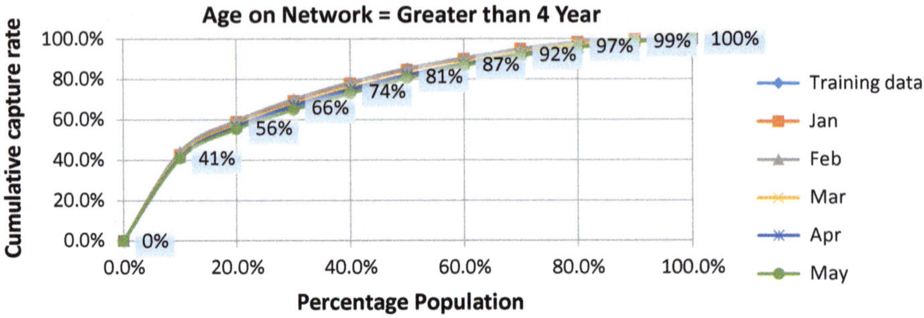

Fig. 11 Cumulative capture rate (ROC) curve at segment level: AON greater than 4 years

References

Bansal, S. (2015). *What's eating DTH operators*. LiveMint, May 07, 2015.

Breiman, L. (2001). Random forests. *Machine learning*, *45*(1), 5–32.

Hastie, T., Friedman, J., & Tibshirani, R. (2013). *The elements of statistical learning* (1st ed., pp. 359–361). New York: Springer.

Kamalraj, N., & Malathi, A. (2013). A survey on churn prediction techniques in communication sector. *International Journal of Computer Applications* (0975 – 8887), *64*(5).

Lazarov, V., & Capota, M. (2007). *Business analytics course*. TUM Computer Science.

Portela, S., & Menezes, R. (2009). *Modelling customer churn: An application of duration models*. Portugal: Department of Quantitative Methods, ISCTE Business School.

Rajeswari, P. S., & Ravilochanan, P. (2014). An empirical study on customer churn behaviour of indian prepaid mobile services. *Middle-East Journal of Scientific Research, 21*(7), 1075–1082.

Sharma, A., & Panigrahi, P. K. (2011, August). A neural network based approach for predicting customer churn in cellular network services. *International Journal of Computer Applications* (0975 – 8887), *27*(11).

The Indian Telecom Services Performance Indicators. Telecom Regulatory Authority of India, January–March 2016.

Tsai, C.-F., & Lu, Y.-H. (2010). Data mining techniques in customer churn prediction. *Recent Patents on Computer Science*, *3*, 28–32.

Applying Predictive Analytics in a Continuous Process Industry

Nitin Merh

Abstract In this paper an attempt is made to develop data driven models on pilot data set for predicting fault in machines of continuous process industry on various selected attributes using techniques of Multiple Linear Regression Model (MLR), Regression Tree (RT) and Artificial Neural Networks (ANN). Association rules are also derived from the available data set. Efforts are also made to predict total shutdown time of machines. These machines are used for manufacturing components machined for Heavy Commercial Vehicles (HCV), Light Commercial Vehicles (LCV), Multi Axle Vehicle (MA) and Tractors. To check the robustness of models a comparison is made between the results derived from various techniques discussed above. Performance evaluation is done on the basis of the errors calculated between the actual and predicted values of down time. Based on actual and predicted results various error scores are calculated to evaluate best model and check robustness of the models under study. Training and validation of the model is done using datasets collected from a manufacturing unit located at Pithampur industrial area near Indore, Madhya Pradesh, India. In the current paper an association is also developed between the attributes and occurrence of the fault. The developed model will be used on the bigger data set which will help the stakeholders of the organization for smooth functioning of the unit and for better governance in the organization. XLMiner is used for model development and simulations. After analysis results show that ANN, RT and Association Rule techniques are capable of capturing the data set.

Keywords Artificial neural network (ANN) · Regression tree (RT)
Association rule

N. Merh (✉)
Information Technology, SVKM'S NMIMS (Deemed to be University),
Indore 452001, Madhya Pradesh, India
e-mail: nitinmerh0812@gmail.com

A. K. Laha (ed.), *Advances in Analytics and Applications*, Springer Proceedings
in Business and Economics, https://doi.org/10.1007/978-981-13-1208-3_10

1 Introduction and Literature Review

Fault prediction mainly deals with the fault that is likely to happen in the system on the basis of past and current states of the system. Fault prediction has attracted considerable attention across the world due to the growing demand for higher operational efficiency, safety in industrial systems and scheduling of shutdowns.

Modeling, data mining and machine learning are among the few areas of study for predictive analytics. Various statistical techniques are used for their implementation that analyze past data to make future prediction. Prediction helps organization in making right decision at right time by right person as there is always time lag between planning and actual implementation of the event. In a continuous work-flow or continuous process all outputs are treated similar. In such a case the process itself is divided in separate operations. Each unit flows among these operations, individually. In such kind of system the manufacturing of the standard products is carried out at a fixed rate. The mass production is carried on continuously for stock in anticipation of demand.

Principal component analysis (PCA) is a multivariate technique that analyzes data sets in which several inter-correlated quantitative variables exist. PCA is a mathematical technique which tries to find a set of uncorrelated variables among several correlated variables. Main goal of PCA is to extract important information from data set and to represent it as a set of new uncorrelated variables.

The PCA aims at reducing the number of variables of the dataset which define the dimensionality of the system, but the original variability in the data is retained and the complexity is reduced.

Thus PCA mainly explains the variance- covariance factor of a high dimensional system using only a set of few linear combinations of actual component variables.

Multiple Linear Regression (MLR) is one of the several prediction techniques used. It is applied on the dataset to understand the relationship between response and predictor variables or prediction of the response based on input variables. MLR uses following linear model for the selected dataset:

$$Y = \alpha_0 + \alpha_1 X_1 + \alpha_2 X_2 + \cdots + \alpha_m X_m + \varepsilon$$

where Y is the dependent or response variable. X_i : for $i = 1$ to m; represents the independent or predictor variables, α_0 is intercept term and ε is the random error. α_i: for $i = 1$ to m are regression coefficients.

The input variables and predicted variables are main components of a decision tree. The nodes of the tree represent a test performed on an input variable and the prediction variables are the terminal nodes of the decision tree. A regression tree can be viewed as an adoption taken from decision trees.

Learning and pattern recognition problems can be solved by the use of Artificial Neural Networks (ANN). The learning process of ANN helps to find meaningful patterns in data (Afolabi and Olude 2007). Approximation of unknown functions, with no assumptions being made for the distribution of data, can be done by ANN, that to for a desired accuracy (Sexton and Sikander 2001). Approximation of both types, linear and non-linear functions can be done by ANN, resulting in achievement of good performance. Because ANN learn and follow a non-parametric approach, they have gained popularity (Dacha 2007; Rimpley 1996).

The current study aims to find the predictive models for fault detection in the machines of continuous process industry. Selected predictive model/s will help stakeholders to take right decision at right time. It will also help in scheduling planned shutdowns and selecting key attributes responsible for forced or unscheduled shutdowns.

Indore CNC Pvt. Ltd. is a manufacturing unit located in Pithampur, Indore is selected as the organization under study. It is manufacturer of gear boxes which are supplied to heavy commercial vehicles, light commercial vehicles, multi axle vehicle and tractors manufactures located in Madhya Pradesh and other parts of the country. Due to unscheduled shutdowns/breakdowns inventory management, manpower planning and finances have suffered a lot. Scheduled delivery of the finished product to the clients is also hampered, which creates a bad name to the organization.

Samantha and Al-Balushi (2003) and Kankar et al. (2011) have demonstrated use of ANN for diagnosing faults in the manufacturing of rolling element bearings. The inputs which are used for ANN are time domain vibration signals of all bearings normal or defective used in the rotating machinery.

Artificial Intelligence (AI) has the capability to learn and acquire knowledge from facts, data and principles, which is then applied to a process. This capability of AI is used in engineering applications thus attracting many researchers and practitioners.

Key objective of the study is to construct predictive models for predicting main attributes of fault detection in continuous process industry. Next reason for undertaking this study is to compare trends and results of actual and predicted value generated by various models and finding the best model understudy, to find out key factors responsible for the unscheduled shutdowns and to prescribe actions to be taken to reduce unscheduled shutdowns, to find out which types of errors occur together. Last objective is performance evaluation of models by statistical methods and by calculating and comparing various errors.

Methodology Used

In the current study an effort is made to develop models using predictive methods like Regression Tree (RT) and Artificial Neural Network (ANN) for predicting fault detection in a continuous process industry. Primary data for analysis is collected from Indore CNC located at Pithampur industrial area near Indore, Madhya Pradesh (M.P.).

At the first stage data is preprocessed, transformed, missing values are handled, outliers are identified and handled, data normalization and principal components are selected. After transformation of the data various selected techniques are used for predicting the fault and models are created.

Convergence, robustness and model evaluation has been done on the basis of the simulation results obtained by XLMiner. After the development of models using MLR, NN and RT the comparison of various forecasting errors have been calculated.

2 Data

Input data for all methods for developing predictive modeling is collected from manufacturing unit located at Pithampur, Indore, M.P., India for a period during 01.04.2015 to 30.09.2015. The collected data is used for the pilot study and on the basis of the results and inferences generated same can be applied for the larger dataset. Total 155 days sample is collected during period of 01.04.2015 to 30.09.2015. Few days were dropped due to holidays, shutdowns or when data was not generated. Data was collected for Tongtai-1 CNC machine used for manufacturing gear boxes. Time loss data was collected under various heads and attributes.

Pre-processing and Normalization of Data
Final dataset has been prepared after removing attributes having 0 values (No time loss), unary values. Attributes selected from Table 1 for developing model are as follows (Table 2).

Random partitioning has been done on the data set where 60% data is for training and 40% for validation.

For association rule analysis data set is converted in binary format where zero (0) represents non-occurrence of time loss and one (1) represents occurrence of time loss.

3 Data Analysis and Results

Main objective of the study is to find suitable predictive data driven model using various techniques. Neural Networks, Regression Tree, Multiple Linear Regression and Association Rule Mining are used to fit the data for developing the model. After developing models from above mentioned techniques, model with best results can be selected for final deployment on the bigger data set generated by not only Tongtai-1 machine but also on Taknio 86C and Hyundai machines which are installed in Indore CNC. A comparison can also be made on the performance on all machines based on the results generated.

Results of data analysis are as follows:

Dependent variable—Total (It is derived from calculating total down time occurred during a day).

Independent variables—Air Pressure Low Loss Time, APC Loss Time, ATC Loss Time, Magazine Problem Loss Time, Operator Door Problem Loss Time, Electrical Problem Loss Time, Power Cut Loss Time, Tool Broken Loss Time, Tool Fall Down Loss Time, Tool Grinding Loss Time, Coolant Rust Problem Loss Time, Fixture Work Loss Time, Setting Time Loss Time, House Keeping Loss Time, Insert Change Loss Time, Offset Given Loss Time, Spindle Chips Cleaning Problem Loss Time, Tool Proving Time Loss Time, Gauges Problem Loss Time, Inspection Time Loss Time, Rework Loss Time, Tools Not Available Loss Time, Insert Not Available Loss Time,

Table 1 Dimension wise attributes for Tongtai-1 machine

Mechanical	Electrical	Electronics or system	Age of machine	Human intervention	Training or efficiency	Quality	Planning
Air pressure low	Electrical problem	Sensor problem	Running time tool broken	Coolant rust problem	Bore adjust	Casting problem	Coolant problem
APC problem	Power cut	Servo alarm	Spindle over load problem	Data not filled		Excess load	Hydraulic oil low
ATC problem	Lubrication alarm	Spindle drive problem		Fixture work	House keeping	Gauges problem	Insert not available
B Axis clamping delay		X Axis drive problem	Tool broken	Operator late coming	Insert change	Inspection time	Load delay
Conveyor not working		Y axis drive problem	Tool fall down	Setting time	Lathe insert change	Lathe inspection	No helper
Coolant leakage		Z axis drive problem	Tool grinding		Machine cleaning	PDI work	No operator
Crane problem					Offset given	Rework	Production meeting
Magazine problem					Spindle chips cleaning problem	Tools not available	Single pallet
Mechanical problem					Tool proving time		Stud problem
Oil leakage							Tool problem
Operator door problem							
Telescopic cover							

Load Delay Loss Time, No Operator Loss Time, Stud Problem Loss Time, Servo Alarm, Spindle Drive Problem.

Variables dropped due to invalid inputs/unary values—Casting Problem Loss Time, X Axis Drive and Timeloss.

Results:
Artificial Neural Network:
Following parameters of NN were used for designing NN model (Tables 3, 4 and 5).

Table 2 Final selected attributes for Tongtai-1 machine

1	Air pressure low loss time	16	Offset given loss time
2	APC loss time	17	Spindle chips cleaning problem loss time
3	ATC loss time	18	Tool proving time loss time
4	Magazine problem loss time	19	Casting problem loss time
5	Operator door problem loss time	20	Gauges problem loss time
6	Electrical problem loss time	21	Inspection time loss time
7	Power cut loss time	22	Rework loss time
8	Tool broken loss time	23	Tools not available loss time
9	Tool fall down loss time	24	Insert not available loss time
10	Tool grinding loss time	25	Load delay loss time
11	Coolant rust problem loss time	26	No operator loss time
12	Fixture work loss time	27	Stud problem loss time
13	Setting time loss time	28	Servo alarm
14	House keeping loss time	29	Spindle drive problem
15	Insert change loss time	30	X axis drive

Table 3 Artificial neural network parameters

Parameters	
If input variables are normalized	Yes
Type of network architecture	Manual
Initial weights used/seed value	12,345
Number of hidden layers	1
Number of nodes in hidden layer 1	25
Number of epochs	30
Step size for gradient descent	0.1
Weight change momentum	0.6
Error tolerance	0.001
Weight decay	0
Cost function	Sum of squares
Activation function used at hidden layer	Standard
Activation function used at output layer	Standard

Table 4 Summary report of ANN training data set

Total sum of squared errors	Root mean square error	Average error
37887.84548	20.18406	4.336432

Table 5 Summary report of ANN validation data set

Total sum of squared errors	Root mean square error	Average error
423250.3804	82.62336	10.84254

Table 6 Regression tree parameters

Parameters	
If input variables are normalized	Yes
Minimum number of records in a terminal node	9
Maximum number of levels displayed in tree drawing	7
Do you want to draw full-grown tree?	Yes
Do you want to draw best-pruned tree?	Yes
Do you want to draw minimum-error tree?	No
Do you want to draw user-specified tree?	No

Primary investigation of the errors shows that NN model is capable to capturing the data set. It indicates that this model can be used on the bigger data set. Error is very less and most of the data set points are predicted correctly.

Regression Tree:

Following parameters of RT were used for designing regression tree (Tables 6, 7 and 8).

Good number of rules are generated with respect to validation pruned tree which will help in making decision regarding fault detection. Following are two key rules generated from regression tree:

Rule 1:

IF (sprindledriveproblem $\leq$ 620 AND (settingtimelosstime $\leq$ 53) AND (rework-losstime $\leq$ 220) AND (toolsbrokenlosstime $\leq$ 54.83) AND (inspectiontimeloss $\leq$ 12.50) AND (nooperatotlosstime $\leq$ 35) Then Down Time = 69.73.

Table 7 Summary report of training data using fully grown regression tree

Total sum of squared errors	Root mean square error	Average error
162489.2	41.79947	5.11896E−15

Table 8 Summary report of validation data using fully grown regression tree

Total sum of squared errors	Root mean square error	Average error
409415.2	81.26174	8.017533753

Table 9 Input data for association rule

Details of input data set	
Number of transactions in input data	155
Number of columns in input data	30
Number of items in input data	30
Number of association rules	12
Minimum support for association	16
Minimum confidence percentage	50.00%

Rule 2:

IF (sprindledriveproblem $\leq$ 620 AND (settingtimelosstime $\leq$ 53) AND (rework-losstime $\leq$ 220) AND (toolsbrokenlosstime $\leq$ 54.83) AND (inspectiontimeloss $\leq$ 12.50) AND (nooperatotlosstime > 35) Then Down Time = 169.62 (With sub tree beneath).

From the above regression tree rules its can be concluded that down time can be reduced or avoided if necessary maintenance preventive measures can be taken. Other rules can also be derived and interpreted to reduce or avoid down time.

Association Rule:

Inputs:

See Table 9.

After applying association rule on the data set total twelve rules were generated. Following are the rules generated (Table 10).

Above table indicates that Consequents (C) are Servo Alarm and Insert Change Loss Time with various combinations of Antecedent (A). Prior maintenance of machine and man power training measures to be when any alarm is generated from any attribute so that machine down time can be avoided.

4 Findings and Interpretation of Results

Artificial Neural Networks—Model generated using given set of parameters, validation error and actual and predicted chart indicates that model is capable of capturing the inferences in the pilot data set (Tables 4 and 5 and Fig. 1). Hence we can conclude that it can be used on a larger dataset.

Regression Tree—Using this technique it is observed that machine is down primarily due to attributes in the order starting from root node spindle drive problem, setting time loss time, power cut loss time, rework loss time, tool broken loss time, inspection time loss, no operator loss time and insert changing loss time (Tables 7 and 8, Figs. 2 and 3). In case of reducing the unscheduled shut downs these attributes are to be controlled, maintenance and manpower is scheduled.

Table 10 Association rule

Confidence %	Antecedent (A)	Consequent (C)	Lift ratio
50.00	Tool grinding loss time	Servo alarm	1.192308
50.00	Tool grinding loss time and insert change loss time	Servo alarm	1.192308
51.61	Insert change loss time and inspection time loss time	Servo alarm	1.230769
83.78	Inspection time loss time	Insert change loss time	0.947919
88.89	Setting time loss time	Insert change loss time	1.005677
88.89	No operator loss time	Insert change loss time	1.005677
89.23	Servo alarm	Insert change loss time	1.009545
89.29	Offset given loss time	Insert change loss time	1.010167
92.00	Power cut loss time	Insert change loss time	1.040876
94.12	Inspection time loss time and servo alarm	Insert change loss time	1.064835
94.44	Tool grinding loss time	Insert change loss time	1.068532
94.44	Tool grinding loss time and servo alarm	Insert change loss time	1.068532

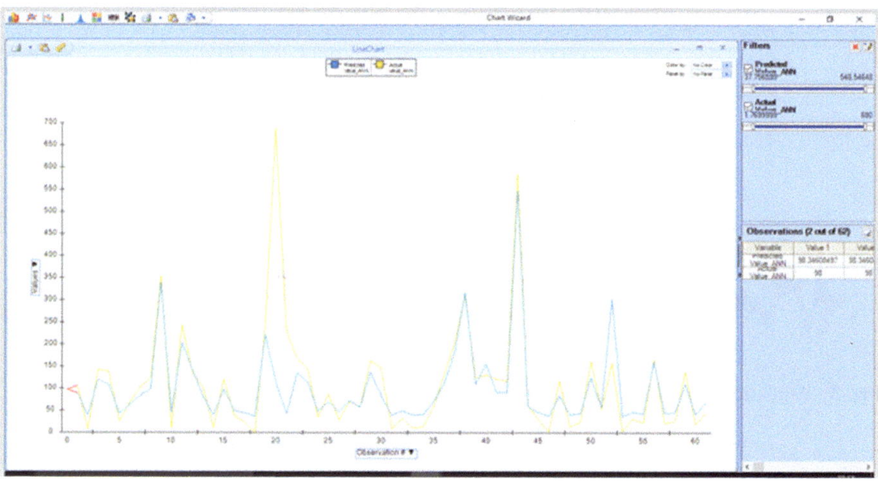

Fig. 1 Validation score actual and predicted values of total down time

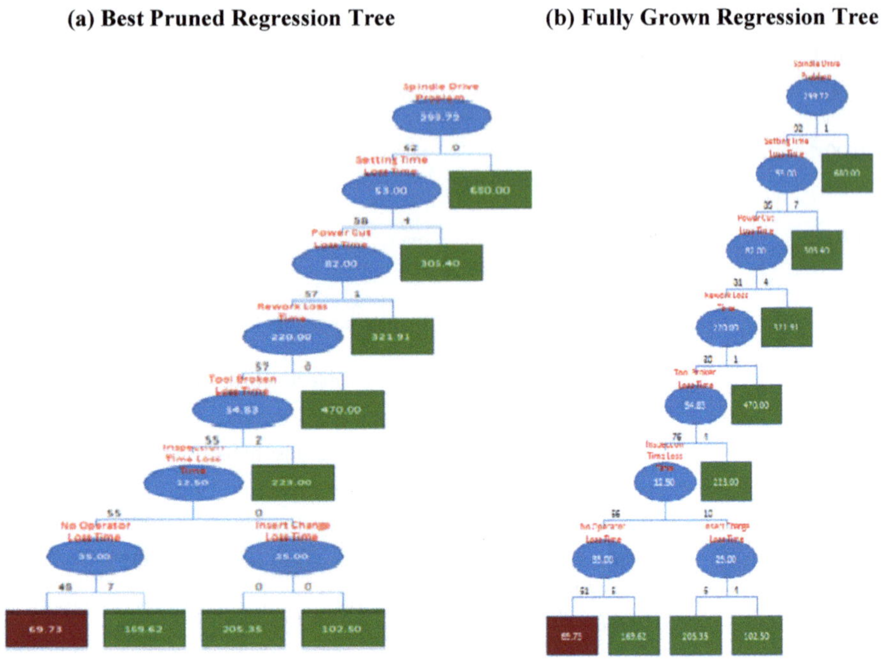

Fig. 2 **a** Best pruned regression tree, **b** fully grown regression tree

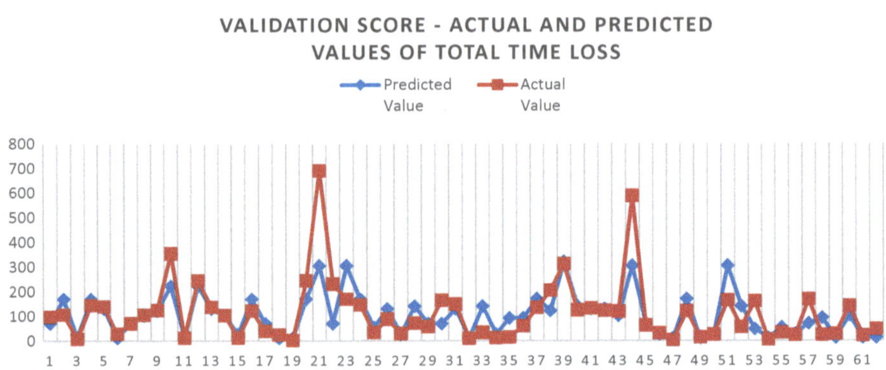

Fig. 3 Validation score—actual and predicted values of total down time using RT

Association Rule—After observing various rules generated by association rule with different antecedents servo alarm and insert change loss time are the consequents with a confidence percent between 50.00 and 94.44%. In all cases except one lift ratios are higher than 1 which indicates that rules can be accepted for decision making. This indicates that which type of loss time (error) are kept together in a basket (Table 10).

5 Conclusion

Applying various predictive techniques mentioned above it is observed that RT, NN and association rule are capable to predict and generate some meaningful results but MLR modeling technique was not able to predict due to over fitting problem and large set of unary values. It reflects that the data collected for the pilot study is not sufficient but same techniques will certainly generate good results when applied on larger data sets.

References

Afolabi, M. O., & Olude, O. (2007). Predicting stock prices using a hybrid Kohonen self organizing map (SOM). In *Proceeding of 40th International Conference on Systems Sciences* (pp. 1560–1605).

Dacha, K. (2007). Casual modeling of stock market prices using neural networks and multiple regression: A comparison report. *Finance India, 21*(3), 923–930.

Kankar, P. K., Sharma, S. C., & Harsha, S. P. (2011). Fault diagnosis of ball bearings using continuous wavelet transform. *Applied Soft Computing, 11,* 2300–2312.

Rimpley, B. D. (1996). *Pattern recognition and neural networks*. Cambridge University Press.

Samantha, B., & Al-Balushi, K. R. (2003). Artificial neural networks based fault diagnostics of rolling element bearings using time domain features. *Mechanical Systems and Signal Processing, 17,* 317–328.

Sexton, R. S., & Sikander, N. A. (2001). Data mining using a genetic algorithm—Trained neural network. *International Journal of Intelligent Systems in Accounting, Finance and Management, 10*(4), 201–210.

Part III
Machine Learning Applications

Automatic Detection of Tuberculosis Using Deep Learning Methods

Manoj Raju, Arun Aswath, Amrit Kadam and Venkatesh Pagidimarri

Abstract In this paper, we present a deep learning based approach for automatically detecting tuberculosis manifestation from chest X-ray images. India is the country with the highest burden of tuberculosis. A chest radiograph in symptomatic patients is used to diagnose active tuberculosis. This screening method is ideally done at the primary health care centres where a clinician is available and sometimes through mobile X-ray unit. The major challenge for this method of screening is timely reporting and further follow-up of patient for initiation of treatment. We built multiple convolutional neural networks, the state-of-the-art deep learning algorithm, to build the model for automatic tuberculosis diagnosis. We classified the chest X-rays into two categories, namely, tuberculosis presence and tuberculosis absence. The dataset used to train the model contained 678 images, having 340 normal chest X-rays and 338 chest X-rays with tuberculosis manifestation. The validation dataset contained 235 images, which observed a sensitivity of 84.91% and a specificity of 93.02%. This demonstrates the potential of convolutional neural networks to automatically classify chest X-rays in real time.

Keywords Convolutional neural network · Machine learning · Artificial intelligence · Pattern recognition · Computer vision

M. Raju (✉) · A. Aswath · A. Kadam · V. Pagidimarri
Enlightiks Business Solutions, Bangalore, India
e-mail: manoj.raju@enlightiks.com

A. Aswath
e-mail: arun.a@enlightiks.com

A. Kadam
e-mail: kadam.amrit@enlightiks.com

V. Pagidimarri
e-mail: venkatesh@enlightiks.com

A. K. Laha (ed.), *Advances in Analytics and Applications*, Springer Proceedings in Business and Economics, https://doi.org/10.1007/978-981-13-1208-3_11

1 Introduction

Tuberculosis (TB) is an infectious bacterial disease which spreads through air and originates from a bacterium named Mycobacterium tuberculosis. It spreads when an uninfected person has a prolonged exposure with an infected person. Major manifestations of TB are persistent cough, coughing up blood, fatigue, weight loss, loss of appetite, etc. As per World Health Organization's (WHO's) Global TB Report 2016, new TB cases rose to 10.4 million worldwide, comprising 56% men, 34% women and 10% accounted for children. India reports more than a quarter of the world's TB cases and deaths (World Health Organization, WHO 2016).

Mycobacterium TB is prone to develop resistance to disinfectants when the patient is not cured first time. The continuous increase in the multidrug resistant TB (MDR-TB) patients is due to limited and expensive treating options. WHO estimated 480,000 new MDR-TB cases and in addition, 100,000 people with rifampicin-resistant TB were also diagnosed (World Health Organization, WHO 2016). India, China and the Russian Federation accounted for 45% of the 580,000 cases. The main reason for multidrug resistance to emerge is due to mismanagement of TB treatment.

A supreme laboratory environment possessing contemporary diagnostics is a prerequisite for the early and precise detection of TB and drug resistance. Many countries lack these provisions, which has led to rapid increase in the number of infected patients. The obvious test for TB detection is the identification of mycobacterium TB in a clinical sputum or pus sample. This test may take several months to identify the bacteria in the laboratory, and hence does not solve the purpose in most of the situations. Another test used for TB detection is sputum smear microscopy. Sputum smear microscopy has been the elementary method for pulmonary TB diagnosis in low-and-middle-income countries. It is a straightforward, speedy and an economical technique focusing on areas with a very high prevalence of the organism. However, sputum smear microscopy has considerable drawbacks when the density of bacteria is lesser than 10,000 organisms/ml. It also underperforms in extra-pulmonary TB, paediatric TB and in patients coinfected with human immunodeficiency virus (HIV) and TB (Perkins 2000). Another test used to detect the presence of TB is the Mantoux skin test, which is cost-effective but highly unreliable (Murtagh 1980). Few molecular tests, such as nucleic acid amplification test (NAAT), are available which provide quick diagnosis of TB (~100 min) but are expensive and unaffordable to common man.

Continuous increase in TB patients has created a requirement for early and a cost-effective TB screening. An automated system to examine X-ray images aids TB screening process of sizable population at various remote locations. Hence, a reliable system for TB detection using chest X-ray (CXR) images would be a great achievement towards quick and trouble-free TB diagnostics.

This work presents a deep learning based approach using convolutional neural network (CNN) for automatically detecting TB manifestations in CXR images. CNNs are variants of multilayer perceptron (MLP) that responds to small blocks of the image which helps in exploiting the spatial correlation between neurons of adjacent layers. Deep convolutional neural networks (Simonyan and Zisserman 2014; Szegedy et al. 2015) are continuously marking their presence in the challenging ImageNet competition (Russakovsky et al. 2015) as well as other visual recognition tasks. We explored and implemented two networks, namely,

- Deep residual learning approach described by He et al. (2016a), for which a variant was provided by He et al. (2016b).
- Oxfordnet architecture proposed by Simonyan and Zisserman (2014).

2 Related Work

The emergence of digital image processing and machine learning techniques have yielded new direction to computer-assisted screening. Computer vision has evolved in a way that medical imaging has attracted increasing attention in recent years. Many researches have been carried out in computer-aided medical imaging analytics using CXR. Few of the related studies are listed here.

Jaeger et al. (2014) proposed an automated TB screening system using support vector machines (SVMs). They used a lung segmentation based approach for processing CXR images before feeding to the SVM. The datasets used were from two hospitals, one each from United States (US) and China. They observed an accuracy of 78.3% and 84% on the US hospital and China hospital datasets, respectively. Jaeger et al. (2012) proposed another system by combining a lung shape model, a segmentation mask and an intensity model to achieve a better segmentation of the lung. They achieved an area under the ROC curve of about 83%, validated on data compiled within a TB control program.

Van Ginneken et al. (2002) presented an approach to detect manifestations in frontal CXRs. The objective of this method is to detect unusual signs in the lung texture. It was found by segmenting lungs and subdividing them into overlapping regions of various sizes.

Van Cleeff et al. (2005) established the performance of CXR in TB suspects. Cleeff also compared the cost-effectiveness of using sputum microscopy proposed by Ziehl–Neelsen (ZN) succeeded by CXR with an alternative pathway using CXR followed by ZN.

The remainder of the paper is organized as follows. Section 3 presents an overview of the dataset, hardware and software configuration, architecture of the CNN and the training strategies used in this work. Section 4 presents the results and discusses them. Section 5 concludes the work and proposes possible future work.

3 Methodology

3.1 Data, Software and Hardware

For our study, we used CXR image dataset collected from three hospitals, namely,

- Montgomery County, Maryland and provided by National Library of Medicine (World Health Organization, WHO 2016). The dataset contained 138 posteroanterior CXRs, of which 80 were normal and 58 were abnormal containing TB manifestations.
- Shenzhen Hospital in Shenzhen, Guangdong Providence, China and provided by National Library of Medicine (World Health Organization, WHO 2016). Shenzhen Hospital is one of the most reputed hospitals in China for treating infectious diseases and is one of the two teaching hospitals of the University of Hong Kong. The dataset comprised of 662 CXRs, containing 326 normal and 336 abnormal CXRs with TB manifestations.
- Dataset 3 comprised of 113 images obtained from a medical college in India and annotated by a radiologist. This dataset comprised of 63 normal CXR and 50 CXR with TB manifestation.

Figure 1 shows CXRs of patients who are not infected with TB, and Fig. 2 shows CXRs of patients infected with TB. The images ranged up to 3000 × 3000 pixels, and it takes very long time to train on a CPU. We used two GPUs, namely, NVIDIA GTX980TI containing 2816 cores and an Amazon EC2 instance having NVIDIA GPU with 1536 cores, to train the CNNs. On the software side, we used Python, NumPy, scikit learn and Theano (www.deeplearning.net/software/theano/), in combination with the cuDNN library. To preprocess the images, we used the Python Imaging Library (PIL).

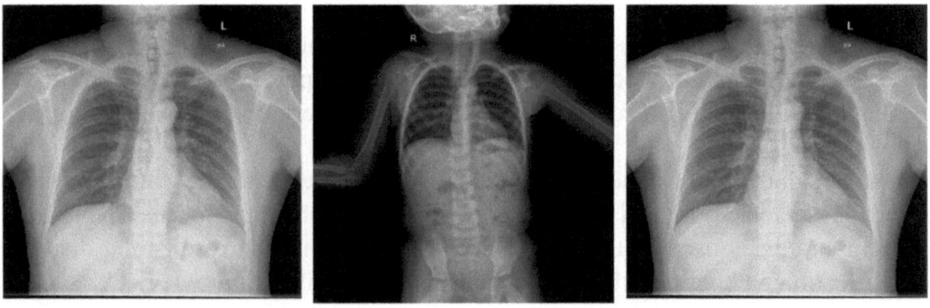

Fig. 1 CXRs of patients without TB

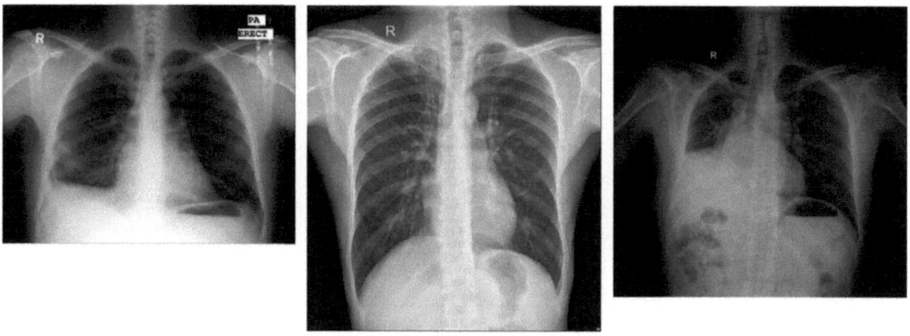

Fig. 2 CXRs of TB patients

3.2 TB Diagnosis

As a part of data preprocessing, images from the dataset are intensified at the edges. The edge-enhanced images are cropped by identifying the foreground from background. The cropped images are resized to 128×128, 256×256, 512×512 and 1024×1024 pixels, and their pixel values were normalized.

3.2.1 Method 1: Deep Residual Network

Building deeper convolutional nets are not as simple as stacking more layers as it leads to hindrance in the convergence. The work implements a deep residual learning based approach described by He et al. (2016). The preprocessed images are passed to the CNN which consisted of three residual blocks. Residual learning is a technique by which each network layer fine-tunes the input from previous layer by adding a learned residual to the input (refer Fig. 3). The intuition behind residual learning is that it is more challenging to fit an identity mapping by stacking nonlinear layers than to bring the residual to zero. The greatest benefit of this is during backpropagation, where the gradient of higher layer can directly pass to lower layer (skip connection as shown in Fig. 3). This also solves the vanishing gradient (Bengio et al. 1994) problem.

The activation functions, ReLU (Krizhevsky et al. 2012) and Batch Norm (Fig. 3), are used as pre-activations, in contrast to post-activation of weighted layers. Batch norm (Ioffe and Szegedy 2015) allows to use much higher learning rates and be less mindful about weight initialization. It acts as a regularizer and eliminates the use of dropout (Srivastava et al. 2014). X_l is the input to the CNN at layer 'l', and X_{l+1} is the output from the layer 'l'.

The parameters/weights for the network were initialized using orthogonal weight initialization. A depth of 10 for each residual block was used, which in total forms 60 convolutional layers. The network also contains a convolutional layer after the input layer and a fully connected layer as the last layer, which aggregates to a 62-layered

Fig. 3 Residual block

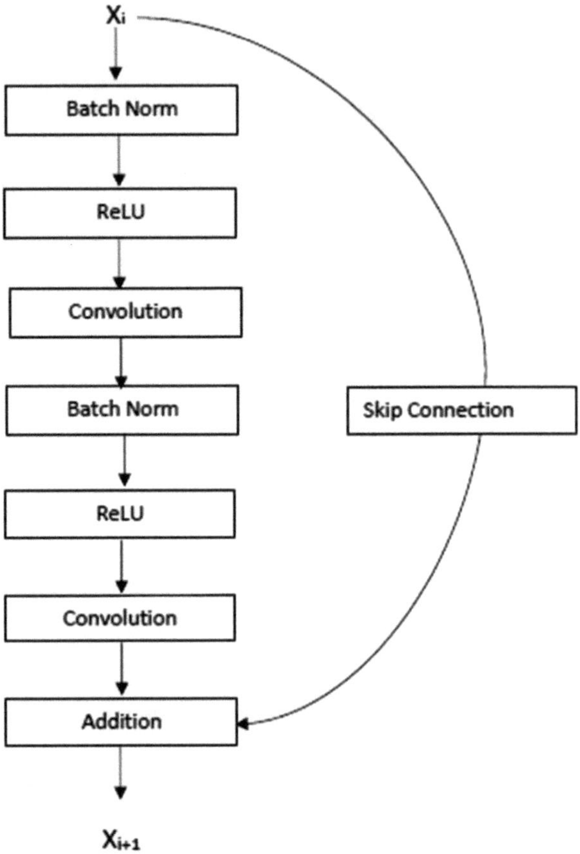

CNN. A batch normalization, ReLU and an average pooling (LeCun et al. 1990) were performed on the output of the final residual block before passing to the fully connected layer.

The network was trained with various filter sizes of 3×3, 4×4, 5×5 and 7×7. At the first convolutional layer, eight filters were used and the filters were gradually increased up to 32 at higher layers. To reduce over fitting, L2 regularization of 0.0001 was applied at all layers. The network was trained over 750 epochs with a batch size of 16 and cross-entropy objective function to calculate loss. Learning rate was continuously decreased from 0.01 to 0.0001 as the number of epochs grew. Stochastic gradient descent (SGD) optimization with a Nesterov momentum of 0.9 was used to minimize the loss. For data augmentation, the images were randomly rotated between $-45°$ and $45°$, randomly translated between -40 and 40 pixels, and random stretching between 1/1.3 and 1.3 over each epoch.

3.2.2 Method 2: Oxfordnet (Variant)

The preprocessed images were passed to the CNN, whose network architecture is shown in Fig. 4. The problem was treated as a regression problem with output threshold at 0.5, to calculate sensitivity and specificity of TB detection. The input to the network holds the raw pixel values of the image. The weights were randomly initialized using orthogonal weight initialization. The network was trained over 750 epochs with a batch size of 48 and mean square error objective function to calculate train and validation loss. The network training was initiated with a learning rate of 0.003 and was constantly decreased as the number of epochs grew and ended up

Layer	No of Filters	Filter Size	Stride	Output Size
Input				448x448
Convolution	32	5x5	2	224x224
Convolution	32	3x3	1	224x224
Maxpool		3x3	2	111x111
Convolution	64	5x5	2	56x56
Convolution	64	3x3	1	56x56
Convolution	64	3x3	1	56x56
Maxpool		3x3	2	27x27
Convolution	128	3x3	1	27x27
Convolution	128	3x3	1	27x27
Convolution	128	3x3	1	27x27
Maxpool		3x3	2	13x13
Convolution	256	3x3	1	13x13
Convolution	256	3x3	1	13x13
Convolution	256	3x3	1	13x13
Maxpool		3x3	2	6x6
Convolution	512	3x3	1	6x6
Convolution	512	3x3	1	6x6
RMSpool		3x3	3	2x2
Dropout				
Fully connected	1024			
Feature pool	512			
Dropout				
Fully connected	1024			
Feature pool	512			
Fully connected	1			

Fig. 4 Method 2, CNN architecture

Fig. 5 Blend network

No	Layer	Output size
1	Input	4096
2	Fully connected	32
3	Feature pool	16
4	Fully connected	32
5	Maxout	16
6	Fully connected	1

with a learning rate of 0.00003. A ReLU of 0.01 was used after every convolutional and fully connected layer. L2 regularization with a factor of 0.0005 is applied to every layer. RMSpool (Yang et al. 2009) performs a downsampling operation by calculating the root-mean-square (RMS) value of the local region. A dropout of 0.5 was used which helps to prevent overfitting. Feature pool layer was used with a pool size of 2, which acts as an activation function. Output of the RMS pool layer was used as features for the blend network (Fig. 5) which was trained again to improve the prediction results. To increase the quality of the extracted features, feature extraction was repeated up to 50 times with different augmentations per image, and the mean and standard deviation of each feature were used as input to the blending network. The same data augmentation steps which were performed in previous method were followed.

In the blend network, all features were normalized to have zero mean and unit variance and were used to train a simple fully connected network. A ReLU of 0.01 was introduced after each fully connected layer. L2 regularization with a factor of 0.005 was applied to every layer. The batch size used was 128 with a mean-squared-error objective.

4 Results and Discussion

4.1 Results

4.1.1 Method 1

After training the network with multiple filter sizes, we observed that 3×3 filters produced better prediction results. The confusion matrix for the same is shown in Table 1. The sensitivity and specificity for TB prediction are observed to be 82.08% and 93.80%, respectively.

Table 1 Confusion matrix (method 1)

		Predicted	
		TB	Non TB
Actual	TB	87	19
	Non TB	8	121

Table 2 Confusion matrix (method 2)

		Predicted	
		TB	Non TB
Actual	TB	90	16
	Non TB	9	120

4.1.2 Method 2

In this network, we observed a combination of 3×3 and 5×5 filters produced the best prediction results. The confusion matrix for the same is shown in Table 2. The sensitivity and specificity for TB prediction are observed to be 84.91% and 93.02%, respectively.

4.2 Discussion

Out of the resized images, the network started overfitting with images of size larger than 512×512 pixels. Also, the lesions in the images were not well detected with low-resolution images, due to the misinterpretation of noise and lesions. Thus, we finalized the network with images of size 512×512 pixels. Although both networks were trained using filters of size 3×3, 4×4, 5×5 and 7×7 for TB diagnosis, it was observed that the network training with smaller filter sizes better suited for identifying TB manifestation on CXRs. Dynamic data augmentation before the blend network was a crucial step in improvement of prediction result. Preprocessing methods other than cropping and resizing did not help reduce the noise during network training. Ensembling of networks with different filter sizes, as well as images with different resolutions, was attempted but did not observe significant improvement in the classification result. The exciting aspect of both methods of automatic TB diagnosis is its ability to classify thousands of images every minute which helps in providing real-time diagnosis.

5 Conclusion

To classify TB images automatically, we used the CXRs collected from multiple hospitals and trained multiple CNN models. As far as we are aware, this is the first work on CXR images for detection of the presence of tuberculosis using a complete CNN-based approach. We also observed a very good classification accuracy (sensitivity, as well as specificity) on the validation CXR images. This demonstrates the potential of CNNs to automatically classify CXRs in real time. The system must be trained on more images from different environments to achieve robustness to the system and usable in a clinical setting. Though the automatic detection of tuberculosis was a good success, we could perform few more preprocessing methods before predicting the CXR images to increase the prediction accuracy. We can also extend the research on X-rays from other bony structures of the body.

References

Bengio, Y., Simard, P., & Frasconi, P. (1994). Learning long-term dependencies with gradient descent is difficult. *IEEE Transactions on Neural Networks, 5*(2), 157–166.

He, K., Zhang, X., Ren, S., & Sun, J. (October 2016a). Identity mappings in deep residual networks. In *European Conference on Computer Vision* (pp. 630–645). Springer International Publishing.

He, K., Zhang, X., Ren, S., & Sun, J. (2016b). Deep residual learning for image recognition. In *Proceedings of the IEEE Conference on Computer Vision and Pattern Recognition* (pp. 770–778).

Ioffe, S., & Szegedy, C. (June 2015). Batch normalization: Accelerating deep network training by reducing internal covariate shift. In *International Conference on Machine Learning* (pp. 448–456).

Jaeger, S., Karargyris, A., Antani, S., & Thoma, G. (August 2012). Detecting tuberculosis in radiographs using combined lung masks. In *Engineering in Medicine and Biology Society (EMBC), 2012 Annual International Conference of the IEEE* (pp. 4978–4981). IEEE.

Jaeger, S., Karargyris, A., Candemir, S., Folio, L., Siegelman, J., Callaghan, F., et al. (2014). Automatic tuberculosis screening using chest radiographs. *IEEE Transactions on Medical Imaging, 33*(2), 233–245.

Krizhevsky, A., Sutskever, I., & Hinton, G. (2012) Imagenet classification with deep convolutional neural networks. In *NIPS*.

LeCun, Y., Boser, B. E., Denker, J. S., Henderson, D., Howard, R. E., Hubbard, W. E., et al. (1990). Handwritten digit recognition with a back-propagation network. In *Advances in Neural Information Processing Systems* (pp. 396–404).

Murtagh, K. A. T. H. L. E. E. N. (1980). Unreliability of the Mantoux test using 1 TU PPD in excluding childhood tuberculosis in Papua New Guinea. *Archives of Disease in Childhood, 55*(10), 795–799.

Perkins, M. D. (2000). New diagnostic tools for tuberculosis [The Eddie O'Brien Lecture]. *The International Journal of Tuberculosis and Lung Disease, 4*(12), S182–S188.

Russakovsky, O., Deng, J., Su, H., Krause, J., Satheesh, S., Ma, S., et al. (2015). Imagenet large scale visual recognition challenge. *International Journal of Computer Vision, 115*(3), 211–252.

Simonyan, K., & Zisserman, A. (2014). Very deep convolutional networks for large-scale image recognition. arXiv:1409.1556.

Srivastava, N., Hinton, G. E., Krizhevsky, A., Sutskever, I., & Salakhutdinov, R. (2014). Dropout: a simple way to prevent neural networks from overfitting. *Journal of Machine Learning Research, 15*(1), 1929–1958.

Szegedy, C., Liu, W., Jia, Y., Sermanet, P., Reed, S., Anguelov, D., et al. (2015). Going deeper with convolutions. In *Proceedings of the IEEE Conference on Computer Vision and Pattern Recognition* (pp. 1–9).

Van Cleeff, M. R. A., Kivihya-Ndugga, L. E., Meme, H., Odhiambo, J. A., & Klatser, P. R. (2005). The role and performance of chest X-ray for the diagnosis of tuberculosis: a cost-effectiveness analysis in Nairobi. *Kenya. BMC Infectious Diseases, 5*(1), 111.

Van Ginneken, B., Katsuragawa, S., ter Haar Romeny, B. M., Doi, K., & Viergever, M. A. (2002). Automatic detection of abnormalities in chest radiographs using local texture analysis. *IEEE Transactions on Medical Imaging, 21*(2), 139–149.

World Health Organization, WHO (2016). Global Tuberculosis Report 2016. http://www.who.int/tb/publications/global_report/en/.

Yang, J., Yu, K., Gong, Y., & Huang, T. (June 2009). Linear spatial pyramid matching using sparse coding for image classification. In *IEEE Conference on Computer Vision and Pattern Recognition, 2009. CVPR 2009.* (pp. 1794–1801). IEEE.

Connected Cars and Driving Pattern: An Analytical Approach to Risk-Based Insurance

SrinivasaRao Valluru

Abstract Insurance companies are witnessing a significant drop in their profit margins particularly in the segment of vehicle insurance due to heavy competition in the industry. Insurance companies are trying to improve their customer base by retaining existing customers and launching new policies with additional benefits. Customers are expecting insurance policies which match to their requirements and at the same time, companies also want to charge more premium for the customers with risky driving behaviour and less for safe driving. Insurance companies are reducing costs with the help of historical risk data and advanced analytics to improve their profits. Insurance companies are capturing real-time vehicle movement data through IoT to monitor the driving behaviour of their customers. By applying advanced analytics on this data, insurance companies can study customers driving pattern to assess the risk involved in it. In this study, we are presenting an analytical approach to categorize driving patterns using advanced machine learning techniques which will lead to risk-based insurance premium. It will help insurance companies to provide personalized services to their customers and in assisting insurance companies in the process of claims approval when an accident took place.

Keywords K-means clustering · Decision tree · Fuzzy forest · Random forest (RF) · Support vector machine (SVM)

1 Introduction

Insurance companies are launching new policies with additional benefits to increase their customer base. Profit margins of insurance companies are under pressure, especially in the segment of vehicle insurance due to heavy competition in the industry. Over the years, vehicle insurance companies are charging premiums based on accumulated data such as age, gender, type of vehicle, etc. Customers are expecting

---

S. Valluru (✉)
HCL Technologies, Hyderabad, India
e-mail: srinivasarao.v@hcl.com; srinivas_stat@rediffmail.com

A. K. Laha (ed.), *Advances in Analytics and Applications*, Springer Proceedings in Business and Economics, https://doi.org/10.1007/978-981-13-1208-3_12

insurance policies which match to their specific requirements and at the same time, companies also want to charge more premium for the customers with risky driving behaviour and less for safe driving. Insurance companies also want to provide premium discounts as a reward for safe driving behaviour of their customers. Insurance companies are monitoring real-time vehicle movement of their customers through IoT. By leveraging this information with the help of advanced analytics, companies can derive better insights on their customer's driving patterns.

The objective of the study is to provide an analytical approach to classify driving behaviour which will lead to risk-based insurance premiums. Study of the driving pattern will also help insurance companies to quickly determine the fault when accidents took place which is a key factor in claims settlement. In our study, driving patterns are categorized into three groups such as Risky, Potential Risky, and Safe. We give below an analytical approach to categorize the driving pattern.

Vehicle movement information is captured through sensors connected to the cars. K-means clustering technique (Zaki and Meira 2014) is applied to form homogeneous groups of similar driving patterns. Number of Clusters in K-means is decided using Elbow rule technique and Hierarchical clustering (Zaki and Meira 2014). Studied the properties of each cluster through Decision Tree (Zaki and Meira 2014) and labelled each cluster with the help of domain experts. Labelled data has been divided into 'Train and Test' using Stratified Random Sampling (Cochran 1977). Feature selection technique Fuzzy Forest (Conn et al. 2015) is used to select the important features for model building. A classifier is built using Support Vector Machine (Hastie et al. 2009) (SVM) to classify driving behaviour and monitored the performance of classification model post-deployment.

2 Our Approach

2.1 Data

Data comprises of vehicle movement information captured through multiple sensors connected to the cars. A total of 23 features related to vehicle movement, engine condition, driving speed, weather condition, and other engine performance measures are collected along with timestamp. Captured second-wise data is aggregated to minute level using mean for continuous variables and mode for categorical variables so that all 23 features information is available minute wise. Applied pre-processing steps outlier removal and data normalization to clean the data.

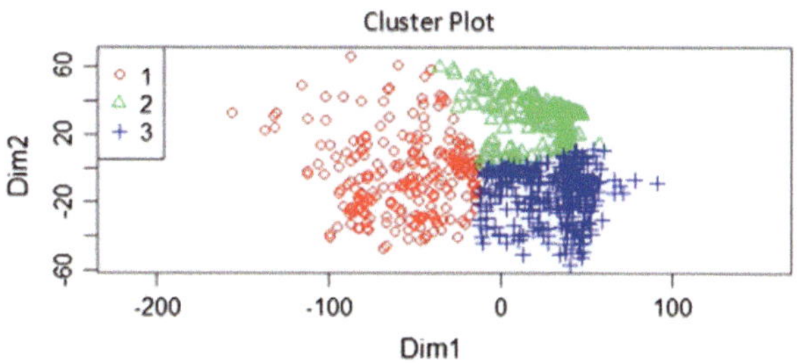

Fig. 1 Two-dimensional view of the clusters formed through K-means clustering

2.2 *Clustering*

The captured data is not a labelled data, so we need to derive categories from the data. We can study the inherent patterns of the data by applying unsupervised machine learning technique like clustering. To form homogeneous groups of similar driving pattern, clustering technique K-means (Zaki and Meira 2014) is applied. Number of clusters (K) is decided through the technique of Elbow rule and by looking at the dendrogram of Hierarchical clustering. The Elbow method looks at the total Within-Cluster Sum of Squares (WSS) as a function of the number of clusters: Plot the graph with number of clusters (K) on X-axis and cluster WSS on Y-axis. The location of a bend (elbow) in the plot is considered as the optimal number of clusters. The number of clusters formed are three ($K=3$). Figure 1 represents the two-dimensional view of clusters formed through K-means. Each colour and shape represents a cluster.

Here, Dim1 and Dim2 are the two-dimensional coordinates of multidimensional data derived using distance matrix and classical Multidimensional Scaling (Gower 1966) (MDS). It can be observed from Fig. 1, three clusters are well formed and observations in each cluster are non-overlapping with other cluster members.

2.3 *Cluster Labelling*

In order to derive target variable, we need to study the cluster properties and label each cluster. We applied rule-based technique Decision Tree (Zaki and Meira 2014) with cluster number as response variable, and all other features as independent variables to study the cluster properties. The rules formed will provide concrete information related to the behaviour of each cluster. Table 1 represents some of the rules extracted from each cluster.

After going through the rules formed for each cluster, with the help of domain experts labelled clusters as Risky, Potential Risky, and Safe Zones. The observations falling under 'Risky' cluster are of risky driving behaviour and the records falling under 'Safe' cluster are following safe driving pattern. Each record is labelled with cluster name, which is the target variable with three levels Risky, Potential Risky, and Safe.

Table 1 Rules extracted through Decision Tree to study properties of each cluster

Cluster	Rule
Cluster 1	V3 ≥ 36.42 and V9 ≥ 17.98
Cluster 2	V3 ≥ 36.42 and V9 < 17.98
	(V3 ≥ 1.044 and < 36.42) and (V16 ≥ 22.32 & < 39.28)
Cluster 3	(V3 < 1.044) and (V16 ≥ 22.32 and < 39.28)
	V3 < 36.42 and V16 ≥ 39.28

2.4 *Sampling*

Data is divided into Train and Test in the ratio of 70:30 using stratified random sampling (Cochran 1977). Train data set is used for model building and Test data set is used for model evaluation. Stratified sampling is used to give a fair representation of each category in the training data.

2.5 *Feature Selection*

All features may not be relevant for model building and presence of irrelevant features might reduce the model's predictive power, which ultimately leads to low accuracy. Feature selection (Guyon and Elisseeff 2003) technique 'Fuzzy Forest' (Conn et al. 2015) was used on Train data to select a subset of features which are important. Fuzzy Forest is one of the best feature selection techniques which provide unbiased variable importance rankings when the features are correlated. Fuzzy Forest using 'R' software identifies five variables V3, V9, V14, V16, and V21 as important variables to build a classifier. These key variables are related to Speed, Torque, Pedal Position, Brake, and Weather.

2.6 *Building Classifier*

Using the features selected through Fuzzy Forest, we built Support Vector Machine (Hastie et al. 2009) (SVM) and Random Forest-based (Breiman 2001) classification model using Train data. Model performance has been evaluated on Test data and the accuracies are reasonably good. Support Vector Machine with Radial kernel and optimum parameters (gamma, cost, and degree) has an accuracy of 98.2%, which is slightly higher than Random Forest 96%. SVM Classifier is used for categorizing the driving patterns of new data as Safe, Potential Risky, and Risky.

2.7 Analytical Results

Classifier based on Support Vector Machine is used to predict the driving patterns of new data. Figure 2 represents the period-wise predicted results of a single vehicle's driving pattern.

Here, period represents a week, and each bar shows the percentage of time vehicle was driven in Safe, Potential Risky, and Risky categories. For Period3, the percentage of vehicle driven in Risky zone is 27.2% and in Period4, the vehicle was driven more in Potential Risky and less in Safe zone.

Figure 3 represents the cumulative percentage of time a vehicle was driven in each category of Safe, Potential Risky, and Risky.

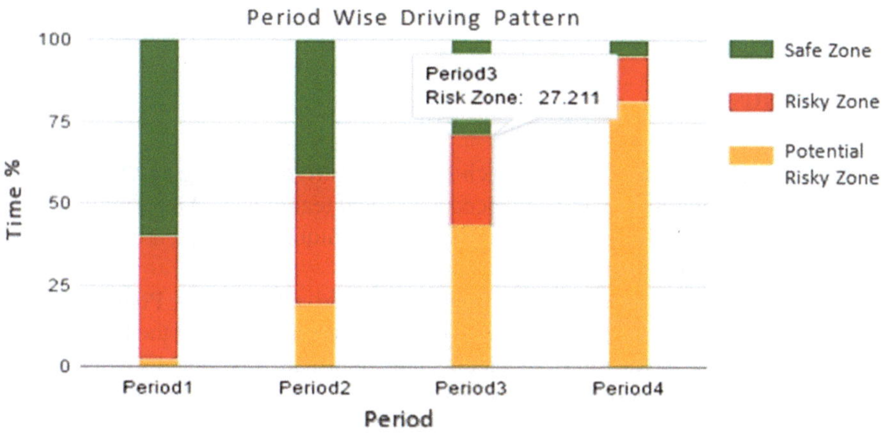

Fig. 2 Period wise predicted driving pattern of a vehicle in Risky, Potential Risky and Safe Zone

Fig. 3 Cumulative percentage of time a vehicle was driven in Potential Risky, Risky and Safe Zone

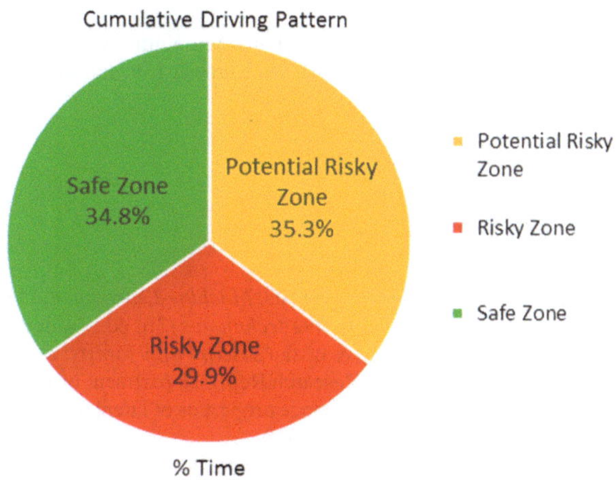

Over the periods of time, the vehicle was driven more in Potential Risky zone (35.3%) and less in Risky zone (29.9%) categories.

2.8 Performance Monitoring

Once the model has been deployed, we need to monitor the performance of the model continuously to make sure that model is performing as per expectations and there is no degradation in model prediction power. For this, collecting predicted results of the model at regular intervals and examining whether the rules extracted earlier are still valid or not for each category. If there is any variation in the predicted results, we need to refit or rebuild the model by considering latest data.

3 Conclusion

In this study, we presented an approach to classify drivers driving pattern as Risky, Potential Risky, and Safe using advanced analytical techniques. Along with Support Vector Machine, we tried another classification technique Random Forest. And when compared, SVM gave better results.

By quantifying the percentage of time vehicle driven in Risky, Potential Risky, and Safe zone, insurance companies can fix the premium based on the risk assessed with each category. At the time of accident claim approval, this analysis will assist the insurance companies by providing a clear picture on the driver's driving condition.

This analysis will aid insurance companies in providing certain services which are specific to their customers. It will ultimately lead to customer loyalty and healthy revenues.

Acknowledgements I sincerely thank Mr. Kumar G. N. for his continuous support and timely inputs. I would also like to thank Mr. Kamesh J. V. and my colleagues who encouraged me during this journey.

References

Breiman, L. (2001). Random forests. *Machine Learning, 45*(1), 5–32.

Cochran, W. G. (1977). *Sampling techniques* (3rd ed.). Wiley.

Conn, D., Ngun, T., Li, G., & et al. (2015). *Fuzzy forests: extending random forests for correlated, high-dimensional data* (Research Rep.). Department of Biostatistics UCLA.

Gower, J. C. (1966). Some distance properties of latent root and vector methods used in multivariate analysis. *Biometrika, 53,* 325–338.

Guyon, I., Elisseeff, A. (2003). An introduction to variable and feature selection. *Journal of Machine Learning Research.*

Hastie, T., Tibshirani, R., & Friedman, J. (2009). *The elements of statistical learning* (2nd ed.). Springer.

Tan, P. N., Steinbach, M., & Kumar, V. (2014). *Introduction to data mining* (2nd ed.). Pearson Education.

Zaki, M. J., & Meira, W. (2014). *Data mining and analysis: fundamental concepts and algorithms*. Cambridge University Press.

Part IV
Human Resource Analytics

Analytics-Led Talent Acquisition for Improving Efficiency and Effectiveness

Girish Keshav Palshikar, Rajiv Srivastava, Sachin Pawar, Swapnil Hingmire, Ankita Jain, Saheb Chourasia and Mahek Shah

Abstract Large IT organizations every year hire tens of thousands of employees through multiple sourcing channels for their growth and talent replenishment. Assuming that for each hire at least ten potential profiles are scrutinized and evaluated, the Talent Acquisition (TA) personnel ends up processing half a million-candidate profiles having multiple technical and domain skills. The scale and tight timelines of operations lead to possibility of suboptimal talent selection due to misinterpretation or inadequate technical evaluation of candidate profiles. Such recruitment process implementation due to manual, biased, and subjective evaluation may result in a lower job and organizational fit leading to poor talent quality. With the increased adoption of data and text mining technologies, the recruitment processes are also being reimagined to be effective and efficient. The major information sources, viz., candidate profiles, the Job Descriptions (JDs), and TA process task outcomes, are captured in the eHRM systems. The authors present a set of critical functional components built for improving efficiency and effectiveness in recruitment process. Through multiple real-life case studies conducted in a large multinational IT company, these components have been verified for effectiveness. Some of the important components

G. K. Palshikar · R. Srivastava (✉) · S. Pawar · S. Hingmire · A. Jain · S. Chourasia · M. Shah
Tata Research Development and Design Centre, Pune 411013, India
e-mail: rajiv.srivastava@tcs.com

G. K. Palshikar
e-mail: gk.palshikar@tcs.com

S. Pawar
e-mail: sachin7.p@tcs.com

S. Hingmire
e-mail: swapnil.hingmire@tcs.com

A. Jain
e-mail: ankita7.j@tcs.com

S. Chourasia
e-mail: saheb.c@tcs.com

M. Shah
e-mail: shah.mahek@tcs.com

A. K. Laha (ed.), *Advances in Analytics and Applications*, Springer Proceedings in Business and Economics, https://doi.org/10.1007/978-981-13-1208-3_13

elaborated in this paper are a resume information extraction tool, a job matching engine, a method for skill similarity computation, and a JD completion module for verifying and completing a JD for quality job specification. The tests performed using large datasets of the text extraction modules for resume and JD as well as job search engine show high performance.

Keywords e-Recruitment · Talent acquisition · Resume information extraction Job matching · Skill similarity · HR analytics

1 Introduction

The HR analytics is an active area of research and talent acquisition, in particular, has garnered significant attention of late. The HR journals as well as IT applications and systems journals have encouraged the analytics and technology enablement of the HR processes. Schiemann (2014) gives importance to the selection process and interview process to focus on measuring effectively the candidate alignment and engagement to the organization's goals values and culture. The failure results in a misfit and can be improved using the technology-enabled recruitment processes as well as by evaluating candidates on the right set of parameters related to the organizational culture, values, and environment. Parthasarathy and Pingale (2014) inform that the use of e-recruiting and web-enabled functions has brought the collaborative approach in talent acquisition and management. He also emphasizes that the online experience of web browser access with interactive user-interfaces, social networking enabling collaboration, and participation of community is essential. He emphasizes that the efficiency metrics such as days-to-hire are popular but now recruitment effectiveness in terms of the quality of talent hired is also being targeted for measurement.

Srivastava et al. (2015) provide several predictive analytics based point solutions to address TA needs such as predicting joining delay, selection likelihood, offer acceptance likelihood, and other similar solutions to improve effectiveness of the TA processes. Dutta et al. (2015) highlight the importance of the quick decision-making, innovative methods of talent acquisition, and focused metrics for the function of the HCL TA Group (TAG) as it is gearing up for strategic recruitment. HCL has realigned TAG by implementing change initiatives for its members using aggressive SLAs with the business stakeholders. They are yet to start use of data mining for insights and text mining for efficiency improvement. Faliagka et al. (2012a, b) describe an approach for ranking job applicants for recruitment in web-enabled systems. Their proposed system implements candidate ranking, using objective criteria which are made available from the applicant's LinkedIn profile. The candidate's personality features are also extracted from her social activity using linguistic analysis. The Faliagka et al. (2012a) use Analytic Hierarchy Process (AHP) for ranking profiles, whereas we have devised functions for similarity and matching score computation which are based on the acquired domain knowledge. The Faliagka et al. (2012b) use text mining of LinkedIn for creating profile and use linguistic analysis for inferring personality characteristics. In our work, we are extracting attributes from candidate's resume using Resume Information Extractor (RINX) as well as planning to combine information from multiple online and social platforms for the technical and domain skills using extraction tools. Bui and Zyl (2016) give insight into how gamification, a new technique, can be used for acquiring talent. The findings suggest that gamification platforms can be used to align the prospective employee's interests and identity with the organization. It can also act as a personalization tool which may make employee onboarding smoother.

Edmundson (1969) has proposed simple rule based techniques for automated extraction of structured information from the unstructured textual data. Mooney and Bunescu (2005) have applied knowledge extraction from the unstructured text using text mining. With increase of machine learning and natural language processing techniques, Téllez-Valero et al. (2005) and other researchers tried to solve this problem of automatic extraction. On resume documents different extraction techniques are used to make candidate selection process (Tomassetti et al. 2011) easier and automatic.

2 Analytics for Talent Acquisition

Talent Acquisition (TA) is a fundamentally important function within a company's HR function, responsible for recruiting high-quality workforce, and important for the successful operations and growth. The recruitment fulfills demand pipeline of futuristic domains and technical skill requirements in shortest possible time frames, at lowest costs, using diverse channels and from multiple locations. Leading IT service providers which have strength of more than 200,000 Full-Time Employees (FTEs) every year hire 40,000–50,000 employees for their growth and talent replenishment. These organizations work in multiple industrial domains requiring combinations of technical and domain skills for executing projects worldwide. The experienced candidates are sourced from recruitment agencies, online job portals, and social collaboration platforms such as LinkedIn™[1]. The fresh graduates are engaged and groomed through social, community portals run by these organizations who are subsequently screened for hiring as FTEs. Assuming that for each hire at least ten potential profiles are shortlisted by the TA personnel, it requires half a million-candidate profiles collected for scrutiny and proportionate efforts for the complete recruitment process. These shortlisted candidate profiles are further screened by the requestor group who selects, for example, five out of ten profiles for the interviews and onward selection process as depicted in Fig. 1.

The talent selection which results from inadequate evaluation or misinterpretation of candidate profiles would result in lower job and organizational fit resulting in poor talent quality. These losses can be attributed to the large volume of candidate profiles being screened and profiles mostly being textual leading to subjective and biased interpretations by the personnel involved in shortlisting and selection. An important

[1] LinkedIn™ is a trademark owned by LinkedIn™ recruiting services.

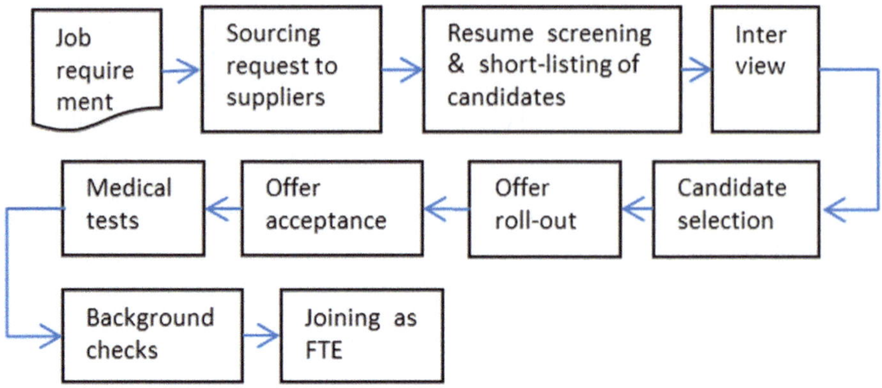

Fig. 1 A typical recruitment process flow

issue is inadequate technical or domain knowledge held by the TA personnel who have HR background. Other process constraints such as fulfilling a talent requirement within a given time limit or requirement being in bulk for a quick ramp-up would put additional pressure on the recruitment personnel leading to inadvertent oversights during the selection. Also, a JD being a crucial talent requirement document any oversight such as noninclusion of an important supplementary skill may lead to suboptimal fit, utilization of available talent as well as possible effort and time overrun in a project. To avoid these situations, an intelligent JD completion recommender can be developed which does an automatic review and suggests best skills or roles or task names likely missing from a JD.

With the increased adoption of technologies enabling analytics, the recruitment processes are also being reimagined. With the increase in number of skills being used across knowledge-based industries as well as the increase in available skilled candidates, it is an imperative for TA function to adopt analytics based on the data mining, text mining, machine learning, and statistical methods (Hastie et al. 2008; Tomassetti et al. 2011) to incorporate intelligence and automation for improving effectiveness and efficiency.

The major information sources, viz., candidate profiles, the JDs and TA process outcomes such as shortlists, interviewers, evaluations, and selected profiles, are available in electronic form. These data are captured in the eHRM (Bondarouk et al. 2009; Strohmeier 2007) systems deployed for automating workflows and tasks of recruitment process such as shortlisting, interviewing, offer generation as well as early engagement of the selected candidates to increase probability of acceptance of job offer and early joining. In the commonly used workflow automation systems, the inputs of candidate profile, JD details, and various decision support activities are completely manual. The primitive matching available in the job portals uses only a few attributes such as experience range, role, and skills. Most of the data being textual, the word similarity-based matching is elementary and is prone to suboptimal solutions. There are several manual processing steps which are effort intensive, repetitive, and strenuous, and these lead to the end-user dissatisfaction due to the low-quality hires. Some of the important automation opportunities are listed as follows:

- Candidate has to *manually re-enter* details of her profile to facilitate structured storage in the employer database through a web interface, though her resume already has all details in it.
- Sourcing personnel has to *manually derive* additional latent attributes from the candidate profile, such as quality of education, skill level, as well as soft skills such as communication abilities, leadership potential, etc. to help in shortlisting.
- Sourcing personnel has to *shortlist* a list of matching candidates for a JD using a rich set of latent quality attributes, in addition to experience and held skills.
- Continuing the previous point, *automated matching* would require user to understand the salient parts of the *given textual JD* to form a query to retrieve best matching candidate profile.

In addition to the possible automation of these abovementioned manual tasks, there are several improvements possible by use of the data and text mining techniques (Ronen et al. 2006). For example, using information retrieval we can improve the accuracy of the automated matching of a JD to candidate profiles for increasing effectiveness of the talent acquisition function. A few of such improvement opportunities are listed as follows:

- Deriving and using *skill similarity* for shortlisting candidates for a JD
- Suggestions to *improve a submitted JD* specification for better job fitment.

In this paper, we present a set of components built during multiple real-life case studies for improving the efficiency and effectiveness of TA function in a large multinational IT company. The list of important functionalities implemented as components is as follows:

- A tool for resume information extraction, RINX;
- A search and matching engine for job description and similar queries, RINX-SE;
- A module to derive the richer set of quality attributes from resume and social media;
- A JD completion module for suggesting possible improvements in a JD;
- A module for computing skill similarity.

Earlier, extraction task was performed using gazette-based matching in which entities are marked in the text if that particular word or sentence is present in the gazette. This "strict" pattern-based matching does not handle variations, i.e., minor spelling mistakes, new evolving patterns, etc. leading to weak matching results. These proposed components have been developed and tested with large sample size of ~10,000 resumes and 200 JDs for matching. The extraction results for resume and JD are evaluated using the *F-measure*.[2] The *F-measure* achieved for extraction of various entities from JD is above 0.8, and from resume it averages around 0.75.

[2] *F-measure—harmonic mean of precision and recall. F-measure* $= \frac{2 * precision * recall}{precision + recall}$.

The rank correlation measures such as *Kendall's τ (tau)*[3] in job matching and skill similarity achieve value of 0.71 and 0.94, respectively, compared to earlier pattern-based matching.

3 Data Mining and Text Mining-Based Solution Components

The following subsections provide description and salient features of the developed components.

3.1 Resume Information EXtractor (RINX)

The *resume* or *curriculum vitae* (CV) or *bio-data* document contains vital information about the education, skills, experience, and expertise of a person. A resume contains a summary (or profile) of the entire work experience of a person. Typical information elements in a resume are personal details (e.g., name, current address, phone number, e-mail, etc.), work experience profile (e.g., period, organization, designation, etc.), educational qualifications (e.g., degree, branch, year of passing, college, score, etc.), and project worked on (e.g., duration, role, services delivered, technologies used, etc.). One significant way to improve the operational efficiency of people who need to use resumes is to automatically extract all the relevant information elements from each resume in the given set. For automated resume content extraction, Resume Information eXtractor (RINX) tool has been developed. RINX uses NLP techniques (Aggarwal and Zhai 2012) such as parsing (using *Google TensorFlow syntaxnet*) and Part-of-Speech (POS) tagging (kindly refer Brown Corpus manual at http://clu.uni.no/icame/manuals/BROWN/INDEX.HTM for a complete list of POS tags), named entity recognition for enriched text generation to facilitate robust pattern specification for the targeted information extraction as well as RINX includes gazette-based look-up.

As an example the sentence "`My role as a Business Intelligence Expert is responsible for Managing Business Intelligence Architecture`" from a resume is enriched for extraction using NLP tools as follows:

[3] *Kendall's τ considers rank ordering among entities. It is the ratio of difference between number of concordant pairs and discordant pairs to the total number of ranked pairs possible.*

<S><NP><PRP$>My </PRP$><NN>role </NN></NP><IN>as </IN><NP><DT>a </DT><nmod_role><NNP>Business </NNP><NNP>Intelligence </NNP><NNP>Expert </NNP></nmod_role></NP><VP><VBZ>is </VBZ><JJ>responsible </JJ><IN>for </IN><NP><NNP>Managing </NNP><NN>Business </NNP><NNP>Intelligence </NNP><NNP>Architecture </NNP></NP><PUNCT>. </PUNCT></VP></S>

In this enriched text of the sentence from resume, the Noun Phrases (NPs) `My Role` and `Managing Business Intelligence Architecture` are eliminated since none of the hypernym trees for nouns `Role` and `Architecture` contain any word from the given list {*creator*, *expert*, *specialist*, *specializer*, *planner*, *person*, *individual*}. The NP `a Business Intelligence Expert` is selected as a "Role" value because it contains the role indicating noun *expert* from the list. The same technology of entity extraction from resume is used to automatically convert the textual JD into structured form.

3.1.1 Extraction of "Service Line"

A "Service Line" is a domain-specific (often technical) task or activity carried out by humans, with or without tools. A "Service Line" is typically mentioned as part of the project or job description which represents the actual work done or the service provided by the resume author. A "Service Line" is typically indicated by a Noun Phrase (NP), such as user interface design. Figure 2 shows a few examples of the ways in which service lines are mentioned in resumes.

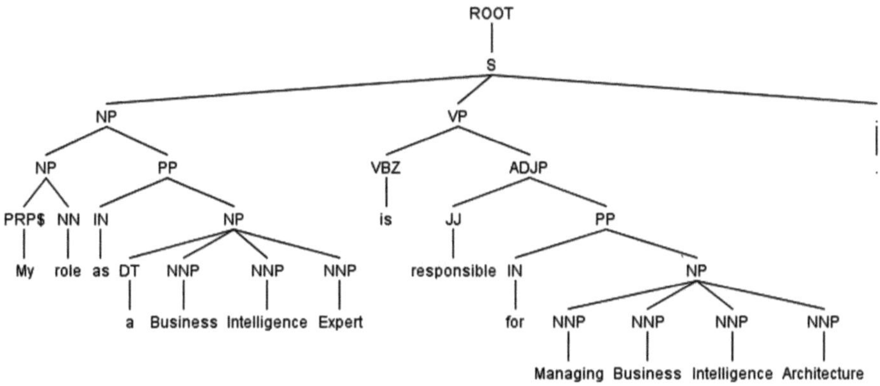

Fig. 2 Parse tree for an example sentence

```
Main responsibilities include Development, testing,
implementation and Design of the modules offshore, developed at
client interaction, handling client escalations etc.
Development, supervision and migration of maps as per the
specification.
Execution of Unit Test Cases and Functional Test Cases.
```

Fig. 3 Examples of "service line"; values (underlined)

"Service Line" values are usually domain-dependent; e.g., "Service Line" in IT and finance domains are different. For higher flexibility, we want to minimize the use of gazettes (i.e., lists) containing known values for "Service Line". Moreover, we want to standardize different "Service Line" values, so as to facilitate retrieval/comparison in later stages (e.g., when matching candidates to specific job requirements). Lastly, we aim for at least 90% overall accuracy for extracting "Service Line" values. All this makes extraction of "Service Line" challenging (Fig. 3).

Algorithm 1 Service Line extraction

algorithm Service_Line
input set of N sentences $\{S_1, S_2, ..., S_N\}$ from given resume
input k // max no. of word senses to consider (default $k = 2$)
input m // max no. of 'Service Line' values per project
    // description (default $m = 7$)
input L_1; // list of prohibited nouns: work, scope etc.
input L_2; // list of known 'Service Line' values: walk-through,
    // documentation, review etc.
output L // set of service lines selected for given resume
for each sentence S_i and its parse tree T_i **do**
  for each lowest-level NP (say X) in the sentence S_i do
    if X occurs under a VP node containing a copula verb **then**
      continue;
    if X occurs under SINV node **then continue**;
    if X contains comma or 'and' **then** // a list of nouns
      treat each noun in X as a separate NP // e.g., if X = "design
        // and development" then treat "design" and
      // "development" as two separate NPs
    if X does not contain at least one noun (NN, NNS, NNP etc.)
      then continue;
    if X fails any one of the tests specified in function check_np
      then continue;
    $L = \emptyset$; // L = list of 'Service Line' values found in sentence S_i
    for each noun Y in X **do** // parent node of Y in parse tree T_i
                  // has label NP, NNS, NNP etc.
      if $Y \in L_1$ **then continue**; // prohibited noun

```
        if Y′ is similar to a value in L2 then // known value
          Add Y′ to L; continue; end if
        if Y ∈ WordNet (e.g., budgeting) or Y has a related noun
        then
          obtain root Y′ of Y // testing→test, budgeting→budget
          Let H = {H1, H2, …, Hk} denote the set of top k
          hypernym trees for Y′ obtained using WordNet;
          // use first k senses
          if no Hi in H contains one of {"speech act", group
            action", "action", "change", "activity"} followed by
            "human activity" then continue; // not a human action
          Remove articles and prepositions from X;
          Add X to L; // X is a possible value for 'Service Line'
          break; // finished checking X; go to next NP
        end if
      end for
    end for
  end for
Remove duplicate (or very similar) entries in L, if any;
// Keep at most m NPs in L;
Select NPs which contain nouns from the list of known 'Service Line' values; if the number of such
NPs is less than m then keep other NPs in order of appearance in the project description;
return(L)
```

Function check_np is used to filter out NPs which cannot be a possible value for "Service Line" entity.

Algorithm 2 For checking Noun Phrase (NP)

```
Boolean check_np(X) // check if given NP X is acceptable
// e.g., 'Oracle Retail 12 Applications' or 'Tally 4DOT5'
  if X contains a number then return(0);
  // e.g., 'Raymark Merchandising Systems' is ruled out because
  // Raymark is not in WordNet.
  if X contains a word not present in WordNet then return(0);
  if X contains an acronym then return(0); // GIS, I.B.M.,
  // 'GE Consumer Finance' etc.
  if X contains a cue word for an organization then return(0);
  // e.g., systems, company, incorporated, inc, bank.
  if X contains a word having underscore then return(0);
  // e.g., 'SQL_Server'
  // Following strong conditions can be turned off by user
  if X contains more than 2 nouns then return(0);
  // e.g., 'Sales Order Management System'.
  if X contains ALL words with first letter capitalized
    then return(0); // e.g., 'Warehouse Management System'.
return(1); // OK
```

3.1.2 Extraction of Other Entities

RINX extracts a large number of entities from resumes: name, address, phone number, e-mail address, gender, date of birth, career profile (e.g., period, organization, and designation), educational qualifications (e.g., degree, branch, year of passing, college, score, etc.), and project worked on (e.g., duration, role, services performed for a project, technologies used, etc.). We describe extraction of a few of these entities now.

Period of a Project

The date range, if mentioned in a project description, usually indicates the period of that project. It always consists of two dates. The individual date is mentioned in various formats. We designed several complex patterns using regular expressions to convert a date in a commonly occurring format to a standard date format. A few examples of the common formats used for mentioning period of projects are "`Jan 06–Till date`", "`August 04-Dec 05`", and "`Oct 2001-Dec 2002`".

Technology and Tools

Within a given business domain, a technology refers to an organized body of formal knowledge and techniques that help the organization in providing better service or in producing better products. A technology is often implemented (and made useful in practice) as a machine, device, or a computer program. In IT domain, a technology is often indicated in the form of software tools (e.g., the database technology is indicated by tools such as Oracle, Sybase, and SQL Server). The names of tools mentioned in project descriptions in a resume are considered for deriving competencies, as these are the technologies used by the associate in actual work. Some examples of the technologies from IT domain are databases, graphics, programming languages, compilers, image processing, data mining, and model-driven development. The following list gives sample statements from resumes, which contain mentions of technology and tools.

(a) `Solution Environment: WINDOWS-2000/XP`
(b) `Tools: Eclipse 3.0, MS Visual Source Safe 6.0`
(c) `Support for Queue implementation in Message Driven Beans in MasterCraft 6.5.2 and 7.0`
(d) `Support for Weblogic 9.0 appserver in MasterCraft 6.5.2.`

RINX has complex patterns (as well as extensive gazettes of known tool names) to recognize technology and tools.

3.1.3 Experimental Results

Accuracy of the algorithm is measured on 160 free-form resumes of candidates in which entities are manually annotated. The resumes with manually annotated entities are part of the "Gold copy" and are used for the verification purposes. Only

Table 1 Extraction accuracy of RINX

Entity	Recall	Precision	F-measure
Date of birth	0.99	0.95	0.97
E-mail	0.98	0.93	0.95
Phone	0.96	0.95	0.95
Project date range	0.92	0.9	0.91
Project role	0.95	0.7	0.8
Project service line	0.97	0.95	0.96
Job duration	0.74	0.94	0.83
Employer	0.94	0.91	0.93
Degree name	0.86	0.9	0.88
Degree marks	0.71	0.93	0.81
Degree specialization	0.86	0.9	0.86
Degree university	0.77	0.89	0.82
Year of degree	0.74	0.94	0.83

information technology domain resumes have been selected for testing. The entities like date of birth, e-mail, phone numbers, project dates, service line, role, education qualifications, and career profile were extracted using RINX, wherein the extraction results were post-processed, e.g., cleaning with a negative list and then manually compared with the content of resume. The results consisting of precision, recall, and F-measure are provided in Table 1.

Please note that the extraction accuracy of entity 'role' is further improved by deriving it from the extracted service lines, as majority of service lines grouped together imply a role.

3.2 RINX Search Engine (RINX SE)

A resume search engine, for matching JDs or similar queries to the resume profiles (Patil et al. 2012), is developed to incorporate domain-specific knowledge to enhance the quality of search results. Given a search query for "`Microsoft`" as previous employer, a domain agnostic search engine would not be able to distinguish between resumes mentioning `Microsoft based skills` or `Microsoft` as an employer. As resume repository is first processed using RINX to extract structured information qualifying "`Microsoft`" as an employer eliminates error possible with a domain agnostic search engine. The extracted structured as well as unstructured resume contents are further indexed using Lucene, a text search engine library, for domain knowledge enabled search. Generally, the matching is performed using the technical skills and experience-based attributes specified in a typical JD, which only evaluates the "Job Fit" of a candidate. Our search engine provides domain-driven

Table 2 Skill competency results for selected technologies

Competency	Similarity score
ASP.NET 2.0	1
Microsoft.Net compact framework	0.8009
ASP.NET 4.5	0.7032
Microsoft collaboration tool	0.4762

approximate or elastic matching for attributes such as technical skill, role, specialization, etc. For the enrichment of the profile, we also derive further knowledge from resume, including the attributes such as "Quality of education", "Quality of experience", and "Proficiency Level in a skill", to aid evaluation of "Organizational Fitness". A set of domain-specific matching functions are also developed for ranking query results for shortlisting and expert identification purposes.

3.3 *Skill Similarity Computation*

The most crucial component for matching profiles to a JD is the skill or competency equivalence computation module. Going by the adage that *a word can be understood by the company it keeps,* we model this problem as a comparison of two feature vectors which comprise important topics and associated features of the two skills. For comparing and ranking skills, we have created a corpus of the skill definition documents. These skills are represented as a *tf-idf* [4] feature vector of length equal to the size of the vocabulary. We computed similarity between two competencies using cosine distance between their respective feature vectors.

We use the standard way of identifying similarity between two documents d_1 and d_2 by computing cosine similarity between their vector representations $\vec{V}(d_1)$ and $\vec{V}(d_2)$, where "." denotes the scalar product of the vectors $\vec{V}(d_1)$ and $\vec{V}(d_2)$.

$$sim(d1, d2) = \frac{\vec{V}(d1) \cdot \vec{V}(d2)}{\left|\vec{V}(d1)\right|\left|\vec{V}(d2)\right|}$$

Some example results are presented in Table 2.

[4]*tf-idf: term frequency-inverse document frequency is a measure of importance of a word in a document belonging to a corpus or collection.*

3.4 JD Extraction

Job Description (JD) document is an important document that describes talent requirement in detail including important aspects, namely, technical skills such as "Java" and "Big Data", behavioral skills such as "Communication Skill" and "Team work", service lines such as "Banking" and "Manufacturing", educational qualifications such as "MCA" and "B.E.", location of job, and so on. JDs also include desired expertise levels in skills and experience range. To define expertise level, a JD uses various adjectives such as "Extensively Experienced", "Strong", "Proficient", "Good", "Experienced", and "Understanding". A JD typically also has (a) the profile of the recruiting company in the form of statements "About Company", "Company Strength", "Working Areas", etc. and (b) other elements which are divided into the following categories:

- Technical skills (Technological area) which include attributes such as name, experience (in years), expertise level, mandatory or not, etc.
- Educational qualifications which contain degree name, specialization, subject, area, duration of study, part-time or full-time, etc.
- Service lines represent task involved, experience and expertise level in a task category, etc. Behavioral skills along with a description of expertise level, location, and language(s).

The purpose of extracting content from JD is manifold, and the relevant ones are as follows: (i) the structured attributes from a JD can be used for automated matching and ranking candidate profiles (as described in section on RINX-SE) and (ii) the JD information is useful for verification and validation of a submitted JD for the purpose of completing it in all respects. Given a JD, assuming it can be improved, we describe the recommendation methods targeting completion of JD in this section. The completion of the JD would improve its specification resulting in the identification and shortlisting of better quality candidates for hiring.

3.4.1 Extraction of Technical Skill

A technical skill consists of the knowledge and abilities required to accomplish engineering, scientific, or computer-related work, as well as related specific tasks. A technical skill in IT domain refers to an ability to use a specific IT technology platform, software system, software product, framework, programming language, operating systems such as "Java", "ASP.NET", "SAP", "COBOL", "J2EE", etc. In a JD, the technical skill requirements are specified in multiple forms possible in language. Few examples are as follows:

(a) 3-5-year experience in "Core Java", "Spring".
(b) 3-year experience in "Java", 2 years of experience in "Spring".

(c) `At least 2 years of experience in "Hibernate" or related frameworks.`
(d) `Experience in at least (one/two) technology/frameworks "Hibernate"/"JPA"/"EJB".`
(e) `Knowledge of "Angular JavaScript" is preferred (in this case, the technology is not mandatory).`
(f) `Strong knowledge of "Java".`
(g) `Proficiency in "Java".`
(h) `Must have "Core Java" experience.`
(i) `"Spring" and "Hibernate" are must.`

In examples mentioned above, the technical skills are "`Core Java`", "`Spring`", "`Hibernate`", "`JPA`", and "`EJB`", and keywords like "`Strong`", "`Experience in`", "`Knowledge of`", and "`Proficiency in`" specify the level of expertise required. We analyzed few hundred JDs to identify a list of cue words which are commonly used to specify the expertise levels. The cue words can help in getting intuition which technology is important over other in the current job description profile. Some cue words such as "`Extensive Experience`", "`Strong`", and "`Must have`" indicate that the technologies are of higher importance in the JD and need be assigned higher importance during use in either profile matching or in recommending technical skill to an incomplete JD. Mandatory skills are those which are specifically mentioned in the JD as a "`must have`" or as a "`mandatory skill`". The desired experience in using a skill is mentioned as a numeric value for the number of years as "`2-4`", "`2`", "`2 plus`", "`2-4`", etc. The year as a time unit can be written as "`year`", "`years`", "`yrs`", "`yrs.`", etc.

To extract fields such as skill, expertise level, and experience from JDs, we use the approach as described earlier in section on RINX. A set of examples of the extraction patterns written and applied to the enriched text are given as example. Input sentence is a text input from which entities and its expertise level to be extracted. In the second step, sentence has been enriched using part-of-speech tagging. On this enriched text, handcrafted patterns in the form of regular expressions described in step 3 have been applied to extract the entities with the expertise level. The extraction results for these sentences are shown in Table 3.

Table 3 Sample entity extraction result

S. no.	Expertise level	Skill name	Experience (in years)
1	Experience	SAP Hana Live	5
2	Good	MySQL, Oracle	–

Input Sentence 1: `Minimum 5 year experience with SAP HANA Live content and browser.`

Input Sentence 2: `Good understanding and hands on experience in MySQL or Oracle.`

Enriched Text for Sentence 1: <S><ROOT><S><NP><nsubj_**experience**><NNP>Minimum </NNP></nsubj_experience></NP><NP-TMP><tmod_experience><CD>5 </CD><NN>year </NN></tmod_experience></NP-TMP><VP><VBP>experience </VBP><PP><prep_experience><IN>with </IN><NP><NP><pobj_with><ORGANIZATION><NNP>SAP </NNP></ORGANIZATION><NNP>HANA </NNP><NNP>Live </NNP></pobj_with></NP><ADJP><amod_Live><JJ>content </JJ><CC>and </CC><JJ>browser </JJ></amod_Live></prep_experience></ADJP></NP></PP></VP></S></ROOT><PUNCT>. </PUNCT></S>

Enriched Text for Sentence 2: <S><ROOT><NP><NP><JJ>Good </JJ><NN>understanding </NN><CC>and </CC><NNS>hands </NNS></NP><PP><prep_understanding><IN>on </IN><NP><pobj_on><NN>experience </NN></pobj_on></prep_understanding></NP></PP><PP><prep_understanding><IN>in </IN><NP><pobj_in><ORGANIZATION><NNP>MySQL </NNP></ORGANIZATION><CC>or </CC><ORGANIZATION><NNP>Oracle </NNP></ORGANIZATION></pobj_in></prep_understanding></NP></PP></NP></ROOT><PUNCT >. </PUNCT></S>

Pattern used for Sentence 1:

<CD>([\d\-\s]+)<\/CD>.*?years.*?<\/IN>.*?(<NNP>|<NN>)([A-Za-z\s]+)(<\/NNP>|<\/NN>).*?(Excellent|Understanding|expert|proficient|strong|experience|extensive experience|Knowledge|develop|proficiency|exposure).*?level.*<\/IN>.*?<pobj_in>(.*?)<\/pobj_in>

Pattern used for Sentence 2: (Expert|Strong|Good|Experience|extensive experience|Knowledge|develop).*?(level|skills?)?<\/IN>(.*)(<\/cc_Batch>|<\/[A-Za-z_]+(in|with)>.*?<\/IN>.*?<pobj_[A-Za-z]+>(.*)?<\/pobj_[A-Za-z]+>)

3.4.2 Experimentation and Results

The entities like technical skills, service lines, behavioral skills, location, and language were extracted from the gold copy, wherein the extraction results were post-processed, e.g., standardized using a master list and then manually compared with the content of JDs. The results consisting of precision, recall, and F-measure are provided in Table 4.

Table 4 JD extraction results

Entity	Precision	Recall	F-measure
Technical skill	0.91	0.81	0.85
Behavioral skill	0.96	0.91	0.93
Service line	0.95	0.89	0.92
Location	0.95	0.95	0.95
Language	0.98	0.99	0.98

3.5 JD Completion

The JD completion module generated recommendations for completing the JDs. This can be framed as a problem of mining and learning the most frequent association rules from the training data which are the past JDs. The details of the JD completion module are given in Fig. 4.

As the name suggests, JD completion module tries to *complete the missing information* in case of an *incomplete job description* and makes it more unambiguous in terms of requirements of both technical and behavioral skills. As an example if "`Spring`" technology is present in a JD, then "`Java`" technology is *a must* for the same JD as the Spring framework is built in Java and is used along with Java. There can be many similar examples of such relations as "`JAXB`" (`Java` framework for `XML` binding) and "`Java`" signify use of "`XML`" technology. This problem can be a viewed as identifying the pre-requisite skill names for an advanced technology or frequently used technologies in implementing IT systems. We provide few rules and their examples which guide the completion of a JD:

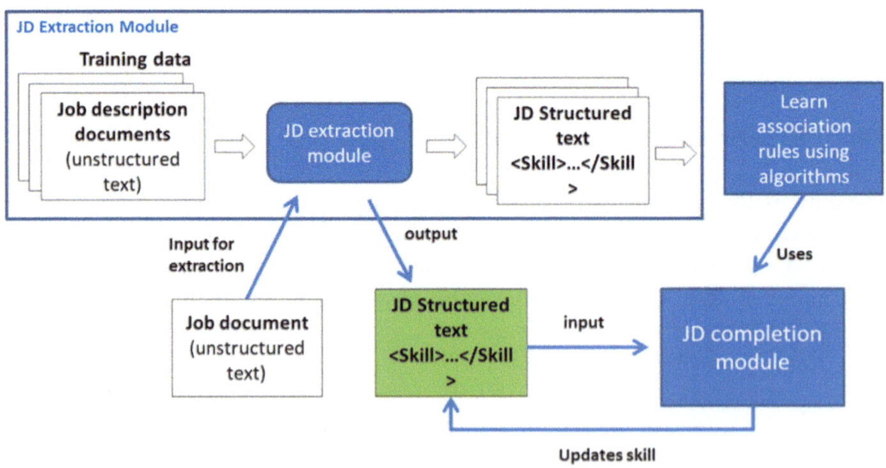

Fig. 4 JD extraction pipeline

1. If "Spring" is present in a JD, then "Java" should also be included, as "Spring" is a framework developed in "Java". Or if "JAX-RS" (Java framework for creating web services) is present, then "Java" and Web services should also be present. Hence, we define "Rule 1" that *if a framework is present then their respective base technology should also be included* in JD.
2. If "Java" is present along with a framework such as "Spring" or "Hibernate", then their IDE's like "Eclipse", "Netbeans" can also be included, as they provide best platform for maintenance of code. Therefore, "Rule 2" can be that *for a framework and root technology, a suitable IDE's should also be included.*
3. If "Oracle" database and "Java" is present, then "JDBC" should also be present, means if two root technologies are present then their connecting framework should also be present. Hence, "Rule 3" will be *if any two root technologies are present, then their connecting framework should also be included*.
4. Few technologies depend upon operating systems such as "Big Data" and "Hadoop" depends on "Linux" operating system. Hence, "Rule 4" will be *if operating system-dependent technologies are present in a JD then include their respective operating systems*.
5. With technologies such as "Java" or IDE like "Eclipse", it is good to have knowledge about version control framework. A candidate has to work in a team and a version control tool is generally used to manage the code versions and collaboration. Hence, "Rule 5" will be a *version control framework recommendation based on technologies present in JD.*

The proposed solution approach uses domain knowledge to lay out a structure or template for a JD, and then uses association rule mining to identify the most likely technology or frameworks to fill the placeholders in the template. The current algorithm is based on co-occurrence in the past data and associated domain knowledge about the composition of a JD.

For the frequent dataset mining, the algorithms used are (a) Apriori and (b) Frequent Pattern (FP)—Growth. For the association rule mining, the algorithms applied are (a) TopK Rules nonredundant (TNR) (Fournier-Viger and Tseng 2012) and (b) closed rule discovery using FP close (Grahne and Zhu 2005) for frequent dataset mining. *TopK Rules* algorithm finds Top"K" rules from the training dataset based on the value of confidence; here, we have used another variant of TopK rules, i.e., *TNR (TopK nonredundant rules)* which removes the redundant rules from the association rule list produced in TopK rule algorithm. In this example of a redundant rule, the rule {a} -> {c, f} is redundant given another rule, {a} -> {c, e, f}.

3.5.1 Experimental Results

For completing a JD, dataset of nearly 300 JDs of "`Java`" and "`SAP`" has been used for training, and 56 JDs for testing. We have used the following experimental approach to test the mining techniques, for recommending five best matching technology names to a JD using Apriori and FP-Growth algorithm:

1. Generate set of frequent patterns(T_{FP}) using training dataset of JDs using an algorithm.
2. Given a test JD, which contains technology set T = $\{tech_1, tech_2, tech_3, tech_4, tech_5\}$, remove two technologies randomly, denoted by $T_{\mathrm{rem}} = \{tech_3, tech_5\}$, and the set of technology names remaining is denoted by $T_{\mathrm{left}} = \{tech_1, tech_2, tech_4\}$.
3. In T_{left}, create all different combinations of technology set and put in T_{set}, e.g., $\{tech_1, tech_2\}$, $\{tech_1, tech_2, tech_4, \ldots\}$, and so on, starting with the longest subsequence.

   a. Find all the sequences from FP algorithm which have all technologies associated with the test JD mentioned. For example, in sequence $\{tech_1, tech_2, tech_4\}$, two sequences such as $T_{list} = \{\{tech_1\ tech_2, tech_4, tech_8, tech_5\}$ and $\{tech_1, tech_2, tech_4, tech_5\}\}$ contain all three technologies.
   b. Make a list of all technologies which are not part of T_{left} but present in T_{list} and count their occurrence. The resultant list of such technologies would be $\{(tech_8, 1), (tech_5, 2)\}$.
   c. Add these technology names to recommendation list (T_{Recomm}).
   d. Repeat from step one with relatively smaller sequences until T_{Recomm} contains at least *five recommendations*.

4. If the technology added to T_{Recomm} matches with any one of the T_{rem}, we mark it as a hit (1) or else a miss (0). The hits are counted for all JDs in the test set as N_{Hits}.
5. Let N_{nRec} represent the count of the JDs for which there are no recommendations and N_{JD} is total number of JDs, then

$$Precision@5 = \frac{N_{Hits}}{N_{JD} - N_{nRec}} * 100$$

The accuracy of the results is measured using Precision@5, which implies the correct matches among the top 5 recommendations. The correct match during testing can be for one or two randomly removed technology names. The results of the correct recommendations against two of the removed technologies are termed as "Horizon = 2" in Table 5.

Table 5 JD completion results

S. no.	Algorithm and setup variables	Hit	Total count	No of Reco.	Precision@5, Horizon = 2
1	Apriori, FP Growth	18	56	–	32.14
2	TNR (top k rules nonredundant) Confidence = 0.8, TopK rules = 2000	12	56	–	21.42
3	TNR (top k rules nonredundant) Confidence ≥ 0.8, TopK rules = 5000	17	56	21	48.71
4	TNR (top k rules nonredundant) Confidence ≥ {0.9–0.4}, TopK rules = 5000	21	49	12	56.75
5	TNR (top k rules nonredundant) Confidence = {0.9–0.4}, TopK rules = 5000	24	55	16	61.53
6	TNR (top k rules nonredundant) Confidence = {0.9–0.4}, TopK rules = 5000	31	49	8	**75.6**
7	Closed rule using FP close	17	54	13	41.46

4 Conclusion and Future Work

In this paper, we focus on the domain-driven text and data mining components to build point solutions to deploy analytics at appropriate juncture in recruitment process. Each such component is designed to address a specific business issue related to efficiency, quality, or cost in TA-related processes. In this paper, we outline text and data mining-based components to build important and mandatory capabilities for enabling efficient and intelligent recruitment.

The future work would aim to assess behavioral attributes to gain deeper insights into the candidate profiles. Also, we would apply information fusion to create comprehensive profile by harvesting data and inputs from social networking and collaboration platforms. The continuous improvement using learning algorithms for better measurement of technology similarity as well as for updating gazettes and ontologies would be pursued. This would be necessary as the evolving technological and business landscape would require continuous mining of the information sources to maintain the readiness of the deployed solutions and capabilities.

References

Aggarwal, C. C., & Zhai, C. X. (2012). Mining Text Data (1st ed.). Springer.

Bondarouk, T., Ruël, H., Guiderdoni-Jourdain, K., & Oiry, E. (2009). Handbook of Research on E-Transformation and Human Resources Management Technologies—Organizational Outcomes and Challenges. IGI Global.

Bui, H. Q., & Zyl, L. T. V. (2016). Talent acquisition gamified: Insights from playing the game at PwC Hungary, Master thesis, Lund University, School of Economics and Management.

Dutta, D., Mishra, S., Manimala, M. J. (2015). Talent acquisition group (TAG) at HCL technologies: improving the quality of hire through focused metrics (p. 22). IIMB-HBP. http://research.iimb.ernet.in/handle/123456789/6698.

Edmundson, H. P. (1969). New methods in automatic extracting. *Journal of the ACM, 16,* 264–285.

Faliagka, E., Tsakalidis, A., & Tzimas, G. (2012a). An integrated e-recruitment system for automated personality mining and applicant ranking, *Internet Research*, **22**(5), 551–568.

Faliagka, E., Ramantas, K., Tsakalidis, A., & Tzimas, G. (2012b). Application of Machine Learning Algorithms to an online Recruitment System. In *ICIW 2012: The Seventh International Conference on Internet and Web Applications and Services* (pp. 216–220). IARIA.

Fournier-Viger, P., & Tseng, V.S. (2012). Mining top-K non-redundant association rules. In *Proceedings of 20th International Symposium on Methodologies for Intelligent Systems (ISMIS 2012)* (pp. 31–40). Springer, LNCS 7661.

Grahne, G., & Zhu, J. (2005). Fast algorithms for frequent itemset mining using fp-trees. *IEEE Transactions on Knowledge and Data Engineering, 17*(10), 1347–1362.

Mooney, R. J., & Bunescu, R. (2005). Mining knowledge from text using information extraction. *ACM SIGKDD Explorations Newsletter, 7*(1), 3–10.

Parthasarathy, M., & Pingale, S. (2014). Study of talent acquisition practices—A review on global perspective. *International Journal of Emerging Research in Management & Technology., 3*(11), 80–85.

Patil, S., Palshikar, G. K., Srivastava, R., & Das, I. (2012). Learning to rank resumes, FIRE 2012: In *Proceedings of the 4th Annual Meeting of the Forum on Information Retrieval Evaluation*. ISI Kolkata, India.

Ronen, F., & James, S. (2006). The Text Mining Handbook: Advanced Approaches in Analyzing Unstructured Data (Vol. 1). Cambridge University Press.

Schiemann, W. A. (2014). From talent management to talent optimization. *Journal of World Business., 49*(2), 281–288.

Srivastava, R., Palshikar, G. K., & Pawar, S. (2015). Analytics for Improving talent acquisition processes. In *International Conference on Advanced Data Analysis, Business Analytics and Intelligence, ICADABAI 2015*. IIM, Ahmedabad, India.

Strohmeier, S. (2007). Research in e-HRM: review and implications. *Human Resource Management Review, 17*(1), 19–37.

Téllez-Valero, A., Montes-y-Gómez, M., Villaseñor-Pineda, L. A. (2005). Machine learning approach to information extraction. *Computational Linguistics and Intelligent Text Processing*, 539–547.

Hastie T., Tibshirani, R., & Friedman, J. (2008). Elements of Statistical Learning (2nd ed.). Springer.

Tomassetti, F., Rizzo, G., Vetro, A., Ardito, L., Torchiano, M., & Morisio, M. (2011). Linked data approach for selection process automation in systematic reviews. In *15th Annual Conference on Evaluation & Assessment in Software Engineering (EASE 2011)* (pp. 31–50).

Assessing Student Employability to Help Recruiters Find the Right Candidates

Saksham Agrawal

Abstract India produces lakhs of engineers every year, but employability has been a concern for educators and recruiters alike. Cost of technical recruitment has been driven extremely high due to inconsistent quality of talent available. Objective assessment of students has helped, but recruiters still use simple cutoffs followed by subjective methods like group discussions and personal interviews which are inherently unscalable and expensive. In the present work, we looked at most recent employment data for almost 50,000 fresh engineering graduates and established precise objective relationships between students' scores across several dimensions, and the students' employability as measured by actual job offers made to them by employers. Through regression modeling, we were able to develop a composite score that identified employable candidates with five times greater precision than any individual score. We then went a step further to cluster employers on the basis of their ability to choose the stars from among a large pool of candidates, to identify employers who can be advised to further optimize their recruitment spend.

Keywords Engineering · Education · Employability · Regression modeling
Clustering

1 Introduction

Each year, India produces more than 16 lakh engineering graduates, products of over 3500 colleges (Mohanty 2016). Unsurprisingly, there is a vast variance among the graduates and the quality of training they receive. Private engineering colleges make up about 85% of total capacity (Sharma 2014). Therefore, the government's role in improving quality of technical training, at least for the vast majority of supply, has been limited to enforcing standards. As a result, AICTE plans to pare down the number of engineering seats by 40% to about 10 lakh graduates a year (Livemint 2015).

S. Agrawal (✉)
Kanpur, India
e-mail: saksham.agrawal@gmail.com

A. K. Laha (ed.), *Advances in Analytics and Applications*, Springer Proceedings in Business and Economics, https://doi.org/10.1007/978-981-13-1208-3_14

As per NASSCOM, only 26% of all engineering graduates are fit to join the workforce (Nasscom 2014). The top 15–20 institutes might be able to place almost all of their students, but outside of that, employability drops to well below 10%. By the time a recruiter visits a 50th ranked engineering college for campus placements, she ends up assessing more than 50 candidates for each offer she rolls out.

Education entrepreneurs have been working on this specific problem over the last decade. The most fruitful approach has been the development of standardized tests. These tests measure each student on dimensions that are more relevant than what education boards measure, and standardize it better than individual colleges can. They also collect data about the educational institutes and companies as a natural part of the operations. This has created the opportunity to use this large amount of data to algorithmically reduce the cost of placements.

2 Research Objectives

Recruiters collect an array of objective (e.g., tests and cutoffs) and subjective (e.g., group discussions and interviews) data before offering a job. Our objective was to see how students' test scores could be used to assess employability and predict salaries, *prior to* conducting any other recruitment activity. Following specific objectives were set:

1. Establish how well each test score predicted a candidate's employability.
2. Determine objectively the salary a candidate could expect to earn based on test scores.
3. Identify the relationships, if any, between available jobs and the candidates they select, remembering that while many students will not make the cut for any given job, several jobs will be overlooked by a student since it does not pay their expected salary.
4. Cluster the recruiters based on their ability to effectively identify good candidates from their applicant pools.

3 Data Analysis

3.1 Approach

Establishing relationship between test scores and employability: We want to establish the role of test scores of a student and the chance that the student will get a job offer. In this analysis, the test scores, as well as any information available to a potential recruiter, become the independent variables, while the outcomes, i.e., whether a student got a job, and for what annual salary, become the dependent variables. Appropriate regression methods have been applied to derive meaningful relationships.

Establishing relationship between a pool of jobs and a pool of candidates: We want to understand better, which kinds of candidates are selected for which kinds of jobs (employers). This becomes a segmentation problem. We determined the most "informative" aspects of the data through a principal component analysis to reduce the number of variables, and then applied a number of clustering methods to this reduced set of "features" to segment the employers by their ability to select superior candidates from their pool of applicants.

3.2 Data Collection

We collected data for 47,197 final-year engineering students who, during the recruitment season of 2015–16, made 61,181 applications to 73 companies. These applications resulted in 767 job offers to 722 unique candidates.

For each student, the following data were available:

Student Profile—Degree, specialization, and college.

Generic Assessment Data—Class X marks, class XII marks, and undergraduate grade point average.

Specific Assessment Data—Scores of specialized tests, including those in English, quantitative, analytical, domain, and coding expertise.

In addition, we combined to this the following:

Employer Profile—For each job, the type of the job profile and offered salary.

Application and Offer Data—Binary indicators of whether a student applied to a job, and whether a student received an offer from a job, as well as the identifier of the company that the student got an offer from.

3.3 Assumptions, Constraints, and Mitigation

1. Overall employment rate is only 1.5%. While this makes the employability question more urgent than ever—who wants to interview 100 students only to reject 98 of them—this made the data more sparse than ideal.
   - *Mitigation*: we oversample the offer data by a factor of 10 prior to regression. To do this, we keep all the 722 records which have an offer. From the remaining 46,475 records, we randomly select 10% of the records. Combining these two gives us a dataset which includes all the positives and has an employment rate of 13.4%, which we use for regression modeling.
2. Graduate percentage was missing for 18% of the pool.

- *Mitigation*: We needed to solve this through missing-value imputation. We supplied the median value of graduation percentage to those for whom it was missing and could not be collected. We did not rely on more sophisticated methods of imputing missing values, such as k-nearest neighbor methods. This was because we did not know the factors because of which the graduation percentage was missing, and therefore could not make any assumptions about the underlying relationship between graduation percentage and other data points. On the other hand, knowing that graduation percentage will be considered independent of other examination scores by our model, and that it has a broadly bell-curve distribution in real life, we went ahead with imputing missing values using the median graduation percentage in our data—70%.

3. Salary contained only 30 distinct values, since it was the one from the job posting made at the campus, rather than the final salary negotiated with a candidate.

    - *Mitigation*: entry-level jobs and salaries, especially for large pools of candidates like these, are fairly standardized. We considered the salaries as given.

4. While we had the "Applied" and "Offered" data, we were missing two steps in between. Information about any filtering in between—which candidates were shortlisted, and which candidates were able to go the interview—was not available.

    - *Potential mitigation*: One way to partially counter these issues would be to add "interview city." The largest factor in determining whether someone goes to an interview or not is whether it is accessible from the college campus. However, this treatment was not done for the present analysis.

5. We assumed that the "Offered" data was "complete", in that the companies hired all their open positions from this pool, i.e., they interviewed as many people as they needed to find the right candidates, which also means that the offer rate of each job is indicative of the efficiency of that recruiter in finding the right candidates.
6. Data included mostly engineering undergraduates and some postgraduates, but also a few MCA candidates who were eligible for the same jobs. We kept them in.

3.4 Exploratory Data Analysis

The software R was used for both data exploration and the subsequent modeling.

Data Preparation
Data is cleaned and quality-checked for accuracy. For example, we ensure that whenever the "offer" data is 1, the "application" data is also 1, and if offer data is 1, then the company that gave the offer is correctly identified. These data quality rules became

important since the data was collated manually and saved hours of work by catching errors early.

Familiarizing with the Data
The overall employability (percentage of students who got a job offer) is 1.5% among the students under consideration, but this is a function of the competitive nature of the jobs on offer. The companies under consideration received an average of 838 applications and rolled out only 10 offers on average. It is also a function of the number of applications each student made—merely 1.3 applications per student.

The salaries offered range from Rs. 1,20,000 lakhs per annum to Rs. 12,00,000 lakhs per annum.

Bivariate Analysis
Bivariate plots (Fig. 1) are developed to see how well each of the scores do in predicting the chance of a student getting an offer and the salary he will fetch, by plotting each of these statistics against equal-sized "bins" of the scores under consideration. These plots give a visual understanding of how the improvement of a certain score affects candidates' chances of success in the job market.

Aptitude and graduation performance turn out to be the best indicators of success, followed by other individual test scores. Graduation percentage is something that is frequently used while hiring. It also indicates, in the Indian context, how sincere a candidate is, since it is the result of 3/4 years of continuous effort.

Basic level of expertise in English (cutting off at nearly 45%) seems to ensure a better average salary. This could be for a number of reasons, including the possibility that these candidates perform better in interviews.

3.5 Data Modeling and Methodology

The following methodologies were used to model and analyze the data:

1. Data partitioning—We separated the data 70:30 to create training and testing partitions. We also compared the key statistics of each group to ensure that partitions were statistically significant.
2. Logistic regression for probability of receiving an offer—We developed a logistic regression model to capture precise and relative contribution of various scores to a probability of getting a job offer for each student.
   Transformations: The following transformations were made to the data based on bivariate analysis:

   (a) Graduation percentage, where missing, was imputed with the median value (70)

$$\text{Graduation.Percent.T} = \begin{cases} \text{Graduation.Percent when available} \\ \text{70, when Graduation.Percent not available} \end{cases}$$

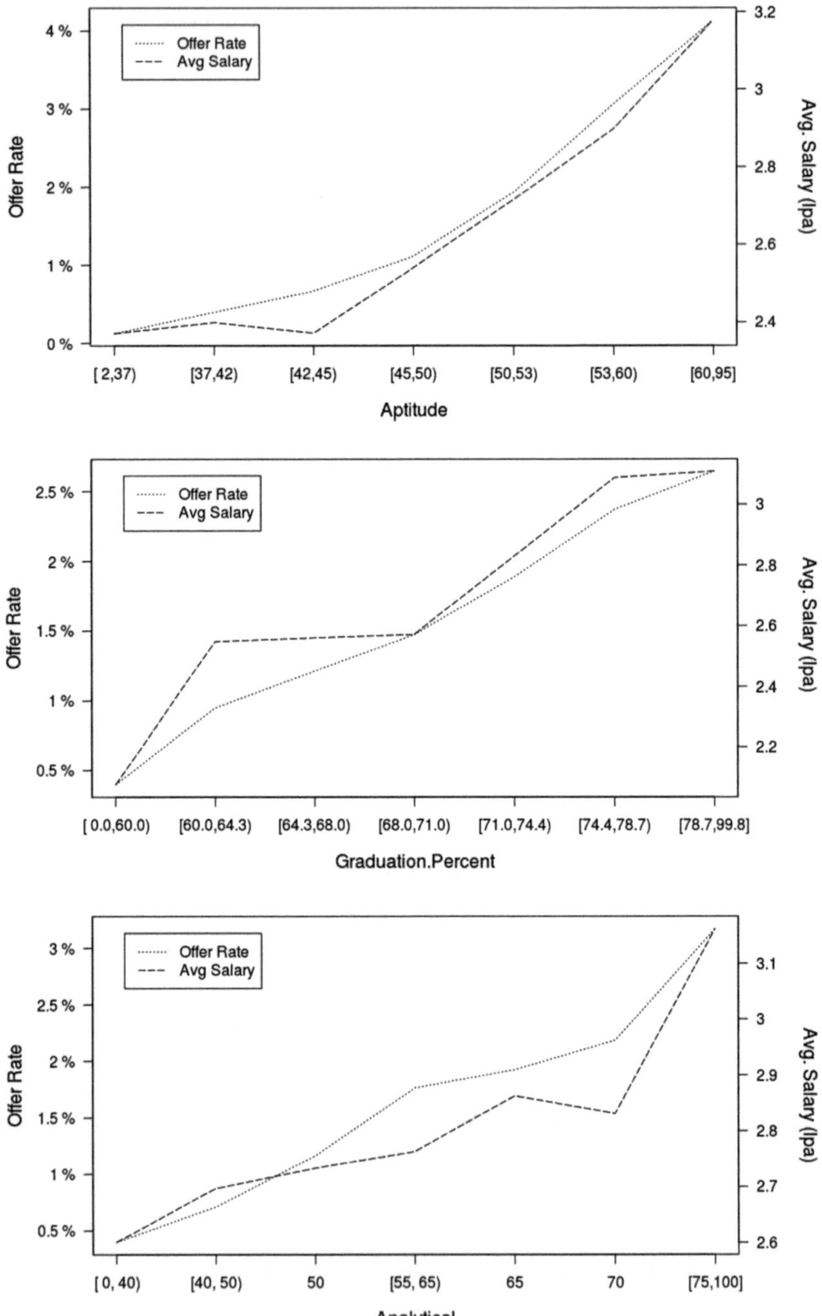

Fig. 1 Bivariate plots showing relationship of certain scores to offer rate (dotted) and avg. salary (dashed)

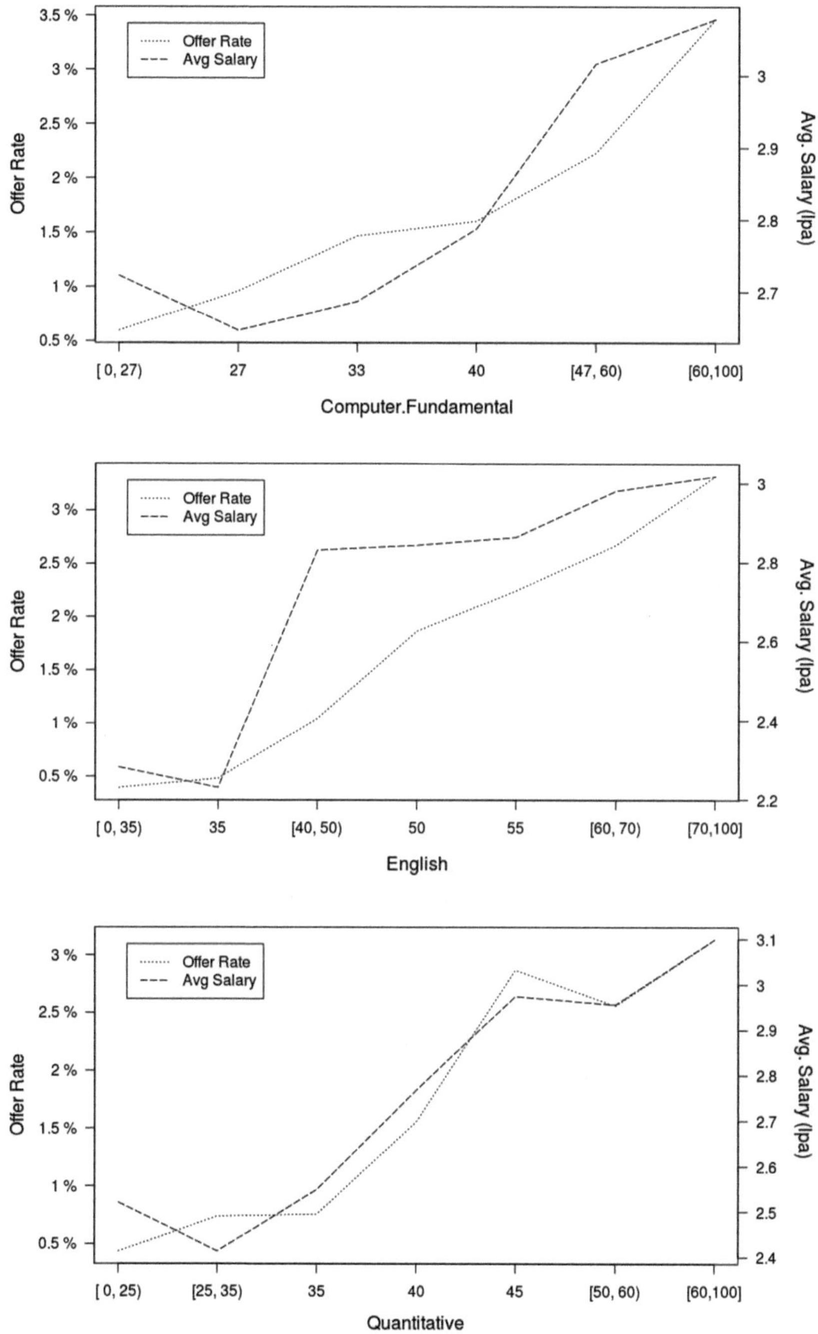

Fig. 1 (continued)

(b) Coding and quantitative scores were converted to binary, using the following transformations:

$$\text{KnowsCoding} = \begin{cases} 0 \text{ if score in Coding} < 10 \\ 1 \text{ if score in Coding} \geq 10 \end{cases}$$

$$\text{KnowsQuant} = \begin{cases} 0 \text{ if score in Coding} < 40 \\ 1 \text{ if score in Coding} \geq 40 \end{cases}$$

This transformation ensures that we are not overburdening these two variables with a predictive power that they do not possess.

English, analytical, and aptitude scores show multi-collinearity, and a stable model is derived by selecting the first two and dropping the third.

3. Linear regression for salary—Similar to above, we developed a linear regression model to estimate the predicted salary for each student. The salary is capped at Rs. 5,00,000 per annum since there are only three offers above this salary, which are discounted as outliers.
4. Preparing employer-level data—Next, we move on to analyze and model the various jobs and employers in the data. For this, we combine the student-level data used for previous analysis to create the employer-level data. For each job, we consider the "applicant pool" capturing the features of the students who applied to each job, as well as the "offer pool," capturing similar features of the students who got an offer from that employer. Features include mean scores across various dimensions.
5. Principal component analysis and dimension reduction—Once we have aggregated the data in this manner, in order to reduce the dimensionality of this large amount of data, we apply a PCA analysis to identify the number of dimensions.
6. K-means clustering—To further arrive at the number of clusters, we use R's built-in library to compare the "within-cluster" and "between-cluster" variances for number of clusters, ranging from one cluster to ten clusters.

We cluster the various jobs on the basis of the quality of their applicant pool (measured by average scores of the applicants for each job), as well as the quality of the selected candidates (measured by the average scores of the selected candidates for each job). This helps us gain insight into which jobs (employers) were doing a good job of selecting superior candidates, and which ones had greater opportunity of further improving the effectiveness of their recruitment effort. After iteratively developing confidence in the stability of the clustering, the final step is to interpret, visualize, and present the results.

Table 1 Coefficients and p-values of regression model for probability of a candidate to receive an offer

	Estimate	Std. error	z value	p value
Intercept	−9.373117	0.364143	−25.74	<2e−16
Graduation. Percent.T	0.024908	0.004076	6.11	9.94e−10
Analytical	0.015759	0.003215	4.902	9.49e−07
English	0.026684	0.002931	9.104	<2e−16
Computer. Fundamental	0.013338	0.003212	4.152	3.29e−05
KnowsCoding	0.454821	0.103291	4.403	1.07e−05
KnowsQuant	0.911164	0.108181	8.423	<2e−16

3.6 Results

Regression: The coefficients and p-values are shown in Table 1. The four scores (Graduation Percentage, Analytical, English, and Computer Fundamental) are highly significant, and binary variables of whether or not a candidate knows Coding, or has Quantitative skills also, make it to the final model.

For the student, this means that scoring well in Quant, Coding, and English, apart from having good graduation percentage, can significantly increase the chance of a job offer, while class X and XII marks do not seem to matter all that much.

No single test score does an outstanding job of predicting employability. Regression provides a composite score, combining the candidate's individual test scores with their graduation percentage, and can become a significantly better lever to control the shortlist (and hence increase the offer rate) for any company. We call it the "Composite Score" and calculate it as below:

$$\text{Composite Score} = 1000 * 1/\{1+e^{-1\times \text{Y}}\}$$

where

$$\begin{aligned} \text{Y} = -9.37 &+ 0.027 * \text{English} \\ &+ 0.025 * \text{Graduation.Percent} \\ &+ 0.016 * \text{Analytical} \\ &+ 0.013 * \text{Computer.Fundamental} \\ &+ 0.91 * \text{KnowsQuant} \\ &+ 0.45 * \text{KnowsCoding} \end{aligned}$$

Equation: Calculating the composite score; multiplication by 1000 makes the score easier to communicate.

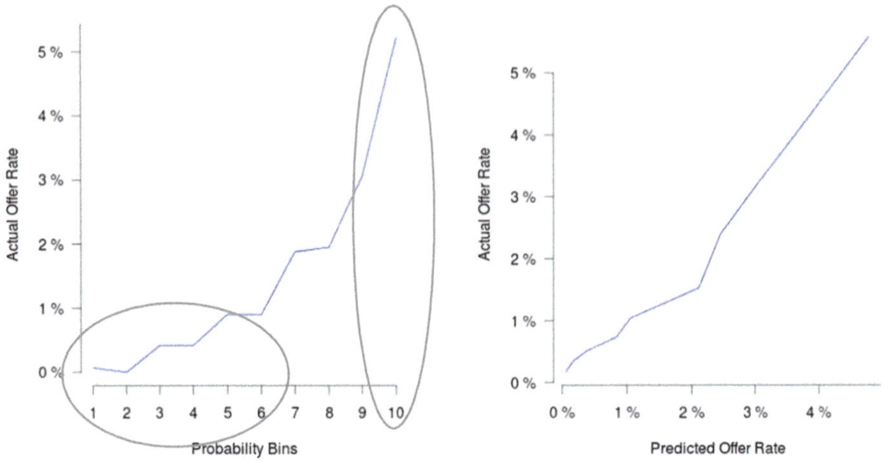

Fig. 2 Composite score using probability of offer: Lorenz curve and actual-versus-predicted charts showing how well a composite score determines employability (X-axis is score, and y-axis is average offer rate)

Composite Score: The composite score combines individual test scores with graduation performance and provides the best prediction of the employability of a student. As shown in Fig. 2, top-10% of students identified using this composite score are up to 12 times as likely as the bottom-60%.

Enhancing the Composite Score: The composite score calculated above makes use only of the estimated probability of getting a job offer. It does not take into account the salary that each candidate can expect to make. This information (salary offered) is also available in the original data and can be regressed linearly with the candidate's scores. In other words, the predicted offer rate can be combined with a predicted salary to further segment the student pool into those with decidedly good scores (high chance of high paying job) versus unemployable (no chance of getting a job).

This approach is further clarified by Table 2. The lowest row (High probability of offer) shows the best 6410 candidates in our data, but we can use the predicted salary to further identify the 4990 "star" candidates from within this pool.

Clustering the Jobs

For each of the jobs in our dataset, we calculate the average scores of all candidates who applied to that job (applicant pool) and the average scores of all candidates who were successful in securing that job (offered pool). This produces a dataset (called jobs data in the rest of this section) of 73 jobs across 20 dimensions (10 scores of the applicant pool and 10 scores of the offered pool).

Principal component analysis of this data showed that out of these 20 dimensions, the top four dimensions capture about 75% of the variance, and the top eight dimensions capture more than 90% (Fig. 3). This gives us confidence that clustering can

Table 2 Segmentation of student pool by both employability and estimated salary

	Low Expected Salary	Medium	High Expected Salary
Low Prob of Offer	**22079** 0.6% 2.33 lpa	**3939** 0.7% 2.37 lpa	**757** 1.1% 2.56 lpa
Medium	**2545** 2.3% 2.57 lpa	**3249** 2.4% 2.66 lpa	**2733** 1.9% 3.10 lpa
High Prob of Offer	**327** 5.2% 2.70 lpa	**1183** 5.9% 2.78 lpa	**4900** 5.9% 3.26 lpa

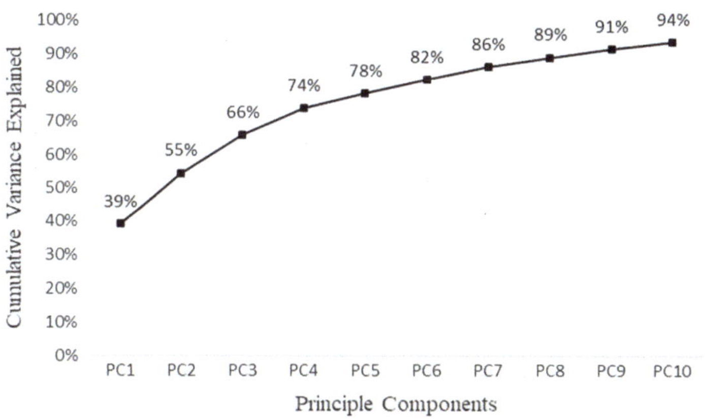

Fig. 3 Principal components of the Jobs data, showing that the top-4 components capture almost three-fourths of the variance present in the dataset across jobs

be performed on this data and gives an idea about the number of clusters we should reasonably hope to segment these 73 jobs—around 4–8.

To further confirm the number of clusters, we also performed a "between-clusters" and "within-clusters" sum of variances across a varying number of clusters. This approach starts with assuming two clusters (n = 2) and calculating the total variance *within* those two clusters, as well as the variance *between* the two clusters. The

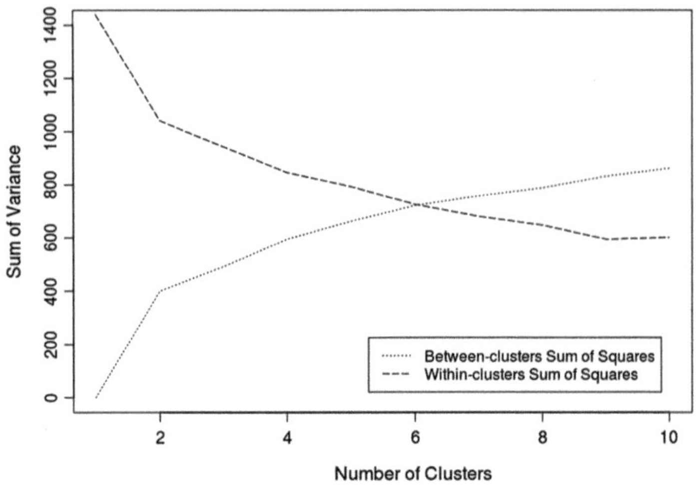

Fig. 4 Between-clusters and within-clusters variances for jobs data, for number of clusters n = 2 through n = 10

Table 3 Defining characteristics of the four job clusters; notice how cluster 1 jobs do significantly better at selecting candidates with higher composite scores than cluster 2 jobs

Cluster ID	Number of jobs	Average salary	Avg # of applications	Avg # of offers	Selection rate (%)	Avg composite score of applicants	Avg composite score of selected candidates
1	14	458,214	1978	28.4	3.7	28.0	57.0
2	23	256,304	616	5.7	1.2	19.5	30.3
3	6	325,167	782	4.8	1.1	18.2	17.4
4	30	204,294	488	7.0	2.1	14.9	22.6

same is repeated while changing the number of clusters n = 2, 3, …, 10 (Fig. 4). This analysis further corroborates our expectation to find between four and eight meaningful clusters within the jobs data.

For our 73 jobs, for number of clusters n = 6, the within-cluster variance is approximately equal to the between-cluster variance. However, two of the clusters are of one job each. For n = 5, the clusters change drastically based on the starting point, i.e., the clustering is not stable. Therefore, n = 4 is selected as the optimal choice for further interpretation.

The characteristics of the four clusters are as shown in Table 3.

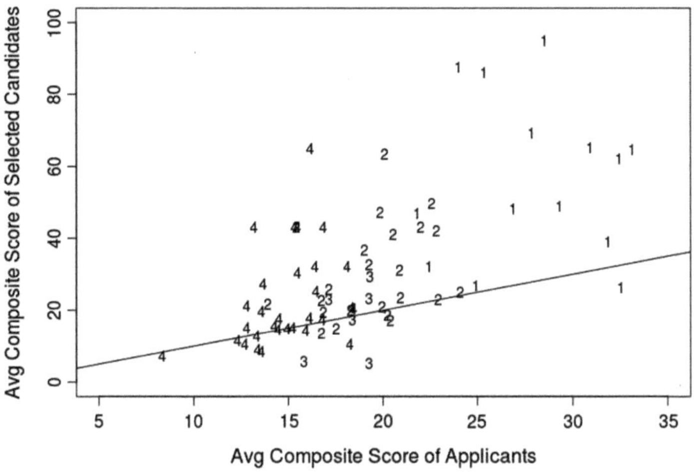

Fig. 5 One of many ways to visualize the job clusters. Every data point represents a job, and labels represent the cluster ID. X-axis shows the average composite score of the applicant pool, while the y-axis shows the same metric for the selected candidates. Notice how cluster 1 jobs do significantly better at selecting good candidates than cluster 3 jobs

Interpreting Job Clusters

The four clusters are visualized in Fig. 5 and can be interpreted as follows:

Cluster 1: Highly Discerning—They are able to recruit high-performance students through their selection process.
Cluster 2: Somewhat Discerning but with Low Offer Rates—These employers can do a better job of reducing the applicant pool even before the selection process starts, by keeping a higher bar of application.
Cluster 3: Non-discerning—These are higher paying and low offer rate jobs, but end up hiring candidates with average scores similar to the applicant pool. They can use significant help in identifying good candidates, and this help will likely come most readily from the testing and recruitment industry.
Cluster 4: Non-discerning and Low-paying—These are bulk-hiring companies, who give out a large number of low-paying employment offers and are satisfied with even a moderately effective recruitment process.

The clusters 2 and 3 indicate opportunity in optimizing selection process (and by extension, return on recruitment costs), through vetting their applicants earlier in the process, as well as by collaborating with recruitment companies to select better candidates more cheaply.

4 Discussion

Standardized scores across relevant dimensions, when combined with the kind of analysis done here, can be used in a variety of ways to reduce the cost of supply of engineering talent. Using test scores on various dimensions can help predict job candidate fit better than any individual score, based on historical data. Also, long-term recruiters can optimize their shortlists on the basis of analyzing the scores of candidates that were successful in the past. Third, since the tests are administered online and remotely, we can measure the same candidate multiple times during his graduation course to further fine-tune the suitability of the student to the role.

Eventually, of course, the role of the recruiter can be automated to the extent that he has to do no more than an elevator conversation with a couple of most promising candidates to achieve the most optimal fit.

References

Atasi Mohanty, D. D. (2016). Engineering education in India: Preparation of professional engineering educators. *Journal of Human Resource and Sustainability Studies*, 92–101.

Livemint (2015). AICTE to cut number of engineering college seats by 600000. Retrieved November 14, 2016, from http://www.livemint.com/Politics/BphkOxYuir6OaYcTrBtldJ/AICTE-to-cut-number-of-engineering-college-seats-by-600000.html.

Nasscom (2014). Transformation roadmap for the Indian technology and business services industries. Retrieved November 14, 2016, from http://www.nasscom.in/NASSCOM-PERSPECTIVE-2020-Outlines-Transformation-Roadmap-for-The-Indian-Technology-and-Business-Services-Industries-56269.

Sharma, N. (2014). Expansion of engineering education in India: Issues, challenges and achievable suggestions. *Journal of Academia and Industrial Research*, 118–122.

Part V
Operations Analytics

Estimation of Fluid Flow Rate and Mixture Composition

Pradyumn Singh, G. Karthikeyan, Mark Shapiro, Shiyuan Gu and Bill Roberts

Abstract Flow rate measurement of oil and gas mixture is the quantification of bulk fluid movement, and it helps in monitoring the oil rig. In this paper, we use acoustic sensor's output which is the result of inside vibrations in pipeline due to fluid movement. Two approaches for estimating the flow rates are discussed: First-order auto-regression and hidden Markov model. Mel-frequency cepstral coefficients are used as features for hidden Markov model. Both approaches show good prediction accuracy.

Keywords Hidden Markov model · Auto-regression · Mel-frequency cepstral coefficients · Nyquist frequency

1 Introduction

One of the key problems in oil production is to determine flow rate of oil/gas mixture in the pipeline. Currently available methods rely on flow meters that are expensive, require installation of dedicated tank separators, and allow to measure the flow rate from a single well at a time for a short period of time. We provided a proof-of-concept using accelerometers with acoustic range bandwidth to estimate flow rates. The key idea is that flow of mixtures inside the pipe generates acoustic range vibrations that can be sensed by an accelerometer magnetically attached to the pipe (http://www.ni.com/white-paper/3807/en/#toc3), and analytic methods can be used to estimate the flow rate from the accelerometer signal. Our approach uses inexpensive sensors that can be installed on pipes in oil wells and provide continuous estimate of flow rates for fluids extracted from the wells.

To simulate oil production conditions, we designed an experiment to collect data at two university laboratories. A closed-circuit pipeline was assembled, and the flow rate and oil/air mixture composition were controlled using an oil pump and air

---

P. Singh (✉) · G. Karthikeyan · M. Shapiro · S. Gu · B. Roberts
Advanced Analytics and Modelling, Deloitte Consulting LLP, New York City, USA
e-mail: pradysingh@deloitte.com; pradyumns3@gmail.com

A. K. Laha (ed.), *Advances in Analytics and Applications*, Springer Proceedings in Business and Economics, https://doi.org/10.1007/978-981-13-1208-3_15

compressor. First, we estimated autoregressive models for each condition. For evaluation, we calculated a normalized log-likelihood score using the AR coefficients. For 63 test conditions, AUC was 0.9676 across five fold. Second, we used a sliding window approach to convert the sensor time series to a sequence of cepstral coefficient vectors for each condition. These cepstral sequences were used as predictors in a Gaussian mixture model and hidden Markov model with an average diagonal accuracy of 92.5% across five fold.

2 Materials and Methods

Acoustic sensors are placed on fluid carrying pipes in a laboratory. The sensors included Class I compliant and noncompliant accelerometers and microphones. A recording is made for 5 s for each condition. We record time series data for 63 different conditions which correspond to a combination of different flow rates and different fluid mixture combinations. Fluid mixture combinations constitute of different ratios of oil, water, air, and gas. Flow rates vary from a minimum of 5–82 gal/min. Air pressure varies from 40 to 50 psi (lb/in.^{2}). Gas amount varies from 5 scf (standard cubic foot) to 6.5 scf.

2.1 Signal Description

Time series data collected from the two sensors were corrupted with high-frequency noise from the acquisition process. In order to get a reasonable estimate of the signals, it was necessary to filter the signal to remove this noise. For this, an implementation of a low-pass filter described in http://www.analog.com/media/en/technical-documentation/dspbook/dsp_book_Ch15.pdf was made.

2.2 Signal Filtering and Down-Sampling

Absolute value of the time series data was considered before applying a low-pass filter described in http://www.analog.com/media/en/technical-documentation/dspbook/dsp_book_Ch15.pdf. A moving average low-pass filter of the form shown in Eq. (1) was implemented:

$$y[m] = \frac{1}{K} \sum_{l=0}^{K-1} x[m+l] \tag{1}$$

Here, K is the window size of filter, x is input signal, and y is output signal.

The cut-off frequency (Oppenheim 1999) for the signal sampled at 25,600 samples/s (henceforth referred to as signal 1) was arrived at approximately 2200 Hz, while the signal sampled at 100,000 samples/s (henceforth referred to as signal 2) had a cut-off frequency of approximately 9000 Hz. Since the signals for each flow rate were assumed to be drawn only from that particular flow rate, a down-sampling procedure was carried out by choosing signals at every 5th sample, which translates into a frequency of approximately 5100 and 20,000 Hz for signal 1 and signal 2, respectively. Both these frequencies obey the Nyquist frequency (Oppenheim 1999) rules, and hence are considered to be appropriate for proceeding ahead with further processing.

2.3 Auto-regression Model: AR(1) Process

An assumed autocorrelation or a linear association between lagged observations in time series data from our sensors provided us with the motivation to adopt an autoregressive model. With experimental evidence, a lag of 1 was arrived at for an $AR(1)$ process of the form $y_t = c + \varphi * y_{(t-1)} + \epsilon_t$, with *yt* being the current signal and $y_{(t-1)}$ being the lagged signal with a time lag of 1. ϵ_t is the white noise with φ and c being the parameters of the model and constant terms, respectively.

2.4 Hidden Markov Model

The signals obtained are continuous in nature and can show properties similar to that of speech signals. Hidden Markov models (HMM) are extensively utilized in speech classification or recognition. Hidden Markov models can be studied in (Baum and Egon 1967; Baum and Petrie 1966; Baum and Sell 1968; Baum et al. 1970; Baum 1972). HMM has been a popular choice in speech recognition and produced extraordinary results when applied to speech signals.

We use hidden Markov models on acoustic data to see if it can be a good choice to classify different conditions, i.e., different flow rates and fluid mixture compositions.

Hidden Markov model has the following parameters:

1. *N*, number of hidden states. In our case, the states would be distinct sounds of the acoustic sensor output.
2. *M*, number of distinct observation symbols in each state. In our case, this would be the number of spectral coefficients per distinct sound.
3. The state transition probability distribution matrix A.
4. The observation symbol probability distribution matrix B.
5. Initial state probabilities π.

Henceforth, we will characterize an HMM with its parameters (A, B, π).

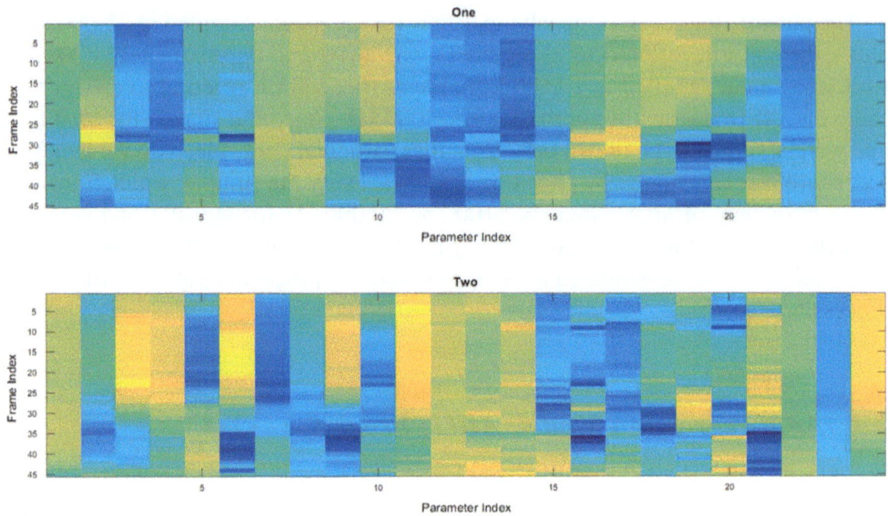

Fig. 1 MFCC heat maps for two different flow conditions

HMMs are categorized into two types: Ergodic hidden Markov model and left-right hidden Markov model (Rabiner 1989). A left–right model is one in which transitions are not permitted from higher states to lower states. We will be using a left–right HMM in our experiments.

2.5 *Mel-Frequency Cepstral Coefficients (MFCC)*

MFCC captures components of acoustic signals which are good for identifying linguistic content (Davis and Mermelstein 1980; Huang et al. 2001). Our purpose here is to use MFCC features from acoustic signals to identify flow rate and mixture composition of fluid. When heard, acoustic sensor produces audio signal which has different characteristics for different conditions. Hence, it will be wise to extract MFCC features of each condition to identify different conditions. A heat map for MFCC features is shown in Fig. 1 for two different fluid conditions, namely, "one" and "two". It can be seen that the heat map for these two conditions behaves differently.

2.6 Relating HMM and MFCC to Flow Rate Classification/Estimation

The motivation for relating HMM and MFCC to flow rate classification lies in the isolated word recognizer using HMM (Rabiner 1989). The goal of such a system would be to estimate the most likely sequence of phonemes (sequence of states) given a sequence of speech signal segments. In our case in place of speech signal, we have acoustic signals from flowing fluids and in place of phonemes, we have equivalent states.

Distinct Observations Distinct observations in the domain of speech recognition are the segments of spoken speech signal. In our case, it is cepstral coefficients of acoustic signals.

Hidden States Hidden states are responsible for the generation of acoustic signals. In the case of isolated word recognizer, we need to create one HMM for every word in the corpus and train it with the utterances of the word to strengthen the mode. In a similar fashion, for flow rate estimation, for every flow rate condition, HMMs will have a unique state. During testing, these HMMs provide us an estimate (via probability score) if a given sequence of test segment matches a sequence of states. Since a sequence of states can be mapped to a particular flow rate condition, HMMs are used to estimate the most likely signal condition of fluid.

2.7 Choice of Hidden Markov Model Parameters and Training the HMM

For each condition, we extract MFCC features and these features are used as continuous observations input to train HMM. Since we have continuous observations to build HMM, we will assume that our input MFCC features are distributed as per Gaussian mixture model with two Gaussian components. We used a grid search for choosing the number of Gaussian components and number of states which gave best model performance. Number of states N in our model is found to be four based on best results. We use left–right hidden Markov model in our modeling approach since it is more intuitive to relate the left–right hidden Markov model with different states of acoustic signal being modeled.

Incorporating all these assumptions, the transition matrix for four states will be of the following form:

$$\mathrm{A} = \begin{bmatrix} a_{11} & a_{12} & 0 & 0 \\ 0 & a_{22} & a_{23} & 0 \\ 0 & 0 & a_{33} & a_{34} \\ 0 & 0 & 0 & a_{44} \end{bmatrix}$$

Algorithm 1 Steps followed in the experimentation for AR(1) process

Step 1: Zero mean the time series and find the absolute value of the time series.
Step 2: Split the data into 80% training and 20% testing segments.
Step 3: The 20% testing segment is further divided into 20 test segments.
Step 4: The AR(1) process is modeled using the training data and tested using the 20 test segments.
Step 5: Calculate the log-likelihood values of each test segment with respect to the training classes.
Step 6: Use these log-likelihood values as pseudo-probability score to construct a ROC curve and compute AUC values from the ROC curve (see Sect. 3).

Algorithm 2 Steps followed in the experimentation for HMM and MFCC

Step 1: Time series data is low-pass filtered using low-pass moving average filter and down-sampled so as to follow the Nyquist criterion and to remove the high-frequency noise.
Step 2: Center the data to zero mean and take absolute value of time series data.
Step 3: Split the time series data into 80% training and 20% test segments.
Step 4: MFCC features are extracted from training data separately for all 63 conditions.
Step 5: For each condition of fluid, we build an HMM.
Step 6: Test data is further divided into 20 segments.
Step 7: MFCC features are extracted for each condition and each segment of test data.
Step 8: Compute the likelihood of each test data segment coming from one of the trained HMMs, i.e., one of the 63 conditions.
Step 9: Assign the test data segment to the condition whose likelihood is highest.
Step 10: Use results from (9) to construct confusion matrix, and hence compute diagonal accuracy.

Also, we chose initial probabilities as $\pi_i = 0 fori \neq 1$, i.e., we assume that the system is in state 1 initially. Taking all these assumptions, we build or train HMMs for each condition.

2.8 Log-Likelihood Estimation for the AR(1) Process

With the parameters φ and c estimated, predictions for the next observation in the time series are computed and the likelihood that a test segment j comes from a particular class i is calculated from a log-likelihood estimate as provided by Hamilton (1994, ch. 5.3, Eq. 5.3.9). For easy reference, Eq. (2) provides this relation.

$$f(y;\theta) = \frac{1}{\sqrt{2\pi\sigma^2}\sqrt{1-\rho^2}}\exp\left\{\frac{y_t - c/(1-\rho)^2}{2\sigma^2/(1-\rho^2)}\right\} \tag{2}$$

where c is the constant, ρ is the $AR(1)$ coefficient, and σ^2 is the error variance. Equation (2) provides likelihood of a test segment coming from a class whose parameters φ and c have already been estimated.

3 Multiclass AUC Computation (In-House Developed)

With likelihood values for every ith test segment calculated with respect to every class j as per Sect. 2.8, there will be a total of n likelihood values for every ith test segment (n is total number of classes). Since we deal with log-likelihood values, there is a high possibility of the values having a wide variation in terms of numerical values. For ease in comparison and for obtaining a normalized score for every (i, j) likelihood pair, we use the relation (Roberts and Ephraim 2015) described in Eq. (3).

$$q_n(j) = \frac{p(h_j)p(y^n \backslash h_j)}{\sum_{i \neq j} p(h_i)p(y^n \backslash h_i)} \tag{3}$$

where $p(h_i)\, and\; p(h_j)$ are prior probabilities of class i and j, respectively, and their values are assumed to be equal for all i and j. $p(y^n \backslash h_i)$ is the likelihood vector arising from ith class; similarly, $p(y^n \backslash h_j)$ is the likelihood vector arising from jth class. $q_n(j)$ is the vector of normalized likelihoods arising from jth class for all test segments.

With the normalized likelihood values for every ith test segment arising from the jth class computed, the vector $q_n(j)$ is taken as the test label vector. A true label vector of 1 is assigned to every ideal correct classification and 0 assigned to the rest. This true label vector along with the test label vector $q_n(j)$ is used to compute a ROC curve from which *multiclass* AUC values are obtained. We call it *multiclass* AUC because it is computed for more than 2 classes.

4 Results

We provide results from the two different approaches. Table 1 shows results for an average *multiclass* AUC for five different folds computed from the AR(1) process. As we can see from the table, both the sensors show similar results. Also, model performs best when there is only oil and air mixture or 100% water.

Once training of HMMs is done for each condition v, for each test segment to be classified among the given 63 conditions, we extract MFCC features and then compute the model likelihood for all 63 HMMs. Finally, we assign this test segment to a condition for which model likelihood is highest.

Each test segment has a true condition associated with it and a confusion matrix with 63 rows and 63 columns is constructed. We achieved a diagonal accuracy of 92.5% from confusion matrix for the HMM/MFCC approach.

Table 1 Average AUC scores under different flow rates

Constituent mixtures	Flow rates	Average AUC score (Sensor 1, Sensor 2)
100% Oil	5 gpm, 8 gpm, 10 gpm, 12 gpm, 15 gpm, 20 gpm, 30 gpm, 50 gpm	0.7878, 0.8389
Oil + Air	10 gpm 5 scf 40 psi, 10 gpm 6.5 scf 50 psi, 15 gpm 6.5 scf 50 psi, 15 gpm 5 scf 40 psi	0.8571, 0.9676
20% Oil + 80% Water + Air	10 gpm 5 scf 40 psi, 15 gpm 5 scf 40 psi, 15 gpm 6.5 scf 50 psi, 10 gpm 6.5 scf 50 psi	0.8498, 0.691
20% Oil + 80% Water	5 gpm, 8 gpm, 10 gpm, 12 gpm, 15 gpm, 20 gpm, 30 gpm, 50 gpm, 70 gpm	0.8582, 0.839
50% Oil + 50% Water + Air	10 gpm-5 scf-40 psi, 15 gpm-5 scf-40 psi,15 gpm-6.5 scf-50 psi,10 gpm-6.5 scf-50 psi	0.9211, 0.8119
50% Oil + 50% Water	5 gpm, 8 gpm, 10 gpm, 12 gpm, 15 gpm, 20 gpm, 30 gpm, 50 gpm,77 gpm	0.8406, 0.8209
75% Oil + 25% Water + Air	10 gpm 5 scf 40 psi, 10 gpm 6.5 scf 50 psi, 15 gpm 6.5 scf 50 psi, 15 gpm 5 scf 40 psi	0.9084, 0.8747
75% Oil + 25% Water	5 gpm, 8 gpm, 10 gpm, 12 gpm, 15 gpm, 20 gpm, 30 gpm, 50 gpm, 63 gpm	0.8975, 0.8449
Water + Air	10 gpm 5 scf 40 psi, 15 gpm 5 scf 40 psi, 15 gpm 6.5 scf 50 psi, 10 gpm 6.5 scf 50psi	0.9299, 0.8672
100% Water	8 gpm, 10 gpm, 15 gpm, 20 gpm, 30 gpm, 50 gpm, 82 gpm	0.8944, 0.9513

gpm gallons per minute
scf standard cubic feet
psi pounds per square inch

5 Conclusion

We achieved an average diagonal accuracy of 92.5% across five different folds for 63 different conditions which correspond to different fluid flow rates and different mixture compositions using MFCC as continuous observations to left–right hidden Markov model. Tuning of HMM parameters is very important while using them for any particular problem. It is evident that HMM with MFCC and auto-regression can be used to estimate flow rates and fluid mixture composition with high accuracy. We also achieved average AUC of ~0.96 with the AR(1) process.

This study shows that it is feasible to use acoustic sensors to classify oil flow rates for possible applications in the oil and gas industry. The implications of this can be understood better if the time taken for these classifications using traditional techniques is taken into account. This is owing to the fact that time constraints would reduce drastically using machine learning techniques combined with acoustic sensors.

References

Baum, L. E. (1972). An inequality and associated maximization technique in statistical estimation for probabilistic functions of Markov processes. *Inequalities, 3,* 1–8.

Baum, L. E., & Egon, J. A. (1967). An inequality with applications to statistical estimation for probabilistic functions of a Markov process and to a model for ecology. *Bulletin of the American Mathematical Society, 73,* 360–363.

Baum, Leonard. E., & Petrie, Ted. (1966). Statistical inference for probabilistic functions of finite state Markov chains. *The Annals of Mathematical Statistics, 37*(6), 1554–1563.

Baum, L. E., & Sell, G. R. (1968). Growth functions for transformations on manifolds. *Pacific Journal of Mathematics, 27*(2), 211–227.

Baum, L. E., Petrie, T., Soules, G., & Weiss, N. (1970). A maximization technique occurring in the statistical analysis of probabilistic functions of Markov chains. *The Annals of Mathematical Statistics, 41*(1), 164–171.

Davis, S., & Mermelstein, P. (1980). Comparison of parametric representations for monosyllabic word recognition in continuously spoken sentences. *IEEE Transactions on Acoustics, Speech, and Signal Processing, 28*(4), 357–36.

Hamilton, J. D. (1994). Time series analysis (Vol. 2, pp. 690–696). Princeton, NJ: Princeton university press.

Huang, X., Acero, A., & Hon, H. (2001). *Spoken language processing: A guide to theory, algorithm, and system development*. Prentice Hall.

Oppenheim, A. V. (1999). *Discrete-time signal processing*. Pearson Education India.

Rabiner, R. (1989). A tutorial on hidden Markov models and selected applications in speech recognition. *Proceedings of the IEEE, 77*(2), 257–286.

Roberts, W. J. J., & Ephraim, Y. (2015). Speaker classification using composite hypothesis testing and list decoding. *IEEE Transactions on Speech and Audio Processing, 13*(2).

http://www.analog.com/media/en/technical-documentation/dspbook/dsp_book_Ch15.pdf.

http://www.ni.com/white-paper/3807/en/#toc3.

Part VI
Analytics in Finance

Loan Loss Provisioning Practices in Indian Banks

Divya Gupta and Sunita Mall

Abstract RBI has started getting stringent on loan loss provisioning in the banks. Today banks are not only maintaining provisions for sub-standard assets, but also they have to keep aside surplus for standard assets. The objective of our study is to find out the factors determining loan loss provisioning in banks in India. Variables considered for our research are loan loss provisions to total assets, loans to total assets, capital adequacy ratio of the banks, bank holding of securities to total assets, earnings before tax to total asset ratio, non-performing assets to total loans, credit growth rate of the bank, and GDP annual growth rate. Our dependent variable is loan loss provisions to total assets. We have taken the data for the past 11 years from 2004–05 to 2014–15. The scope of our study is 46 commercial banks in India, where we have checked the determinants of loan loss provisioning. We have also analyzed whether the loan loss provisioning in Indian banks is procyclical or countercyclical by focusing on the two key variables, i.e., GDP and earnings. The impact of various variables on loan loss provision has been tested and the result indicates that asset size, credit growth, NPA level, earning of the banks, and macroeconomic variable (GDP annual growth rate) have a greater impact on the loan loss provisioning of the banks. The analysis has been done by using OLS regression and dynamic GMM approach. The findings of the study also reveal that Indian banks have shown countercyclical provisioning and the banks tend to increase the provisions with an increase in banks earnings and GDP annual growth rate.

Keywords Loan loss provisioning · Procyclical · Countercyclical · Standard assets

---

D. Gupta
Institute of Management & Information Science, Bhubaneswar, India
e-mail: divya83_g@yahoo.com

S. Mall (✉)
MICA, Ahemdabad, India
e-mail: malsunita@gmail.com; sunita.mall@micamail.in

A. K. Laha (ed.), *Advances in Analytics and Applications*, Springer Proceedings in Business and Economics, https://doi.org/10.1007/978-981-13-1208-3_16

1 Introduction

RBI has started getting stringent on loan loss provisioning in the banks. Currently, Indian banking industry is going through a phase, where profit is deteriorating because of the huge amount of NPAs. Today banks are not only maintaining provisions for sub-standard assets but also they have to keep aside surplus for standard assets. In India, RBI has increased the provisions rate from 0.25 to 0.40% for standard assets in 2005 and for some of the vulnerable sectors like real estate, credit cards, and capital market exposure; the range is from 1 to 2%. These required provisions amount also considers collateral.

The objective of our study is to find out the factors determining loan loss provisioning in banks in India. We have also worked on the individual relationship of loan loss reserves with the GDP growth rate and earnings of the bank. The study has been conducted by considering 26 public sector banks and 20 private sector banks. We have taken the data for the last 11 years from 2004–05 to 2014–15. We have considered those banks which have been in existence from 2004–05 till 2014–15.

The main area of concern for maintaining adequate loan loss provisions is cover up the loss happening for any attenuation in the value of assets and investments of the bank. In India, banks strictly follow RBI guidelines to maintain loan loss reserves. Reduction in the value of loan assets may happen due to default in the payment of loans by the customer. Those assets which cease to generate income for the bank are called as non-performing asset (NPA). This mainly happens when the customer is unable to make the payment of interest due and principal amount on the loan. Credit assets of the bank include loans which are long term in nature and on the other side advances, which are short term in nature. As stated by RBI, the loans where the payment of interest and principal amount are not paid for the period of 90 days from the due date is considered to be non-performing asset. Banks maintain provisions for every loan given by them to ensure to shield up the losses occurring due to default in loans and advances. This is termed as loan loss provisioning.

The loans given by the banks where all dues are paid on time by the customer is known as Standard Asset. Once a default happens and it continues beyond 90 days, it is called as NPA or sub-standard assets. As per RBI directions, in India banks have to maintain provisions for both standard and sub-standard assets. Table 1 explicates the provision norms for banks in India for both standard and sub-standard assets. General provision norms for standard assets are 0.40% for all loans except few categories, where it varies as per the risk factor involved. Table 2 explains the provisions norms for standard assets in India. Once the loan goes to the category of NPA, it is classified into three tiers as sub-standard, doubtful debt, or losses. The classification depends upon the time period since the default. This is as per advancement of a loan from standard loan to loss asset. If a debtor is unable to pay dues for 90 days after the end of a quarter; the loan becomes an NPA. Sometimes these loan dues remain outstanding for a period less than or equal to 12 months; it is termed as sub-standard Asset. Once the loan is classified as sub-standard, banks have to maintain 15% provisions of the outstanding balance for secured loans and in case of unsecured loans, banks have to maintain 25%.

When sub-standard asset remains so for additional 12 months; it would be termed as “Doubtful asset”. This continues till the end of third year. Doubtful debts are further classified into three parts. Under the first category of doubtful assets, the asset remains for 1 year with the provision rate of 25% for the secured loans and 100% for unsecured. Under the second category of doubtful assets with 1–3 years, provisions rate for secured loans is 40% and unsecured is 100%, and the third category of doubtful assets for more than 3 years, rates are 100% for secured and as well as unsecured loans.

When the loan remains unpaid even after it remains sub-standard asset for more than 3 years, it is known as loss asset. Table 1 explains the details of the provision norms in India to be maintained by the banks.

Table 2 explains the provision to be maintained by Indian banks for standard assets.

Table 3 discusses the impact of NPAs on the return on advances of the banks. We have divided the data into three parts, State Bank of India and its Associates, Public sector banks, and private sector banks. The non-performing assets in 2004–05 were high for all the three types of banks which put a strain on bank’s net worth, i.e.,

Table 1 NPA provision rates for assets in Indian banks

Standard assets	0.40% of outstanding balance	
Sub-standard assets (secured)	15% of outstanding balance	
Sub-standard assets (unsecured)	25% of outstanding balance 20% (infrastructure loans)	
Doubtful assets—up to 1 year	25% of secured portion	100% of unsecured portion
Doubtful assets—more than 1 year and up to 3 years	40% of secured portion	100% of unsecured portion
Doubtful assets—more than 3 years	**100%**	100% of unsecured portion
Loss	100% of outstanding balance	

Source www.rbi.org.in

Table 2 NPA provision rates for standard assets

Standard asset	Percentage of provision (%)
Agriculture and SME	0.25
Housing loan above 20 lakhs	1
Personal loans/Credit card loans/Commercial real estate/Loan qualifying as market capital exposure	2
All other advances	0.40

Source www.rbi.org.in

bank's profitability and liquidity and finally result in credit loss. However, high NPA not only reduces bank's profitability but also add burden to the banks' expenses for recovering them. But gradually from 2006 to 2010, the NPA for SBI and its associates have decreased from 3.5 to 3.12%, for public sector banks, it has decreased from 3.8 to 1.9%.

However, for State Bank of India and its associates and public sector banks, it is found that in the period 2011–2015, again the NPA in banks are increasing. But in private sector banks, the NPAs are showing a decreasing trend. So, the problem of NPAs is more in public sector banks, when compared to private sector banks. The NPAs in public sector banks are increasing due to external as well as internal factors which are ineffective recovery tribunal, wilful defaults, natural calamities industrial sickness, lack of demand, defective lending process, inappropriate technology, etc.

Variables considered for our research are loan loss provisions to total assets, loans to total assets, capital adequacy ratio of the banks (CAR), bank holding of securities to total assets, earnings before tax to total asset ratio, non-performing assets to total loans, annual credit growth rate of the bank, and GDP annual growth rate. Variables have been selected on the basis of the literature, we chose the list of explanatory variables in consistency with previous studies, so that readers can compare the provisioning practices in Asia and other countries.

Our dependent variable is loan loss provisions to total assets. The scope of our study is 46 commercial banks in India, where we have checked the determinants of loan loss provisioning by analyzing whether the loan loss provision is procyclical or countercyclical. Procyclical and countercyclical are terms used to describe how an economic quantity is related to economic fluctuations. In this chapter, the dependent variable loan loss provisions are considered to be procyclical if the amount of provisions are reducing during high GDP growth rate and high earnings with the bank. Though the impact of all independent variables on loan loss provis ion as a proporhing total assets' has been tested but the explanatory variables like earning of the banks and macroeconomic variable (GDP annual growth rate) have been more focused on. The literature on the determinants of loan loss provisions suggests that banks may respond differently to the economic upswing or downturn. Some of the studies have worked on the relationship between loan loss provision and GDP growth rate (Bikker and Metztemakers 2005). Few other papers in the literature have worked on the bank-specific factors that impact on the provisioning of the bank see, for e.g., Packer et al. (2014). To the best of our knowledge, there is no study in India which has worked by considering individual variables and aggregate variables simultaneously for determining provision practices by banks. The methodology includes the panel data analysis, which has been carried out by using OLS regression and dynamic GMM approach recommended by Arellano and Bond (1991). The results of the study imply that the factors which majorly impact the loan loss provision are the size of the bank size, earnings, GDP, NPA, and credit growth of the banks. The provisions have no relation to the capital adequacy ratio (CAR). It means capital ratio does not explain the variation in loan loss provisions. Even the results show a positive relationship between earnings of the bank and GDP growth rate with the maintenance of loan loss provisions.

Table 3 Impact of increase of NPA's on the profitability of the banks in India

		2004–05	2005–06	2006–07	2007–08	2008–09	2009–10	2010–11	2011–12	2012–13	2013–14	2014–15
State bank of India and ITS associates	Gross NPA (%)	5.32	3.5	2.59	2.58	2.55	2.76	3.12	4.36	4.42	4	5.1
	Change in return on advances (%)	(5.55)	3.87	8.88	12.75	3.56	(9.6)	(1)	16.06	(3.6)	(3.9)	(1.6)
Public sector banks	Gross NPA (%)	5.36	3.8	2.69	2.06	1.75	1.97	1.97	3.17	3.61	2.87	4.1
	Change in return on advances (%)	(5.85)	(0.97)	8.11	8.08	7.15	(10.69)	(0.11)	13.31	(2.23)	(3.8)	(1.9)
Private sector banks	Gross NPA (%)	5.26	2.43	1.71	2.47	2.91	2.77	2.45	2.08	1.79	1.05	1.5
	Change in return on advances (%)	(13.19)	0.47	9.52	15.29	3.63	(13.3)	(4.84)	14.97	1.42	(2.4)	(2.9)

Source RBI and respective banks' websites
Figures in brackets() represent loss

The rest of the paper is arranged as per the followings. In Sect. 2, we discuss the available literature in detail. Section 3 discusses the research design and sample selection along with the descriptive statistics. Section 4 is devoted for the discussion, analysis, and interpretation of the regression results. Section 5 concludes the paper and indicates research implications.

2 Literature

The literature on loan provisioning is extensively available. The focus of the available research is multifold.

Abbott (1989) examined the concept and forms of provisioning and discussed the recent establishment of international guidelines and their likely effects on the debt analysis. He highlighted that through loan loss provisioning banks transform their bargaining position over debt crises. Research on loan loss provisioning used to focus on if provisions were used by banks to smooth earnings from an accounting viewpoint (Greenawalt and Sinkey 1988). Packer and Zhu (2012) have worked on the determinants of loan loss provision of Asian and Pacific countries. They have worked on a sample of 240 banks from 12 economies. They have found that most of the Asian countries have shown countercyclical loan loss provisioning, but Japan is one of the countries showing procyclical loan loss provision. In their research, they found that banks put aside more provisions when there is high credit risk with the banks and there is a negative relationship between loan growth and the loan loss provisioning.

Zilberman and Tayler (2014) study the interaction between loan loss provisioning rules, monetary policy, and business cycle fluctuations. They remarked that backward-looking provisioning rule leads to macroeconomic instability, whereas forward-looking provisioning systems moderates the anti-inflationary response in the monetary policy rule. It is observed that in most of the industrialized countries, the provisioning system is still backward-looking. It is expected that the banks should set specific provisions related to identified credit losses as past due payments.

Murcia and Kohlscheen (2016) have studied the determinants of loan loss provisions and delinquency ratios of the banks in different emerging market economies. They considered 554 banks for their research, where they found that most banks in emerging market economies react to the aggregate variables and have less relationship with the idiosyncratic variables. The results indicate that earnings of the banks and bank's size have an impact on loan loss provisioning.

Bouvatier and Lepetit (2012) and Agénor and Zilberman (2013) examined the effectiveness of loan provisioning regimes through the theoretical framework and find out the relationship between credit provisioning and monetary policy. Cavallo and Majnoni (2001), Laeven and Majnoni (2003) and Bikker and Metzemakers (2005) revealed that there exists a negative correlation between output and loan loss provisions. On the other hand, the highly countercyclical relationship between the loan loss provisions to loan ratio and the economic activity is found in the study by Clerc et al. (2001), De Lis et al. (2001) and Pain (2003). Some researchers remarked that the level of provisioning has basically a historically procyclical bias, as it is related to contemporaneous problem assets (Borio and Lowe 2001; Bikker and Hu 2002; Laeven and Majnoni 2003). When commercial property prices are rising the provisions are lower. This suggests that provisioning may amplify credit cycles through the collateral channel (Davis and Zhu 2009).

With reference to the commercial banks provisioning cost channel and lending rate decisions are directly related. Raising more loan loss provisions during a downturn can lead to higher risk and borrowing costs as the loan rate is endogenously linked to the risk of default through the risk premium channel. The risk premium is the component in the loan pricing, where it is positively related to the credit rating of the borrower. Jin et al. (2011) documented a strong positive relationship between loan loss provisions and the probability of bank failures for weak banks between 2007 and 2010.

Some of the papers in the literature have empirically examined the short-term borrowing cost. The rules which are related to the loan rate behavior such as monetary policy, non-performing loans, loan loss provisioning, etc., translate into changes in the firms marginal costs, price inflation, and output through borrowing cost channel (Ravenna and Walsh 2006; Chowdhury et al. 2006; Tillmann 2008; De Fiore and Tristani 2013).

3 Research Methodology

This section explains the research methodology, variables, and data used for the research.

3.1 Sample

The study has been conducted with the sample size of 46 banks. We have considered 26 public sector banks and 20 private sector banks. We have taken the data for the past 11 years from 2004–05 to 2014–15. This period has been selected to understand the changes in credit quality of the banks, change in non-performing assets level of the banks and provisions required for NPA in banks before recession and post-recession period. We have considered those banks which have existence from 2004–05 till 2014–15. If any bank got merged with any other bank during this tenure, it has not been included and also banks which have been established after 2004–05 has not been considered. The data has been taken from the RBI site and respective bank's websites.

3.2 Statistical Tool for Analysis of Data

The methodology includes t panel data analysis, which has been carried out using OLS regression and dynamic GMM approach as recommended in Arellano and Bond (1991).

The baseline model specification used in this research is based on the existing literature.

$$\begin{aligned} \text{LLPi, t} = {} & \text{B1LLP i, t} - 1 + \text{B2Loanasset i, t} + \text{B3CAR i, t} + \text{B4GDP i, t} + \text{B5Earnings i, t} \\ & + \text{B6NPA i, t} + \text{B7Liquidity i, t} + \text{B8Loangrowth i, t} + \varepsilon \text{ i, t} \end{aligned}$$

LLP ratio of loan loss provisions to the total assets. Loan asset represents the size of a bank, which is the ratio of loans to total assets, CAR = capital adequacy ratio, i.e., the ratio of capital to assets, GDP = the annual GDP growth rate of India, Earnings = ratio of earnings of the bank before taxes to total assets, NPA = ratio of non-performing assets to total loans. In order to measure the liquidity with the bank, we have taken a proxy by considering the ratio of total investments to total assets of the bank. Loan growth is the median growth rate of the banks in India. For our second objective, we have considered one micro variable (earnings) and one macro variable (GDP) to find the impact of earnings and GDP on the loan loss provisioning of the banks. Through OLS regression, if we get the positive coefficients for both dependent and independent variable in both the cases (Earnings with loan loss provisioning and GDP with loan loss provisioning) then it will represent the countercyclical behavior, which means the banks keep aside more provisions for loan losses in case of higher GDP growth and higher earnings. And, it will be procyclical in case the coefficient of OLS results is negative. If the bank keeps aside more provisions when their earnings are higher is called as “earning smoothing”, which is a common finding in the literature that would act to diminish the financial system procyclicality.

To address the problem of endogeneity of the regressor, which is a common issue in such types of data, we have used GMM approach. Endogeneity is associated with the joint determination between loan loss provisions and the list of bank-specific explanatory variables. The dynamic panel model has been used where the lagged values of explanatory variables are used as an instrument.

3.3 Descriptive Study of the Sample

Table 4 presents the descriptive statistics of independent and independent variables.

Table 4 Descriptive statistics

Variable	Mean	Median	Min	Max	S. dev
LLP	0.898285	0.674444	0.404086	18.5226	1.25224
Loan asset	69.8057	59.9792	0.118683	85.94	67.6894
CAR	13.1995	12.8300	4.86000	22.4600	2.08082
Liquidity	29.9395	28.2744	16.2398	79.5476	7.07537
GDP	7.76545	7.47000	4.47000	9.57000	1.50194
Earnings	1.95718	1.96000	−2.01712	3.92149	0.701687
Loan growth	33.3193	20.1163	−39.7992	65.9	84.484
NPA	2.17209	1.80000	0.01	17.5534	2.26120

4 Analysis

This section deals with the analysis of the paper. We have divided this section of the paper into three parts. The first part explains the determinants of loan loss provisions by OLS method. If the independent variable is correlated with the error term in a regression model then the estimate of the regression coefficient in an ordinary least squares (OLS) regression is biased, and in order to have a check on the same, we have also analyzed our results with the econometric tool Dynamic GMM model. Second part explains the determinants of loan loss provisioning by Dynamic GMM method. The third part focuses on whether the loan loss provisioning in Indian banks is procyclical or countercyclical by focusing on the two key variables, i.e., GDP and Earnings.

4.1 OLS Method

Table 5 indicates the results of OLS regression. This model is significant and the explained dependent variable is 69.64%. This is suggesting that taken independent variables explain the variability of the dependent variables by 69.64%. We found out loan growth (credit growth), NPA (NPA to total loans) and GDP growth rate at 1% level of significance. Loan growth has a negative relationship with the dependent variable LLP and GDP and NPA have a positive relationship with LLP. Earnings and loan assets are significant at 5% level of significance. Earnings have a positive relationship with LLP and loan assets have an inverse relationship with LLP. Loan loss provisioning increases with the increase in GDP and earnings and decreases with the increase in loan asset and credit growth.

The coefficient of NPA has a positive sign indicating the positive relationship with LLP. This suggests when NPAs go up, NPA provisions amount also increases. Although loan assets and credit growth were expected to have a positive relationship

Table 5 Determinants of loan loss provision (OLS method)

Independent variables	Coefficient	t-statistics	Sig. value
Loan asset	−0.385571	–2.5331	0.01170**
CAR	−0.387714	–1.4578	0.14573
Liquidity	−0.0392202	–0.1652	0.86891
GDP	1.29151	4.3350	0.00002***
Earnings	0.402255	5.3887	0.00001**
Loan growth	−0.214681	–3.8595	0.00013***
NPA	0.0864584	3.0307	0.00261***
Adjusted R square	69.64		
Durbin Watson	1.62		
No. of observations	390		

***, **, * indicate significance at 1, 5, and 10% level

with LLP but the results indicate that with the growth in credit and total loans, NPA provisions are not impacted much.

Loan loss provision does not show any relationship with the liquidity position of the banks and capital adequacy ratio. This signifies that change in investment patterns of the bank and capital does not explain the variation in the provisioning of the banks.

4.2 *Dynamic GMM (Generalized Method of Moments)*

Table 6 indicates the results of dynamic GMM model, where LLP is the dependent variable. We have used one-period legged value of LLP called LLP(-1) l as instruments in this GMM model. This is an econometric approach, where variables are taken in first difference to control the bank-specific effects. Loan loss provisions are explained by loan asset, GDP, earnings, NPA, and loan growth. It signifies that bank's size, credit growth, and GDP have a significant impact on the provisioning of the banks. Loan loss provision responds positively to GDP, loan growth, and earnings and NPA and negatively responded to loan asset. It suggests that with the increase in GDP, cr edit growth, and earnings, loan loss provisioning increases. Our results are consistent with Murcia and Kohlscheen (2012).

Most of the results indicate the similarity with the OLS regression results but there are few differences. In case of OLS regression model, the relationship between LLP and loan growth is negative but in dynamic GMM model it is positive. The coefficients of loan asset and credit growth were expected to be positive but both the models have shown the negative relation of LLP with the loan assets.

Table 6 Determinants of loan loss provision (dynamic GMM method)

Independent variable	Coefficient	t-statistics	Sig. value
LLP(-1)	0.0327953	0.5967	0.55071
Loan asset	–5.44602e–05	–2.7203	0.00652***
CAR	–0.0341703	–1.5259	0.12704
Liquidity	0.00822407	–2.3562	0.01847
GDP	0.139968	2.7155	0.00662***
Earnings	0.0892579	1.9446	0.05182*
NPA	0.218216	10.6370	0.00001**
Loan growth	0.00527649	12.9023	0.00001***
Time effect	Yes		
No. of observations	348		
Wald Chi-Square	2302.95		
AB test for AR(2)	0.5564		

4.3 Impact of GDP and Earnings on LLP

One of the objectives of the paper is to test the impact of GDP and earnings on the dependent variable LLP. We have tried to check the determinants of loan loss provisioning by analyzing whether the loan loss provision is procyclical or countercyclical. In our research, we have considered the loan loss provisions to be procyclical if it tends to decrease with the increase in growth rate of GDP and earnings of the bank. We consider loan loss provisions to be countercyclical, if it increases with the increase in GDP and earnings of the banks. Through both the methods of OLS and dynamic GMM, we have seen the result to be consistent. We have found in case of Indian banks, the relationship between loan loss provisioning and GDP and earnings is countercyclical. Loan loss provisions increase with the increase in GDP of India and with the profits of the banks. This result is consistent with previous studies (Packer and Zhu 2012; Murcia and Kohlscheen 2012). It shows a positive reaction between loan loss provisioning with GDP and earnings of the banks.

5 Conclusion

The result of the chapter indicates that the loan loss provisioning in Indian banks is majorly influenced by the volume of loans, credit growth, non-performing assets, GDP growth rate of the banks, and earnings of the banks. We have applied two techniques to find the determinants of loan loss reserves in India. First, we used OLS method and second Dynamic GMM method. The results of both the techniques indicate the same results. We have found in case of Indian banks, the relationship between loan loss provisioning and GDP and earnings is countercyclical. Loan loss provisions increase with the increase in GDP of India and with the profits of the banks.

References

Abbott, G. C. (1989). Loan loss provisioning. *Intereconomics*, 233–240.
Agénor, P.-R., & Zilberman, R. (2013). Loan loss provisioning rules, procyclicality, and financial volatility. Centre for Growth and Business Cycle Research, Working Paper, no. 184, University of Manchester.
Arellano, M., & Bond, S. (1991). Some tests of specification for panel data: Monte Carlo evidence and an application to employment equations. *Review of Economic Studies, 58,* 277–297.
Bikker, J. A., & Hu, H. (2002). Cyclical patterns in profits, provisions and lending of banks and procyclicality of the new Basel capital requirements. *Banca Nazionale del Lavoro Quarterly Review, 55,* 143–175.
Bikker, J., & Metzemakers, P. (2005). Bank provisioning behaviour and procyclicality. *Journal of International Financial Markets, Institutions and Money, 15,* 141–157.
Borio, C., & Lowe, C. (2001). To provision or not to provision. *BIS Quarterly Review*.
Borio, C., Furfine, C., & Lowe, C. (2001). Procyclicality of the financial system and financial stability: Issues and policy options. BIS Papers, no. 1, pp. 1–57.
Bouvatier, V., & Lepetit, L. (2012). Provisioning rules and bank lending: A theoretical model. *Journal of Financial Stability, 8*(1), 25–31.
Cavallo, M., & Majnoni, G. (2001). Do banks provision for bad loans in good times? Empirical evidence and policy implications. In R. Levich (Vol. 18).
Chowdhury, I., Hoffmann, M., & Schabert, A. (2006). Inflation dynamics and the cost channel of monetary transmission. *European Economic Review, 50,* 995–1016.
Clerc, L., Drumetz, F., & Jaudoin, O. (2001). To what extent are prudential and accounting arrangements pro-or countercyclical with respect to overall financial conditions? BIS Papers, no. 1, pp. 197–210.
Cortavarria, L., Dziobek, C., Kananaya, C., & Song, A. (2000). Loan review, provisioning, and macroeconomic linkages. IMF Working Paper, no. 00/195.
De Fiore, F., & Tristani, O. (2013). Optimal monetary policy in a model of the credit channel. *The Economic Journal, 123*(571), 906–931.
De Lis, S. F., Martínez Pagés, J., & Saurina, J. (2001). Credit growth, problem loans and credit risk provisioning in Spain.
Demyanyk, Y., & van Hemert, O. (2011). Understanding the subprime mortgage crisis. *Review of Financial Studies, 24,* 1848–1880.
Drehmann, M., Borio, C., & Tsataronis, K. (2011). Anchoring countercyclical capital buffers: The role of credit aggregates. *International Journal of Central Banking, 7*(4), 189–240.
Foos, D., Norden, L., & Weber, M. (2010). Loan growth and riskiness of banks. *Journal of Banking & Finance, 34,* 2929–2940.
García-Suaza, A., Gómez-González, J., Murcia, A., & Tenjo-Galarza, F. (2012). The cyclical behavior of bank capital buffers in an emerging economy: Size does matter. *Economic Modelling, 29,* 1612–1618.
Gourinchas, P., & Obstfeld, M. (2012). Stories of the twentieth century for the twenty-first. *American Economic Journal: Macroeconomics, 4*(1), 226–265.
Greenawalt, M. B., & Sinkey, J. F. (1988). Bank loan-loss provisions and the incomes smoothing hypothesis: An empirical analysis, 1976–1984. *Journal of Financial Services Research, 1*(4), 301–318.
Jin, J. Y., Kanagaretnam, K., & Lobo, G. J. (2011). Ability of accounting and audit quality variables to predict bank failure during the financial crisis. *Journal of Banking & Finance, 35*(11), 2811–2819.

Jopikii, T., & Milne, A. (2008). The cyclical behavior of European bank capital buffers. *Journal of Banking & Finance, 32,* 1440–1451.
Jorda, O., Schularick, M., & Taylor, A. (2011). Financial crises, credit booms, and external imbalances: 140 years of lessons. *IMF Economic Review, 59,* 340–378.
Kohlscheen, E., & Miyajima, K. (2015). The transmission of monetary policy in EMEs in a changing financial environment: A longitudinal analysis. BIS Working Papers, no. 495.
Leaven, L., & Majnoni, G. (2003). Loan loss provisioning and economic slowdowns: Too much, too late. *Journal of Financial Intermediation, 12,* 178–197.
Murcia, A., & Kohlscheen, E. (2016). Moving in Tandem: Bank provisioning in emerging market economies. BIS Working Paper, no. 548.
Packer, F., & Zhu, H. (2012). Loan loss provisioning practice by Asian banks. BIS Working Paper.
Packer, F., Shek, J., & Zhu, H. (2012). Loan loss provisioning practices of Asian banks. BIS Working Paper, no. 375.
Packer, F., Shek, J., & Zhu, H. (2014). Countercyclical loan loss provisions in Asia. *SEACEN Financial Stability Journal, 3,* 25–58.
Pain, D. (2003). The provisioning experience of the major UK banks: A small panel investigation. Bank of England Working Paper, no. 177.
Ravenna, F., & Walsh, C. E. (2006). Optimal monetary policy with the cost channel. *Journal of Monetary Economics, 53,* 199–216.
Sinkey, J., & Greenwald, M. (1991). Loan loss experience and risk-taking behavior at large commercial banks. *Journal of Financial Services Research, 5,* 43–59.
Zilberman, R., & Tayler, W. J. (2014). Financial shocks, loan loss provisions and macroeconomic stability. Working Papers 124138133, Lancaster University Management School, Economics Department.

Modeling Commodity Market Returns: The Challenge of Leptokurtic Distributions

Arnab Kumar Laha and A. C. Pravida Raja

Abstract In this chapter, we consider modeling leptokurtic daily log-return distributions of three commodities: gold, silver, and crude oil. Three modeling approaches are tried out namely (a) a two-component mixture of normal distributions model, (b) Variance Gamma (VG) distribution model, and (c) Generalized Secant Hyperbolic (GSH) distribution model. The two-component mixture of normal distributions model is found to be a reasonable model for log-returns on gold and crude oil. The VG distribution model and the GSH distribution model are not found to be suitable for modeling log-returns for any of the three commodities considered in this chapter.

Keywords Generalized secant hyperbolic distribution · Variance Gamma distribution · Sampling importance

1 Introduction

Analysis and modeling of financial time series data and forecasting future values of market variables play an important role in quantitative finance. It is common that the distributions of financial asset returns are known to be non-Gaussian. It is often seen that the empirical asset return distributions have high kurtosis, tails thicker than those of a normal distribution, and may exhibit some skewness. The same features can also be seen in the return distributions encountered in the commodity markets. In this chapter, we study the return distributions of three commodities: gold, silver, and crude petroleum with the goal of modeling their return distributions.

The one-period return on a commodity is defined as $R_t = \frac{P_t - P_{t-1}}{P_{t-1}}$, where P_t is the price (including dividends, if any, declared during the period t − 1 to t) of the asset at time t. However, most often for modeling purposes, one works with the "simple

A. K. Laha (✉) · A. C. Pravida Raja
Indian Institute of Management Ahmedabad, Ahmedabad, India
e-mail: arnab@iima.ac.in

A. C. Pravida Raja
e-mail: pravida@iima.ac.in

A. K. Laha (ed.), *Advances in Analytics and Applications*, Springer Proceedings in Business and Economics, https://doi.org/10.1007/978-981-13-1208-3_17

returns" which is defined as $1+R_t = \frac{P_t}{P_{t-1}}$. It is also quite common to use the log-return r_t which is defined as

$$r_t = \ln(1 + R_t) = \log P_t - \log P_{t-1} = p_t - p_{t-1} \tag{1}$$

where $p_t = \log P_t$.

In this chapter, we explore three different approaches to modeling return distributions showing high leptokurtic behavior, namely (a) Generalized Secant Hyperbolic (GSH) distribution model, (b) the mixture of normal distributions model, and (c) the Variance Gamma (VG) distribution model. The three models are discussed in brief in the succeeding paragraphs.

1.1 Generalized Secant Hyperbolic Distribution

The generalizations of the hyperbolic secant distribution were proposed in the past, which seem to be encouraging as a model for financial return data. Vaughan (2002) introduced the Generalized Secant Hyperbolic (GSH) distribution which is symmetric and is able to model both thin and flat tails. The probability density function (p.d.f.) for this family of distributions is given by

$$f_{GSH}(x; t) = c_1(t)\frac{\exp(c_2(t)x)}{\exp(2c_2(t)x) + 2a(t)\exp(c_2(t)x) + 1}, x \in R \text{ where } t \in (-\pi, \infty),$$

$$a(t) = \cos t, c_2(t) = \sqrt{\frac{(\pi^2 - t^2)}{3}}, c_1(t) = (c_2(t)\sin t)/t, \text{ for } -\pi < t \le 0 \text{ and}$$

$$a(t) = \cosh t, c_2(t) = \sqrt{\frac{(\pi^2 + t^2)}{3}}, c_1(t) = (c_2(t)\sinh t)/t, \text{ for } t > 0. \tag{2}$$

The parameter t is called the kurtosis parameter. This family of symmetric distributions can have kurtosis ranging from 1.8 to infinity, and includes the logistic distribution ($t = 0$), and hyperbolic secant distribution ($t = -\pi/2$) as special cases and the uniform distribution on $\left(-\sqrt{3}, \sqrt{3}\right)$ as limiting case as $t \to \infty$.

Fischer and Vaughan (2002) generalized the family of GSH distributions further by introducing a skewness parameter γ. The distributions in this family are called Skew Generalized Secant Hyperbolic (SGSH) distributions.

$$\text{Let } I^+(x) = \begin{cases} 1 \text{ if } x \ge 0 \\ 0 \text{ if } x < 0 \end{cases} \text{ and } I^-(x) = \begin{cases} 1 \text{ if } x < 0 \\ 0 \text{ if } x \ge 0 \end{cases}. \tag{3}$$

Then p.d.f. of the SGSH family of distributions is given by

$$f_{SGSH}(x;t,\gamma)=\frac{2}{\gamma+\frac{1}{\gamma}}\left\{f_{GSH}(x/\gamma)I^{-}(x)+f_{GSH}(\gamma x)I^{+}(x)\right\}$$
$$=\frac{2c_1}{\gamma+\frac{1}{\gamma}}\left(\frac{\exp(c_2x/\gamma)I^{-}(x)}{\exp(2c_2x/\gamma)+2a\exp(c_2x/\gamma)+1}+\frac{\exp(c_2\gamma x)I^{+}(x)}{\exp(2c_2\gamma x)+2a\exp(c_2\gamma x)+1}\right). \quad (4)$$

The SGSH distribution is symmetric for $\gamma = 1$, skewed to right for $\gamma > 1$ and skewed to the left for $0 < \gamma < 1$.

Fischer (2004) studied the weekly returns of the Nikkei during the period July 31, 1983 to April 9, 1995 and concluded that for this data, the SGSH distribution provides an excellent fit.

1.2 Mixture of Normal Distributions

Mixture of normal distributions models provides an interesting alternative way to model leptokurtic return distributions. A random variable X is said to follow an N-component mixture of normal distributions if its p.d.f. is of the form

$$f_X(x)=\sum_{j=1}^{N}w_j\frac{1}{\sqrt{2\pi}\,\sigma_j}\exp\left\{-\frac{1}{2}\left(\frac{x-\mu_j}{\sigma_j}\right)^2\right\},\ \sigma_j>0,\ w_j>0,\ \sum_{j=1}^{N}w_j=1. \quad (5)$$

Wirjanto and Xu (2009) give a selected review of developments and applications of normal mixture models in empirical finance. According to them, increasing attention has been focused on the normal mixture family since many continuous distributions can be approximated very well by an appropriate finite normal mixture family distribution. Kon (1984) compared the application of mixture of normal distribution model with Student-t model and concluded that Gaussian mixture model has substantially more descriptive validity than a Student-t model. Press (1967), Praetz (1972), Clark (1973), Blattberg and Gonedes (1974) give a good history of modeling asset returns with a mixture of normal. Kamaruzzaman et al. (2012) used a two-component mixture normal distribution to fit monthly rates of returns for three indices of Bursa Malaysia Index Series namely FTSE Bursa Malaysia Composite index, Finance Index, and Industrial index.

Venkataraman (1997) studied the Value-at-Risk (VaR) for a mixture of normal distributions using Quasi-Bayesian estimation techniques. He showed that the Quasi-Bayesian Maximum Likelihood Estimation (QB-MLE) when applied to the mixture of normal distributions model provided better estimate of VaR than that based on the normal distribution model.

Bayesian approaches to mixture modeling have been studied in great detail by many researchers (see, for e.g., Diebolt and Robert 1994; Titterington et al. 1985). In this chapter, we use Sampling Importance Resampling (SIR) algorithm for simulating from a general posterior distribution. Alternative approaches include the Gibbs sampler (Gelfand and Smith 1990; Casella and George 1992), Metropolis–Hastings algorithm (Chib and Greenberg 1995), and Markov Chain Monte Carlo (MCMC) method (Geweke 2007; Smith and Roberts 1993).

1.3 Sampling Importance Resampling (SIR) Algorithm

Suppose that θ is the parameter of interest and we wish to obtain a sample of values from the posterior density $\pi\left(\theta/\, data\right) = K\,\pi(\theta)\,L(\theta)$, where $\pi(\theta)$ is the prior density, $L(\theta)$ is the likelihood function, and K is proportionality constant. SIR is a simple approximate method of sampling from $\pi(\theta/\text{data})$ (Rubin 1988; Smith and Gelfand 1992; Albert 1993). Take a sample $\theta_1, \theta_2, \ldots, \theta_m$ from $\pi(\theta)$ and compute the sample weights $w(\theta_i) = \frac{\pi(\theta_i/\, data)}{\pi(\theta_i)} = K\,L(\theta_i)$, $i = 1, 2, \ldots, m$. Then, take a new sample $\theta_1^*, \theta_2^*, \ldots, \theta_n^*$ with replacement from $\{\theta_1, \theta_2, \ldots, \theta_m\}$ with probabilities proportional to $\{w(\theta_1), w(\theta_2), \ldots, w(\theta_m)\}$. The sample $\left\{\theta_i^*\right\}$ is approximately distributed from the posterior density $\pi\left(\theta/\, data\right)$.

1.4 The Variance Gamma Model

Madan and Seneta (1990) introduced the Variance Gamma (VG) Model, a time change of Brownian motion without drift by a gamma process called the symmetric variance process for modeling the stock market returns. Later, Madan et al. (1998) proposed a three-parameter generalization of Brownian motion as a model for the dynamics of the logarithm of the stock price by evaluating Brownian motion with drift at random time given by a gamma process, which allows to control over both skewness and kurtosis of the return distributions. Like normal, inverse Gaussian, hyperbolic, and generalized hyperbolic distribution, VG distribution also belongs to the class of normal variance mean mixture and corresponds to a gamma mixing density.

1.4.1 The Variance Gamma (VG) Distribution

The p.d.f. of the VG distribution is given by

$$f_X(x) = \frac{2\exp(\theta(x-c)/\sigma^2)}{\sigma\sqrt{2\pi}\upsilon^{1/\upsilon}\Gamma(1/\upsilon)}\left(\frac{|x-c|}{\sqrt{2\theta^2/\upsilon+\sigma^2}}\right)^{1/\upsilon-1/2} \mathrm{K}_{1/\upsilon-1/2}\left(\frac{|x-c|\sqrt{2\sigma^2/\upsilon+\theta^2}}{\sigma^2}\right), \quad (6)$$

where $-\infty < x < \infty$, c (the location parameter), $\sigma > 0$ (the spread parameter), θ (the asymmetry parameter), and $\upsilon > 0$ (the shape parameter) are real constants. Here, $\mathrm{K}_\gamma(.)$ is a modified Bessel function of the third kind (Erdelyi et al. 1953; Hurst et al. 1997, p. 107) with index γ, given for $\omega > 0$ by

$$\mathrm{K}_\gamma(\omega) = \frac{1}{2}\int_0^\infty \exp\left\{-\frac{\omega}{2}\left(\upsilon^{-1}+\upsilon\right)\right\}\upsilon^{\gamma-1}\mathrm{d}\upsilon. \tag{7}$$

The fitting of the VG model and some application to analysis of financial market data is discussed in Seneta (2004). Bellini and Mercuri (2011) discuss approximation of the VG distribution with a finite mixture of normal distributions. Loregian et al. (2011) use this approximation for obtaining a formula for valuing the European call option.

2 Modeling Daily Gold Returns

The prices of gold and other precious metals are of high importance in both policy and business circles. Tully and Lucey (2007) investigated the applicability of asymmetric power GARCH model (AP GARCH) introduced by Ding et al. (1993). Mills (2003) studied the statistical behavior of daily gold price data from 1971 to 2002. They found that daily returns are highly leptokurtic. The role of precious metals in financial markets has been analyzed by Draper et al. (2006). They use daily data for gold, platinum, and silver from 1976 to 2004. They found that these metals have low correlations with stock index returns, which suggest that these metals may provide diversification within broad investment portfolios. We examine the daily gold prices (in Indian Rupees per ounce) in the period January 2, 1998 to June 3, 2011, which consists of 3364 trading days obtained from http://www.gold.org/investment/statistics/gold_price_chart/. Figure 1 gives the plots of gold price and daily gold returns in this period.

The summary statistics of the daily gold returns given in Table 1 show that the gold returns are symmetric with high kurtosis. Thus, it is expected that the log-returns will not follow a normal distribution. This is confirmed by an examination of Fig. 2, which gives the probability plot and histogram of log-returns.

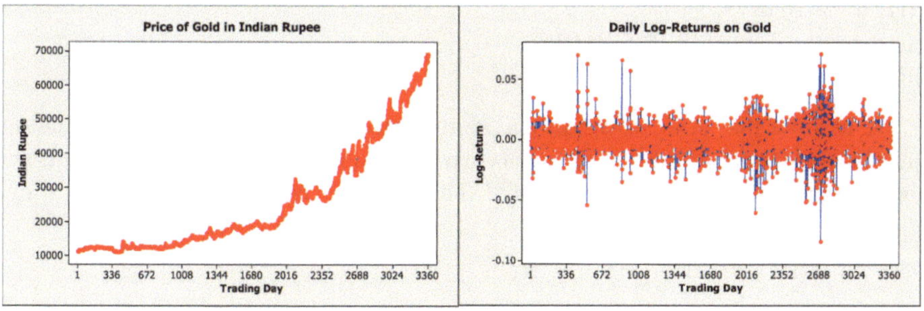

Fig. 1 Time series plot of daily gold price and daily gold returns

Table 1 Summary statistics of the daily gold log-returns (in %)

Minimum	−8.396
Q1	−0.498
Median	0.041
Q3	0.620
Maximum	7.127
Mean	0.054
Std. dev	1.117
Skewness	0.040
Kurtosis	8.129

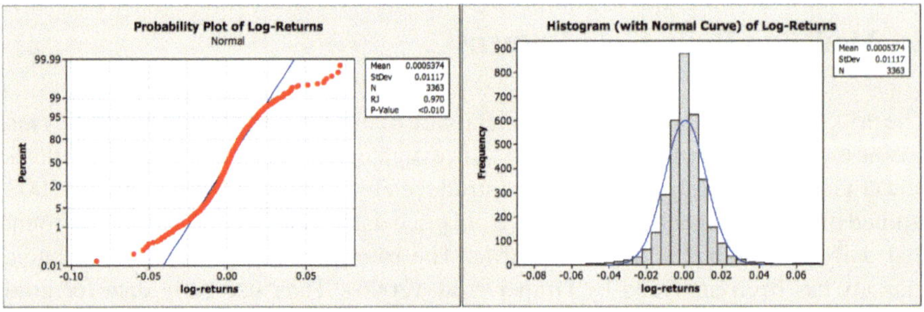

Fig. 2 Probability plot and histogram of log-returns

2.1 Mixture of Normal Model

We assume that the daily log-returns on gold follow a two-component mixture of normal distributions with both components having mean 0. This choice is reasonable because of the unimodal nature of the log-returns as seen in histogram of log-returns with the estimated mean very close to 0. Thus, $r_t \sim \alpha N(0, \sigma_1) + (1 - \alpha) N(0, \sigma_2)$ where $0 < \alpha < 1$. Further, we assume the following priors for the parameters α, σ_1 and σ_2:

(i) $\alpha \sim Beta\,(1,\ 9)$
(ii) $\sigma_1 \sim Gamma\,(8,\ 0.5,\ 0.01)$
(iii) $\sigma_2 \sim Gamma\,(2,\ 0.5,\ 0.01)$

We assume that prior distributions are independent. We obtain 100,000 random samples from the posterior distributions of $\alpha,\ \sigma_1,\ \sigma_2$ by using the SIR algorithm. The histograms of which are shown in Fig. 3. The summary of the posterior distributions of $\alpha,\ \sigma_1,\ \sigma_2$ is given in Table 2.

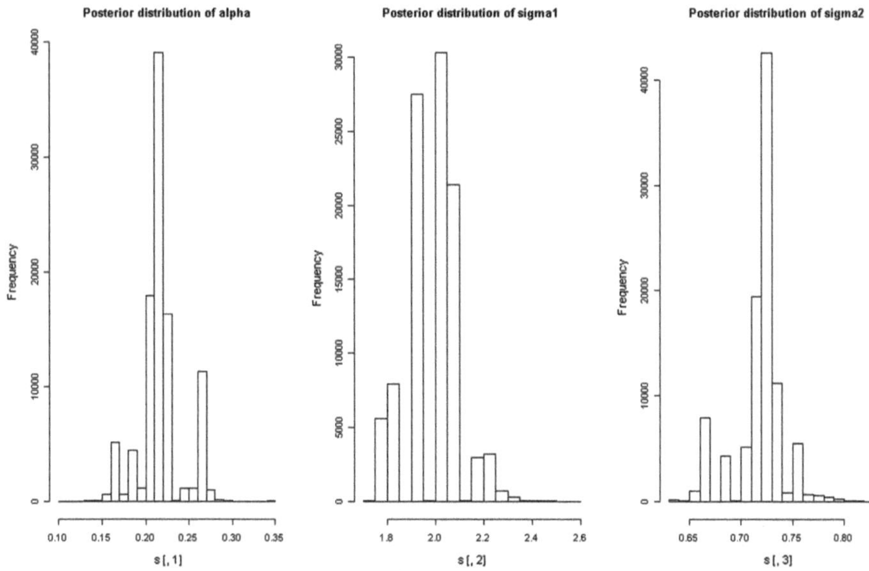

Fig. 3 Histogram of random samples from posterior distributions of α, σ_1, σ_2 for log-returns of gold

Table 2 Summary statistic of the posterior distributions of parameters in the mixture model for log-returns of gold

	α	σ_1	σ_2
Minimum	0.105	1.745	0.635
Q1	0.202	1.937	0.716
Median	0.213	2.029	0.721
Q3	0.224	2.062	0.723
Maximum	0.343	2.589	0.840
Mean	0.216	1.994	0.718
Std. dev	0.024	0.104	0.022

To assess how well the estimated two-component mixture of normal distributions model fits the observed log-returns on gold, we simulated 3363 observations from the estimated distribution (i.e., setting $\alpha = 0.213$, $\sigma_1 = 2.029$, and $\sigma_2 = 0.721$). The histogram of the simulated log-returns is given in Fig. 4 and the summary statistics are given in Table 3 along with a comparison with the actual log-returns.

From the above, we can see that a two-component mixture of normal distributions can be thought of as a reasonable model for log-returns on gold. One can visualize the mixture of normal distributions as follows: On 21% of the days, the market is more volatile and the standard deviation of the log-returns is about 2%. On the remaining days, the market is less volatile and the standard deviation of the log-returns is about 0.7%.

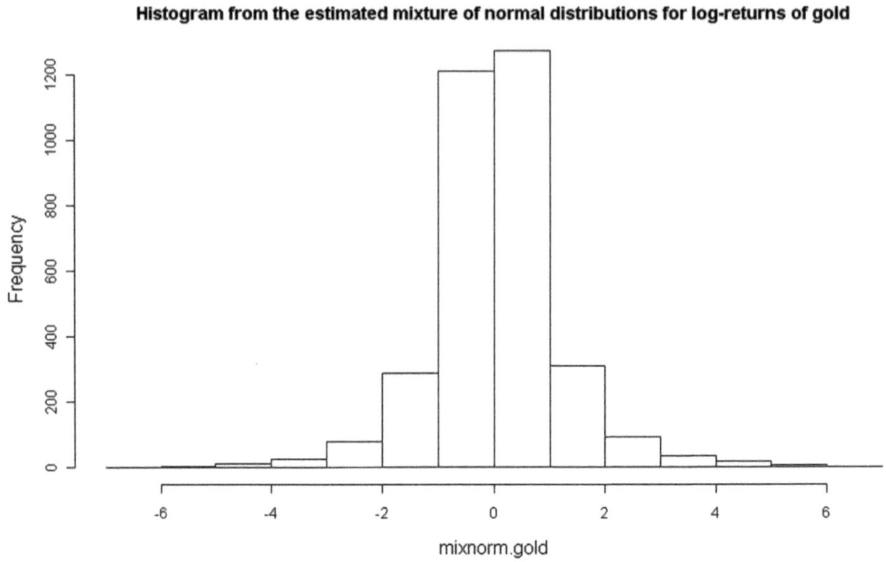

Fig. 4 Histogram from the estimated mixture of normal distributions for log-returns of gold

Table 3 A comparison of summary statistic of the simulated observations from the estimated mixture of normal distributions with the actual daily log-returns of gold (in %)

	Simulation	Actual
Minimum	−6.714	−8.396
Q1	−0.551	−0.498
Median	0.034	0.041
Q3	0.617	0.620
Maximum	6.830	7.127
Mean	0.043	0.054
Std. dev	1.162	1.117
Skewness	0.168	0.040
Kurtosis	7.053	8.129

2.2 *Variance Gamma Distribution Model*

As an alternative, we also fitted the VG distribution to the log-returns of gold. The parameters c, σ, θ and υ were estimated using the "Variance Gamma" package of R software. Table 4 gives the estimated parameter values.

Table 5 gives the comparison of some statistical features of the fitted distribution with that of the actual distribution of log-returns (%) of gold.

From the summary statistics above, we see that the VG distribution is not able to capture the leptokurtic behavior of the log-returns of gold. The Q-Q plot given in

Table 4 Estimated parameter values of VG distribution for log-returns of gold

c	0.031
σ	1.088
θ	0.023
υ	0.910

Table 5 A comparison of summary statistic of VG distribution model fitting for log-returns of gold (in %)

	Fitted	Actual
Mean	0.054	0.054
Variance	1.184	1.248
Skewness	0.057	0.040
Kurtosis	5.732	8.219

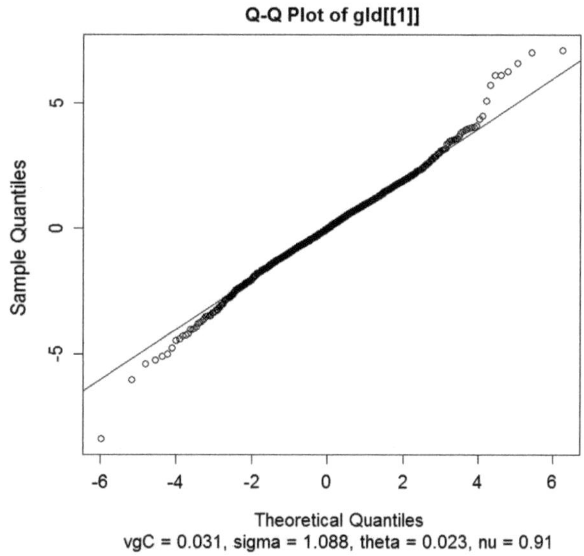

Fig. 5 Q-Q plot of log-returns of gold

Fig. 5 shows that the fit of the VG distribution to the log-returns of gold data is not particularly good near the tails of the distribution.

2.3 *Generalized Secant Hyperbolic Distribution Model*

As a third alternative, we try to fit the GSH(0, 1; t) distribution to the log-returns of gold data. The parameter "t" of the GSH distribution can be estimated as follows: Obtain $\hat{\beta}_2$ from the given data and then estimate the parameter "t" by using expressions (2.3) or (2.4) of Vaughan (2002) as appropriate. Following this procedure, we get the value of "t" as -2.49. To check the goodness-of-fit of the given data, we

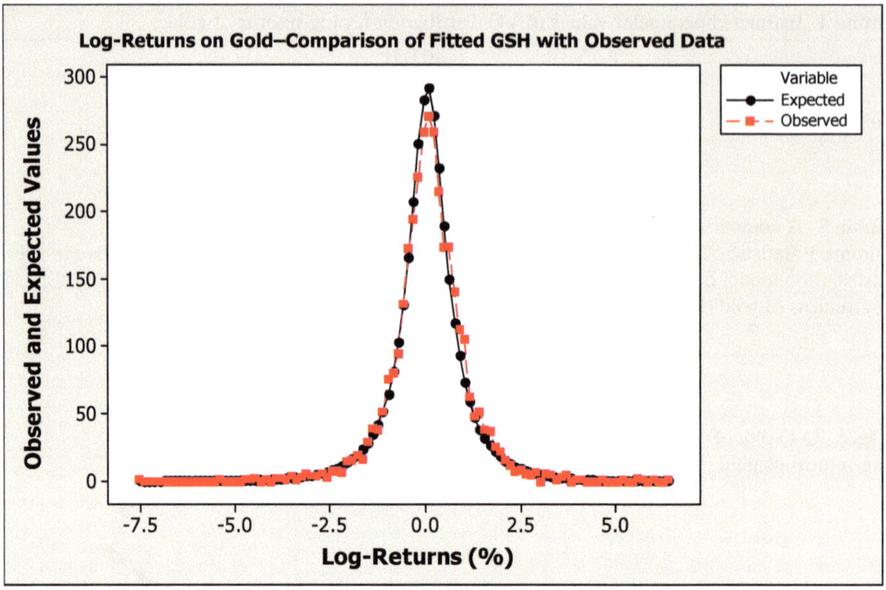

Fig. 6 Log-returns on gold—a comparison of fitted GSH with observed data

use the χ^2 goodness-of-fit test. We obtained a P-value of 0.03 indicating that the GSH(0, 1; t) distribution is not a good fit to the log-returns of gold. However, the plot of the smoothed frequency polygons of the histograms of the actual log-returns (%) data and the expected frequencies derived from the fitted GSH(0, 1, −2.49) distribution, given in Fig. 6, show that the fit is not that bad.

3 Modeling Daily Silver Returns

We examine the daily London Fix silver prices (in USD) in the period January 1, 1998–May 30, 2011, which consist of 3386 trading days obtained from http://www.perthmint.com.au/investment_invest_in_gold_precious_metal_prices.aspx. Figure 7 gives the plot of daily silver returns in this period.

Figure 8 gives the histogram of log-returns and probability plot which exhibits the non-normality of log-returns. The summary statistics of the daily log-returns on silver given in Table 6 show that the silver returns also exhibit high kurtosis as we have seen in the case of gold returns.

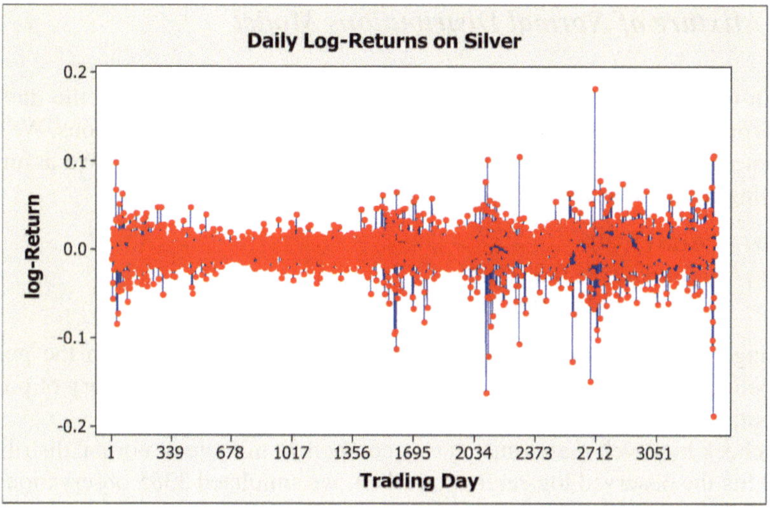

Fig. 7 Daily log-returns on silver

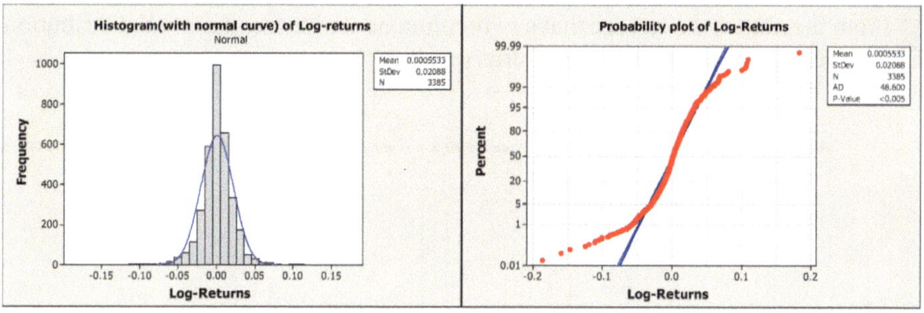

Fig. 8 Histogram and probability plot of log-returns on silver

Table 6 Summary statistic of the daily silver log-returns (in %)

Minimum	−18.693
Q1	−0.878
Median	0.057
Q3	1.104
Maximum	18.27
Mean	0.055
Std. dev	2.088
Skewness	−0.564
Kurtosis	11.738

3.1 Mixture of Normal Distributions Model

Like in the case of log-returns on gold here also, we assume that the daily log-returns on silver follow a two-component mixture of normal distributions. We follow the same approach as described in the case of log-returns of gold. We assume the following independent priors for the parameters α, σ_1 and σ_2:

(i) $\alpha \sim Beta\,(1, 9)$
(ii) $\sigma_1 \sim Gamma\,(8, 0.5, 0.01)$
(iii) $\sigma_2 \sim Gamma\,(2.5, 0.5, 0.01)$

Using SIR algorithm, we obtain 100,000 random samples from the posterior distributions of α, σ_1, σ_2 which can be seen in Fig. 9 and the summary of posterior distributions of α, σ_1, σ_2 is given in Table 7.

To check how well the estimated two-component mixture of normal distributions model fits the observed log-returns on silver, we simulated 3385 observations from the estimated distribution (i.e., setting $\alpha = 0.174$, $\sigma_1 = 4.041$ and $\sigma_2 = 1.361$). The histogram of the simulated log-returns is given in Fig. 10 and the summary statistics are given in Table 8.

From the above, we can see that a two-component mixture of normal distributions is not a very good model for log-returns on silver.

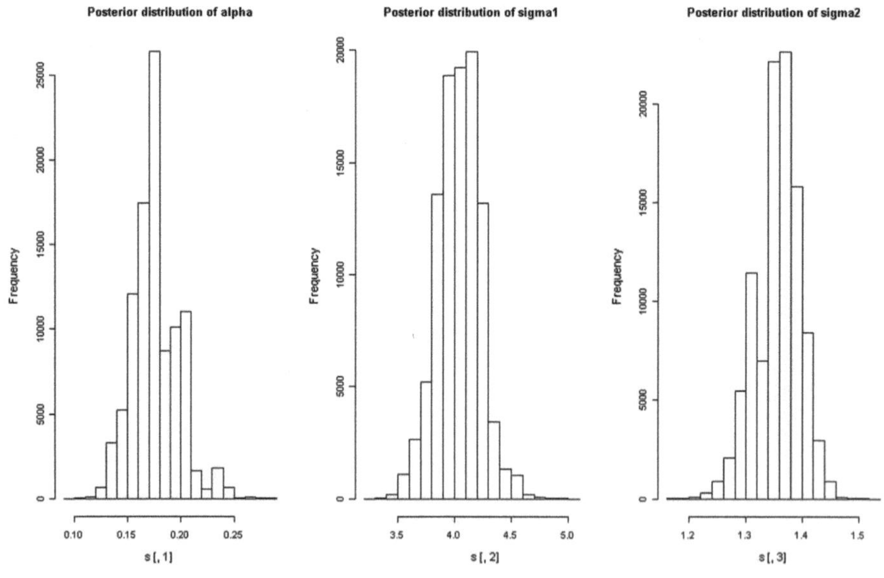

Fig. 9 Histogram of random samples from posterior distributions of α, σ_1, σ_2 for log-returns of silver

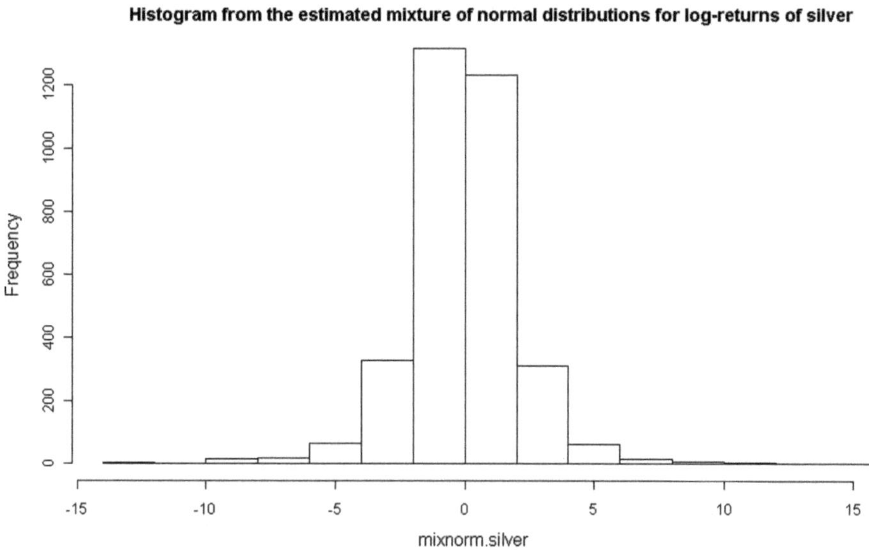

Fig. 10 Histogram from the estimated mixture of normal distributions for log-returns of silver

3.2 Variance Gamma Distribution Model

As in the case of log-returns on gold, we attempted to fit the VG distribution to the log-returns of silver. The parameters c, σ, θ and υ were estimated using the "Variance Gamma" package of R software. Table 9 gives the estimated parameter values. Table 10 gives a comparison of the summary statistics of the fitted VG distribution and the actual log-returns of silver.

We see from Table 10 and the Q-Q plot given in Fig. 11 that the fit of the VG distribution to the log-returns of silver data is not good particularly in the tails of the distribution.

Table 7 Summary statistic of the posterior distributions of parameters in the mixture model for log-returns of silver

	α	σ_1	σ_2
Minimum	0.098	3.245	1.160
Q1	0.163	3.919	1.333
Median	0.174	4.041	1.361
Q3	0.191	4.170	1.381
Maximum	0.287	5.057	1.534
Mean	0.176	4.039	1.357
Std. dev	0.022	0.188	0.040

Table 8 A comparison of summary statistic of the simulated observations from the estimated mixture of normal distributions with the actual daily log-returns of silver (in %)

	Simulation	Actual
Minimum	−12.96	−18.69
Q1	−1.14	−0.878
Median	−0.057	0.057
Q3	1.073	1.104
Maximum	14.320	18.27
Mean	−0.069	0.055
Std. dev	2.120	2.088
Skewness	−0.196	−0.564
Kurtosis	7.960	11.738

3.3 *Generalized Secant Hyperbolic Distribution Model*

Like we did in the case of log-returns on gold, as a third alternative, we try to fit the GSH(0, 1; t) distribution to the log-returns of silver data. Following the same procedure discussed in the analysis of log-returns on gold, we get the value of "t" as −2.74. The χ^2 goodness-of-fit test is seen to have a P-value of 0.000 indicating that the GSH(0, 1, −2.74) distribution is not a good fit to the log-returns of silver data. The plot of the smoothed frequency polygons of the histograms of the actual log-returns on silver (%) data and the expected frequencies derived from the fitted GSH(0, 1, −2.74) distribution, given in Fig. 12, also indicate that the fit is not good.

Table 9 Estimated parameter values of VG distribution for log-returns of silver

c	0.000
σ	2.015
θ	0.059
υ	1.024

Table 10 A comparison of summary statistic of VG distribution model fitting for log-returns of silver (in %)

	Fitted	Actual
Mean	0.059	0.055
Variance	4.064	4.360
Skewness	0.091	−0.564
Kurtosis	6.077	11.738

4 Modeling Daily Crude Oil Returns

Crude oil is one of the most important commodities in the modern world. Ahmad and Asali (2004) studied the behavior of crude oil in the short and long runs. They studied the short-run movements in crude oil prices during business cycles and concluded that the price of crude oil is shock persistent. Dees et al. (2007) described a structural econometric model of the world oil market that can be used to forecast oil supply, demand, and price. They use simulation to show that the model can be used to understand the responses of the world oil market to various types of shocks and changes in OPEC behaviors.

Our crude oil data consists of time series of daily prices (in $/barrel) of WTI-crude oil for the period from January 2, 1998 to May 31, 2011, consisting of 3363 trading days obtained from http://www.eia.gov/dnav/pet/hist/LeafHandler.ashx?n=PET&s=RWTC&f=D. Figure 13 gives the plot of daily crude oil returns in this period.

The summary statistics of the daily crude oil returns given in Table 11 show that like gold and silver, crude oil also exhibits high kurtosis. Figure 14 gives the histogram of log-returns on crude and probability plot, which exhibits the non-normality of log-returns as we have seen in the case of gold and silver.

4.1 Mixture of Normal Model

Like in the case of log-returns on gold and silver here also, we assume that the daily log-returns on crude oil follow a two-component mixture of normal distributions. We

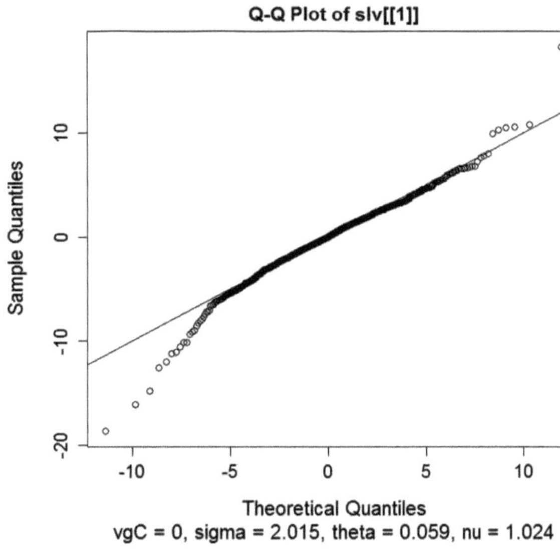

Fig. 11 Q-Q plot of log-returns of silver

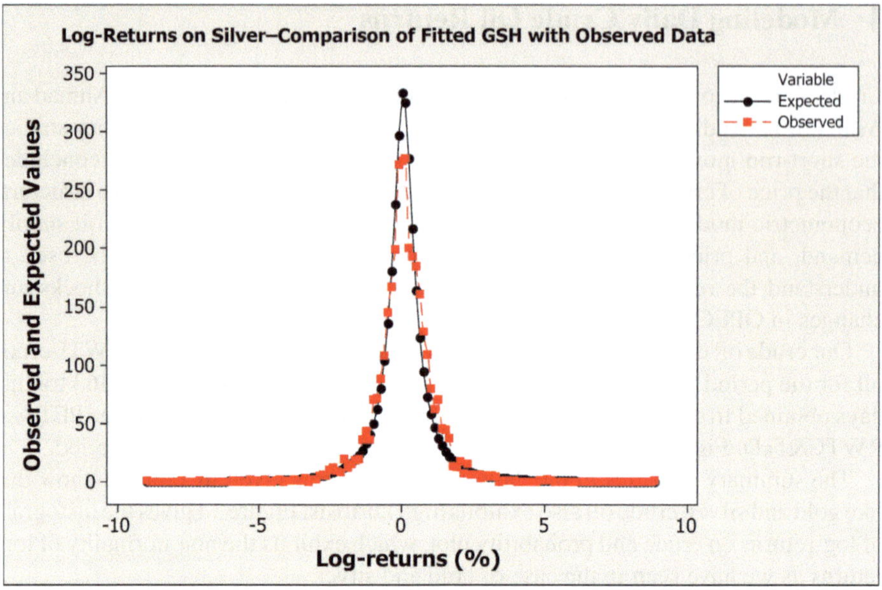

Fig. 12 Log-returns on silver—a comparison of fitted GSH with observed data

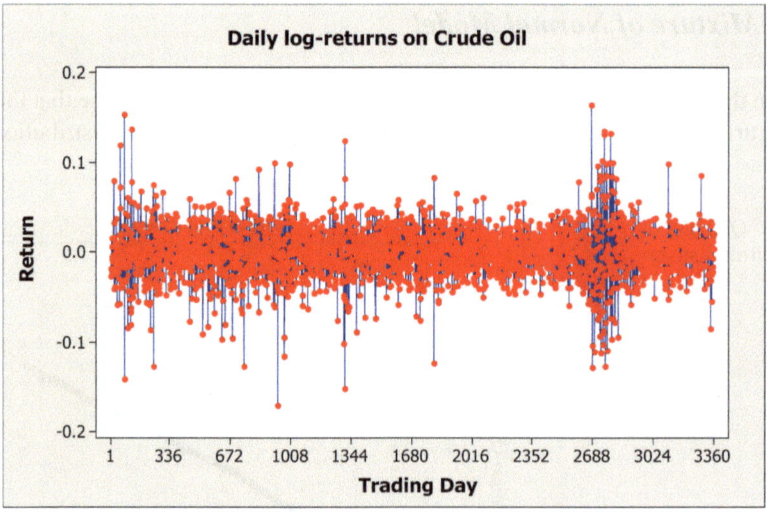

Fig. 13 Daily log-returns on crude oil

follow the same approach as described in the case of log-returns of gold and silver. We assume the following independent priors for the parameters α, σ_1 and σ_2:

(i) $\alpha \sim Beta\,(1,\ 9)$
(ii) $\sigma_1 \sim Gamma\,(8,\ 0.55,\ 0.01)$

Table 11 Summary statistic of the daily crude oil log-returns (in %)

Minimum	−17.092
Q1	−1.332
Median	0.117
Q3	1.528
Maximum	16.414
Mean	0.053
Std. dev	2.664
Skewness	−0.178
Kurtosis	7.453

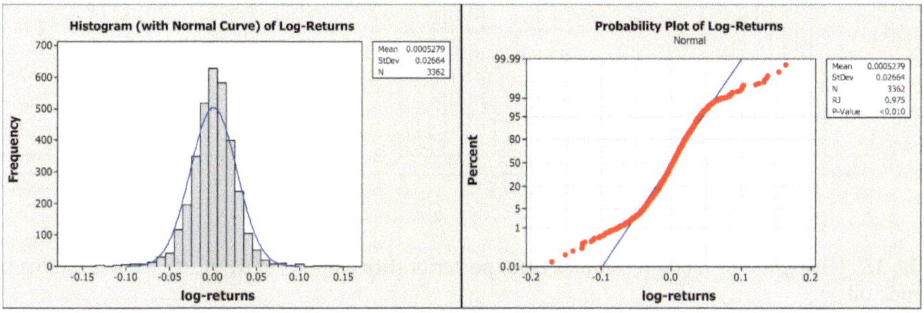

Fig. 14 Histogram and probability plot of log-returns on crude oil

Table 12 Summary statistic of the posterior distributions of parameters in the mixture model for log-returns of crude oil

	α	σ_1	σ_2
Minimum	0.057	4.399	1.838
Q1	0.106	5.433	1.976
Median	0.113	5.629	1.944
Q3	0.124	5.734	2.020
Maximum	0.215	7.635	2.205
Mean	0.114	5.588	1.997
Std. dev	0.017	0.308	0.040

(iii) $\sigma_2 \sim Gamma\,(1,\ 0.55,\ 0.01)$

Using SIR algorithm, we obtain 150,000 random samples from the posterior distributions of α, σ_1, σ_2 which can be seen in Fig. 15 and the summary of posterior distributions of α, σ_1, σ_2 is given in Table 12.

To check how well the estimated two-component mixture of normal distributions model fits the observed log-returns on silver, we simulated 3362 observations from the estimated distribution (i.e., setting $\alpha = 0.113\, \sigma_1 = 5.629$ and $\sigma_2 = 1.994$). The histogram of the simulated log-returns is given in Fig. 16 and the summary statistics are given in Table 13.

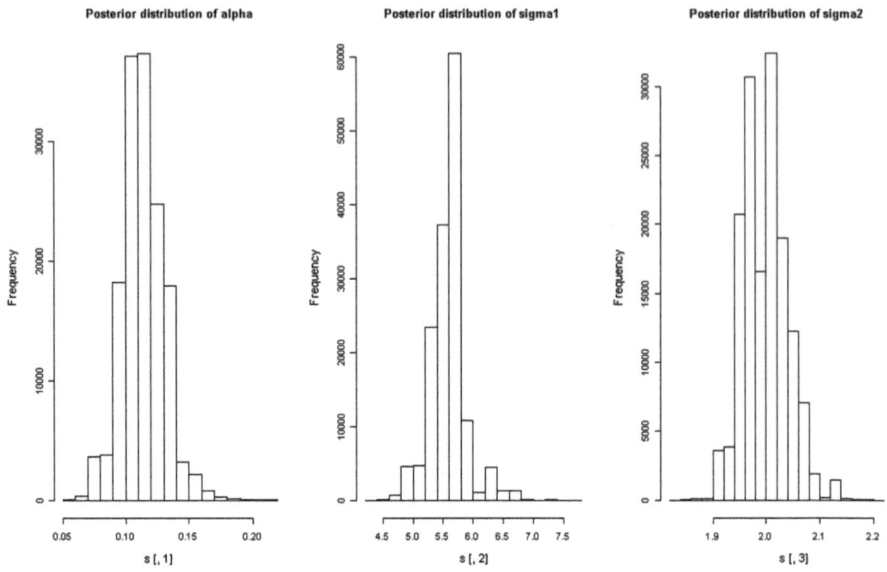

Fig. 15 Histogram of random samples from posterior distributions of α, σ_1, σ_2 for log-returns of crude oil

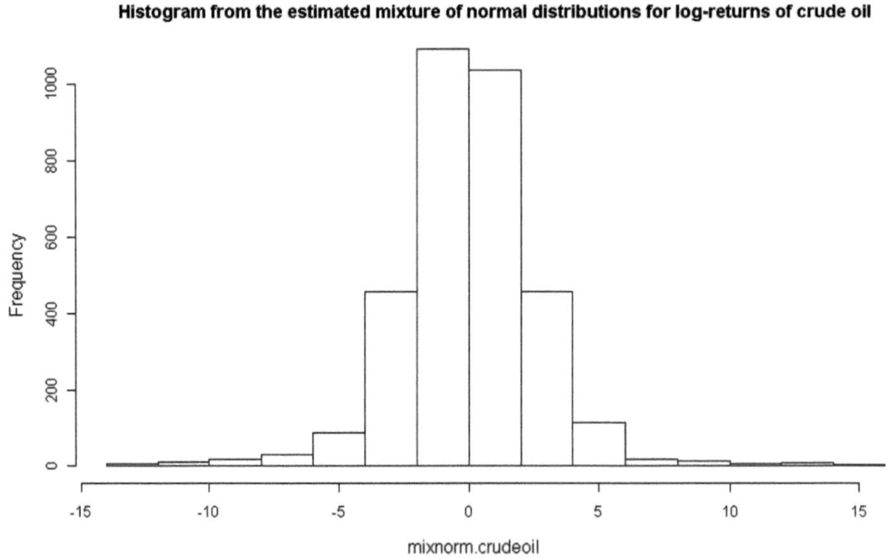

Fig. 16 Histogram from the estimated mixture of normal distributions for log-returns of crude oil

Table 13 A comparison of summary statistic of the simulated observations from the estimated mixture of normal distributions with the actual daily log-returns of crude oil (in %)

	Simulation	Actual
Minimum	−13.760	−17.092
Q1	−1.478	−1.332
Median	−0.041	0.117
Q3	1.467	1.528
Maximum	14.980	16.414
Mean	−0.003	0.053
Std. dev	2.707	2.664
Skewness	0.102	−0.178
Kurtosis	7.473	7.453

Table 14 Estimated parameter values of VG distribution for log-returns of crude oil

c	0.290
σ	2.590
θ	−0.237
υ	0.660

Table 15 A comparison of summary statistic of VG distribution model fitting for log-returns of crude oil (in %)

	Fitted	Actual
Mean	0.059	0.053
Variance	6.743	7.076
Skewness	−0.180	−0.178
Kurtosis	5.000	7.453

From the above, we can see that a two-component mixture of normal distributions is a reasonable model for log-returns on crude oil. One can visualize the mixture of normal distributions as follows: On 12% of the days, the market is more volatile and the standard deviation of the log-returns is about 5.6%. On the remaining days, the market is less volatile and the standard deviation of the log-returns is about 1.99%.

4.2 *Variance Gamma Distribution Model*

As an alternative, we also fitted the VG distribution to the log-returns of crude oil. The parameters c, σ, θ and υ were estimated using the "Variance Gamma" package of R software. The Table 14 gives the estimated parameter values.

From the comparison of the summary statistics given in Table 15 and the Q-Q plot given in Fig. 17, we conclude that the fit of the VG distribution to the log-returns of crude oil data is not good.

Fig. 17 Q-Q plot of log-returns on crude oil

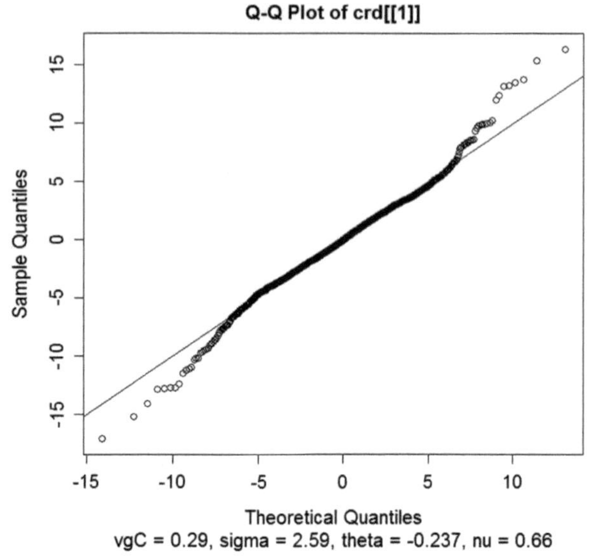

4.3 Generalized Secant Hyperbolic Distribution Model

Like we did in the case of log-returns on gold and silver as a third alternative, we try to fit the GSH(0, 1; t) distribution to the log-returns of crude oil data. Following the same procedure discussed in gold returns, we get the value of "t" is −2.38. To check the goodness-of-fit of the given data, we use the χ^2 goodness-of-fit test. We obtained a P-value of 0.000 indicating that the GSH(0, 1; −2.38) distribution is not a good fit to the log-returns of crude oil. This is also confirmed by an examination of the plot of the smoothed frequency polygons of the histograms of the actual log-returns on crude (%) data and the expected frequencies derived from the fitted GSH(0, 1, −2.38) distribution (Fig. 18).

5 Summary and Conclusions

In this paper, we have attempted to model the log-return distributions of three commodities namely gold, silver, and crude oil each of which exhibit strong leptokurtosis. We try using three different models namely the two-component mixture of normal distributions model, Variance Gamma model, and Generalized Secant Hyperbolic distribution model. We find that the two-component mixture of normal distributions model is a reasonable choice for modeling the log-return distributions of gold and crude oil. This model also lends itself to a simple intuitive explanation. The observations (i.e., log-returns) can be thought to come from two underlying distributions one being for days with less volatility, the other being for days with high volatility, and

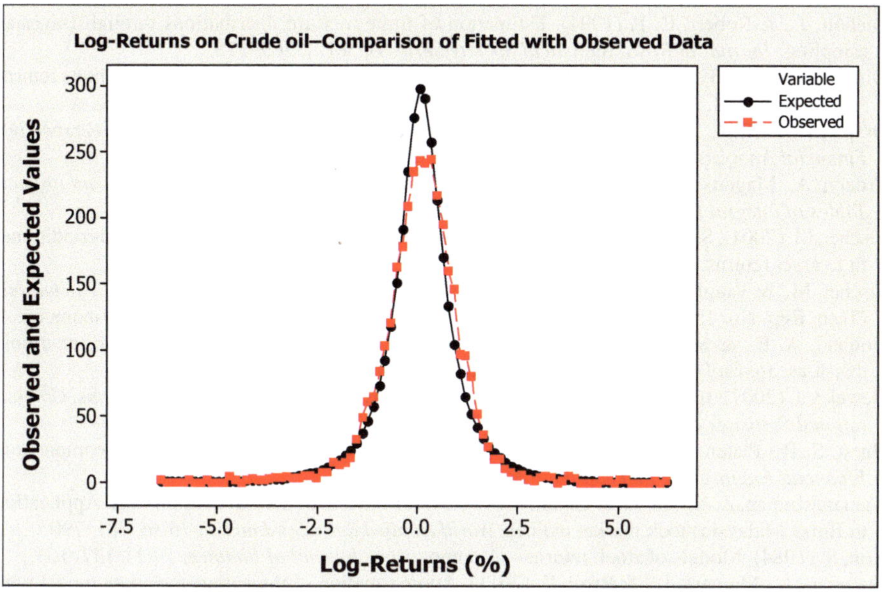

Fig. 18 Log-returns on crude oil—a comparison of fitted with observed data

the mixing proportion indicating the proportion of days with high volatility. None of the three models considered in this chapter give a satisfactory description of the log-return distribution of silver. Thus, we can conclude that a single approach to modeling may not work for all commodities.

References

Ahmad, R. J. N., & Asali, M. (2004). Cyclical behaviour and shock-persistence: Crude oil prices. *OPEC Energy Review, Organization of the Petroleum Exporting Countries, 28*(2), 107–131, 06.

Albert, J. H. (1993). Teaching Bayesian statistics using sampling methods and MINITAB. *The American Statistician, 47*(3), 182–191.

Bellini, F., & Mercuri, L. (2011). Option pricing in a dynamic variance gamma model. *Accepted for publication in Journal of Financial Decision Making*.

Blattberg, R. C., & Gonedes, N. J. (1974). A comparison of the stable and student distributions as statistical models for stock prices. *Journal of Business, 47,* 244–280.

Casella, G., & George, E. (1992). Explaining the Gibbs sampler. *The American Statistician, 46,* 167–174.

Chib, S., & Greenberg, E. (1995). Understanding the metropolis-hastings algorithm. *The American Statistician, 49*(4), 327–335.

Clark, P. K. (1973). A subordinated stochastic process model with finite variance for speculative prices. *Econometrica, 41,* 135–156.

Dees, S., Karadelaglou, P., Kaufmann, R. K., & Sanchez, M. (2007). Modeling the world oil market: Assessment of a quarterly econometric model. *Energy Policy, 35,* 178–191.

Diebolt, J., & Robert, C. P. (1994). Estimation of finite mixture distributions through Bayesian sampling. *Journal of Royal Statistical Society Series B, 56*(2), 363–375.

Ding, Z., Eagle, R. F., & Granger, C. W. J. (1993). A long memory property of stock markets returns and a new model. *Journal of Empirical Finance, 1,* 83–106.

Draper, P., Faff, R., & Hiller, D. (2006). Do precious metals shine? *An Investment Perspective, Financial Analysts Journal, 62,* 98–106.

Erdelyi, A., Magnus, W., Oberhettinger, F., & Tricomi, F. G. (1953). *Bateman manuscript project: Tables of integral transforms* (Vol. 2). New York: McGraw-Hill.

Fischer, M. (2004). Skew generalized secant hyperbolic distributions: Unconditional and conditional fit to asset returns. *Australian Journal of Statistics, 33*(3), 293–304.

Fischer, M., & Vaughan, D. (2002). *Classes of skewed generalized secant hyperbolic distributions* (Tech. Rep. No. 45). Universitat Erlangen-Nurnberg: Lehrstuhl fur Statistik and Okonometrie.

Gelfand, A. E., & Smith, A. F. M. (1990). Sampling based approaches to calculating marginal densities. *Journal of the American statistical Association, 85,* 398–409.

Geweke, J. (2007). Interpretation and inference in mixture models: Simple MCMC works. *Computational Statistics & Data Analysis, 51,* 3529–3550.

Hurst, S. R., Platen, E., & Rachev, S. R. (1997). Subordinated markov models: A comparison. *Financial Engineering and Japanese Markets, 4,* 97–124.

Kamaruzzaman, Z. A., Isa, Z., & Ismail, M. T. (2012). Mixtures of normal distributions: Application to Bursa Malaysia stock market indices. *World Applied Sciences Journal, 16*(6), 781–790.

Kon, S. (1984). Models of stock returns—A comparison. *Journal of Finance, 39*(1), 147–165.

Loregian, A., Mercury, L., & Rroji, E. (2011). Approximation of the variance gamma model with a finite mixture of normals. *Statistics and Probability Letters, 82,* 217–224.

Madan, D. B., & Seneta, E. (1990). The Variance-gamma (V.G) model for share market return. *Journal of Business, 63,* 511–524.

Madan, D. B., Carr, P. P., & Chang, E. C. (1998). The variance-gamma process and option pricing. *European Finance Review, 2,* 79–105.

Mills, T. C. (2003). Statistical analysis of daily gold price data. *Physica A, 338,* 559–566.

Praetz, P. D. (1972). The distribution of share price changes. *Journal of Business, 45,* 49–55.

Press, S. J. (1967). A compound events model for security prices. *Journal of Business, 40,* 317–335.

Rubin, D. B. (1988). Using the SIR algorithm to simulate posterior distributions. In J. M. Bernardo, M. H. DeGroot, D. V. Lindley, & A. F. M. Smith (Eds.), *Bayesian statistics 3* (pp. 395–402). New York: Oxford University Press.

Seneta, E. (2004). Fitting the variance-gamma model to financial data. *Journal of Applied Probability, 41,* 177–187.

Smith, A. F. M., & Gelfand, A. E. (1992). Bayesian statistics without tears: a sampling resampling perspective. *The American Statistician, 46,* 84–88.

Smith, A. F. M., & Roberts, G. O. (1993). Bayesian computation via the Gibbs sample and related Markov chain Monte Carlo methods. *Journal of Royal Statistical Society, Series B, 55,* 3–24.

Titterington, D. M., Smith, A. F. M., & Markov, U. E. (1985). *Statistical analysis of finite mixture distributions*. New York: Wiley.

Tully, E., & Lucey, B. M. (2007). A power GARCH Examination of the gold market. *Research in International Business and Finance, 21,* 316–325.

Vaughan, D. C. (2002). The generalized hyperbolic secant distribution and its application. *Communications in Statistics—Theory and Methods, 31,* 219–238.

Venkataraman, S. (1997). *Value at risk for a mixture of normal distributions: The use of Quasi-Bayesian estimation techniques, economic perspectives*, Issue 2–13 Mar.

Wirjanto, T.S., & Xu, D. (2009). *The applications of mixtures of normal distributions in empirical finance: A selected survey.* Waterloo Economic Series, No. 904.

Part VII
Methodology

OLS: Is That So Useless for Regression with Categorical Data?

Atanu Biswas, Samarjit Das and Soumyadeep Das

Abstract Binary/categorical response data abound in many application areas poses a unique problem; OLS-based model may lead to negative estimate for probability of a particular category and does not provide coherent forecast for the response variable. This unique and undesirable property of linear regression with categorical data impedes the use of OLS which otherwise is the simplest and distributionally robust method. The logit or probit kind of solution is heavily distribution dependent or link function dependent. Failure of such distributional assumption of the underlying latent variable model may cost the estimators heavily and may lead to biased and inconsistent estimates, in general. In this paper, we attempt to fix the inherent problem of linear regression by suggesting a simple manipulation which, in turn, leads to consistent estimates of probability of a category, and results in coherent forecasts for the response variable. We show that the proposed solution provides comparable estimates, and sometimes, with respect to some criterion, the proposed method is even slightly better than the logit kind of models. Here, we consider different underlying error distributions and compare the performances of the two models (in terms of their respective residual sum of squares and also in terms of relative entropy) based on simulated data. It is evidenced that the OLS performs better for many distributions, viz., Gamma, Laplace, and Uniform error distributions.

Keywords Logit model · Ordinary least square · Residual sum of squares
Relative entropy

A. Biswas
Applied Statistics Unit, Indian Statistical Institute, Kolkata, India

S. Das
Economic Research Unit, Indian Statistical Institute, Kolkata, India

S. Das (✉)
Department of Statistics, University of Calcutta, Kolkata, India
e-mail: das.soumyadeep1992@gmail.com

A. K. Laha (ed.), *Advances in Analytics and Applications*, Springer Proceedings in Business and Economics, https://doi.org/10.1007/978-981-13-1208-3_18

1 Introduction

Modeling the relationship between explanatory and response variables is a fundamental activity encountered in Statistics. Ordinary linear regression (OLS) is often used to investigate the relationship between several explanatory variables (predictors) and a single response variable. The OLS is routinely used to find the estimates of the underlying regression parameters. Stigler (1981) rightly mentioned: "The method of least squares is the automobile of modern statistical analysis: despite its limitations, occasional accidents, and incidental pollution, it and its numerous variations, extensions, and related conveyances carry the bulk of statistical analyses, and are known and valued by nearly all."

However, it is generally considered inappropriate to use the OLS in linear regression for modeling categorical (dichotomous) dependent variables. With a dichotomous dependent variable, all the observed dependent data points will fall on either of the two horizontal lines that are parallel, which is a difficult condition to model with the single straight line produced by linear model. More importantly, OLS estimates of the probabilities of the categories may not fall in the permissible interval. The forecast/prediction for the response variable is not coherent; forecast does not conform to the categorical nature of the response variable. This strange and undesirable property of linear regression with categorical data impedes the use of OLS which otherwise is the simplest and distributionally robust method.

Logistic regression analysis is one of the most frequently used statistical procedures that researchers often turn to, and is especially common as a mean to account for the variance in a binary (or categorical) dependent variable (King and Ryan 2002). The technique has become quite popular in social science research due to proponents who have suggested that it is a more appropriate alternative compared to ordinary least squares if a dependent variable is a binary outcome. Peng et al. (2002, p. 259) reported that "research using logistic regression has been published with increasing frequency in three higher education journals: Research in Higher Education, The Review of Higher Education, and The Journal of Higher Education".

In this present paper, we advocate the use of linear regression for categorical data and the use of OLS, and then provide consistent prediction rule for each category. We show that with a simple manipulation, coherent forecast can be obtained without much hassles. We provide the coherent forecast along with forecast distribution. Estimated forecast probability always lies within the permissible range. This forecast distribution is useful to provide standard error and to construct the forecast confidence interval. Here it may be noted that, if the data generating process (DGP) is completely known, logit and probit kind of model is expected to lead to efficient parameter estimates and subsequently may provide the efficient coherent forecasts. However, the affect of misspecification at any level is not completely known, and may lead to biased and inconsistent parameter estimates which may affect the forecast performance.

The motivation of the current work lies in the latent variable based approach that is used to explain the logistic regression; we assume that behind every ordinal categorical variable there exists a latent variable and we can categorize the study variable depending on the value of that latent variable.

Based on the fit of the linear regression on the available data, we set cutoffs to write the probabilities of the categories. So far our knowledge goes, this has not been attempted in the literature. We estimate the parameters using least square method. We then minimize the residual sum of squares (RSS) to estimate the cutoff values. Note that, for categorical data, the RSS is not defined in general as the "value" of a category cannot be defined uniquely. But here, in Sects. 2 and 3, for illustration, we have defined RSS assigning some arbitrary numbers to the categories. Then, in Sect. 4, instead of RSS, we advocate the use of relative entropy (RE), which is sort of a measure of distance between two probability distributions but does not depend on the values of the categories.

We, furthermore, attempt to choose an optimal cutoff point and then convert the predicted values into binary/multinomial outcomes with respect to that optimal point. Here, it may be mentioned that though the predicted probabilities in case of logistic regression fall in (0, 1), we still need to choose an optimal cutoff point and code the predicted values with 1 or 0 with respect to that point. We then compare the efficiency of the two procedures by comparing their forecast performance. We calculated and compared the RSS values (or relative entropy values) along with the AIC (Akaike Information Criteria) values for both the linear regression based model and for the logistic model. For the proposed linear regression based model, we do not exactly use the AIC, instead we use Quasi-AIC (QAIC) as we have not used the MLE's of all the parameters in the formula for AIC; we used the estimates that have been described in this method. Also, we use the "Proportion of Correct Prediction" for comparing two models. We used the method of cross validation by dividing the whole dataset into two parts, namely "training" and "testing". The parameter estimates obtained from the "training" part are used in the "testing" part. In "testing" part, we have assigned the values of the variable into their suitable categories observing their probabilities. "Proportion of Correct Prediction" is nothing but the proportion of newly assigned values that is matched with the values in the "testing" part in terms of their matching categories.

To summarize, in this paper, we attempt to explore the usefulness of the standard linear regression model for obtaining coherent prediction. Extensive simulation study proves the efficacy of the linear regression model over logistic regression. We observe that linear regression model performs as good as logistic model; and in some cases it performs better than logistic model.

The rest of the paper is organized as follows. In Sect. 2, we describe our OLS-based methodology along with the omnipresent logistic regression. A brief description of the choice of the optimal cutoff is also given and the coherent prediction rule is given. In Sect. 3, an extensive simulation study has been carried out to compare the performance of linear regression with that of logistic regression. We also generalize the methodology to multinomial data. We consider category independent measure, namely relative entropy (RE), instead of RSS in Sect. 4. Section 5 concludes the paper.

2 Modeling Categorical Data

2.1 *Logistic Regression*

Let y be a binary random variable and $X^{p\times 1} = (1, x_1, \ldots, x_{p-1})'$ be the corresponding covariate vector. In logistic regression, the model is

$$\text{logit}\,\pi = \text{logit}(P(y = 1)) = X'\beta, \tag{2.1}$$

where $\beta = (\beta_0, \beta_1, \ldots, \beta_{p-1})'$ is a $p \times 1$ vector of parameters. See Agresti (2012) for the relevant details including interpretation of the logistic model.

Let the training data be $\boldsymbol{y} = (y_1, y_2, \ldots, y_m)'$ along with the covariate matrix

$$\boldsymbol{X}^{m\times p} = (X_1, X_2, \ldots, X_m)' = \begin{pmatrix} 1 & x_{1,1} & \cdots & x_{1,(p-1)} \\ 1 & x_{2,1} & \cdots & x_{2,(p-1)} \\ \vdots & \vdots & \ddots & \vdots \\ 1 & x_{m,1} & \cdots & x_{m,(p-1)} \end{pmatrix}.$$

From (2.1), we get

$$\pi_i = P(y_i = 1) = \frac{e^{X_i{}'\beta}}{1 + e^{X_i{}'\beta}}.$$

The log-likelihood function becomes

$$l(\beta) = \sum_{i=1}^{m} \left[y_i(X_i{}'\beta) - \log\left(1 + e^{X_i{}'\beta}\right) \right].$$

One can solve the above equation iteratively by the Newton–Raphson method. Upon convergence, we get $W^{m\times m} = diag(\pi_1(1 - \pi_1), \pi_2(1 - \pi_2), \ldots, \pi_m(1 - \pi_m))$, where we denote $\widehat{\pi}_i$ found in this method by $\widehat{\pi}_i(logistic)$.

As measures of goodness of fit, we obtain

$$\begin{aligned} RSS(logistic) &= \sum_{i=1}^{m} \{y_i - \widehat{\pi}_i(logistic)\}^2, \\ AIC(logistic) &= -2l(\widehat{\beta}) + 2p, \end{aligned}$$

where p is the number of parameters to be estimated.

Also for the purpose of prediction for the test data, i.e., for the data indexed by $i = m + 1, \ldots, n$, we define

$$y_{pi}(logistic) = \begin{cases} 1 & \text{if } \widehat{\pi}_i(logistic) \geq 0.5; \\ 0 & \text{otherwise.} \end{cases}$$

Thus, the proportion of correct prediction is defined as

$$Prop_p = \frac{1}{n-m} \sum_{i=m+1}^{n} I(y_i = y_{pi}(logistic)),$$

where $I(\cdot)$ is an indicator function.

If there are K (>2) categories of the response y, say $C_0, C_1, \ldots, C_{K-1}$ with $P(y_i \in C_k) = \pi_k$ for $k = 0, 1, \ldots, K-1$, we need to take one category as the reference category. Taking C_{K-1} as the reference category, the logistic model is

$$\pi_{(K-1)i} = \frac{1}{1 + \sum_{j=0}^{K-2} e^{X_i'\beta_j}},$$
$$\pi_{ki} = e^{X_i'\beta_k} \times \pi_{(K-1)i}, \quad k = 0, 1, \ldots, K-2.$$

Here $\beta = (\beta_0', \beta_1', \ldots, \beta_{K-2}')'$. Now, the log-likelihood function is

$$l(\beta) = \text{constant} + \sum_{i=1}^{m} \sum_{k=0}^{K-2} I(y_i \in C_k)(X_i'\beta_k) - \sum_{i=1}^{m} \log\left(1 + \sum_{k=0}^{K-2} e^{X_i'\beta_k}\right).$$

Solving this, we get $\widehat{\pi}_{ki}(logistic)$ for $k = 0, 1, \ldots, K-1$. In this case, assuming the representative value of category C_k be k, $k = 0, 1, \ldots, K-1$, we define

$$\widehat{E}_{logistic}(y_i|X_i) = \sum_{k=0}^{K-1} k\widehat{\pi}_{ki}(logistic), \quad i = 1, \ldots, m.$$

Consequently, we have

$$RSS(logistic) = \sum_{i=1}^{m} \left(y_i - \widehat{E}_{logistic}(y_i|X_i)\right)^2,$$

and, as usual,

$$AIC(logistic) = -2l(\widehat{\beta}) + 2p(K-1).$$

For the purpose of prediction, for $i = m+1, \ldots, n$, we define $y_{pi}(logistic) = k$ if $\widehat{\pi}_{ki}(logistic) = \max_{0 \le k \le K-1}\{\widehat{\pi}_{ki}(logistic)\}$ for $k = 0, 1, \ldots, K-1$. As in the binary case, proportion of correct prediction is given by

$$Prop_p = \frac{1}{n-m} \sum_{i=m+1}^{n} I(y_i = y_{pi}(logistic)).$$

2.2 Proposed Ordinary Least Square (OLS) Based Methodology

For the same categorical $\boldsymbol{y}$, we propose to fit (may be brute forcefully) a linear model like $\boldsymbol{y} = \boldsymbol{X}\beta + \boldsymbol{\epsilon}$, where $\boldsymbol{\epsilon}$ follows some distribution with mean vector $\mathbf{0}$ and dispersion matrix $\sigma^2 I_m$. It is well known that $\widehat{\beta}_{OLS} = (\widehat{\beta}_0, \widehat{\beta}_1, \ldots, \widehat{\beta}_{p-1})' = (\boldsymbol{X}'\boldsymbol{X})^{-1}\boldsymbol{X}'\boldsymbol{y}$ and $\widehat{\sigma}^2_{OLS} = \frac{||\boldsymbol{y}-\boldsymbol{X}\hat{\beta}_{OLS}||^2}{m-p}$ by the least square method, assuming full column rank of $\boldsymbol{X}$ (i.e., $rank(\boldsymbol{X}) = p$).

To start with, let us forget the data feature and let the data follow a linear regression with normal error. For binary responses, using a single cutoff $c \in \mathbb{R}$, we would like to model the probability

$$\pi_i(c) = P(y_i \geq c|X_i) = 1 - \Phi\left(\frac{c - X_i'\beta}{\sigma}\right),$$

which is estimated by

$$\widehat{\pi}_i(c) = 1 - \Phi\left(\frac{c - X_i'\widehat{\beta}}{\widehat{\sigma}}\right)$$

Here $\widehat{\beta} = \widehat{\beta}_{OLS}$ and $\widehat{\sigma} = \widehat{\sigma}_{OLS}$ the least square estimate obtained from the Linear model $\boldsymbol{y} = \boldsymbol{X}\beta + \boldsymbol{\epsilon}$.
Moreover, $\widehat{\pi}_i(c)$ is a consistent estimator of $\pi_i(c)$ as the least square estimators are consistent. Then,

$$RSS(c) = \sum_{i=1}^{m} \{y_i - \widehat{\pi}_i(c)\}^2.$$

We find that c for which $RSS(c)$ is minimum, and denote it by c_{opt}. Using $c = c_{opt}$, we define

$$QAIC(c) = -2\sum_{i=1}^{m}\left[y_i \log \widehat{\pi}_i(c) + (1 - y_i)\log\{1 - \widehat{\pi}_i(c)\}\right] + 2(p+2).$$

Then, writing

$$y_{pi}(c) = \begin{cases} 1 & \text{if } \widehat{\pi}_i(c) \geq 0.5; \\ 0 & \text{otherwise;} \end{cases}$$

the proportion of correct prediction is defined as

$$Prop_p(c) = \frac{1}{n-m}\sum_{i=m+1}^{n} I(y_i = y_{pi}(c)).$$

If, for each i, the cutoff depends on the elements of X_i, say $c_0 + \sum_{j=1}^{p-1} c_j x_{ij} = X_i'\boldsymbol{c}$ where $\boldsymbol{c} = (c_0, c_1, \ldots, c_{p-1})'$, then we model the probability

$$\pi_i(\boldsymbol{c}) = P(y_i \geq X_i'\boldsymbol{c}|X_i) = 1 - \Phi\left(\frac{X_i'\boldsymbol{c} - X_i'\beta}{\sigma}\right),$$

which is estimated by

$$\widehat{\pi}_i(\boldsymbol{c}) = 1 - \Phi\left(\frac{X_i'\boldsymbol{c} - X_i'\widehat{\beta}}{\widehat{\sigma}}\right).$$

where like previous case, $\widehat{\beta} = \widehat{\beta}_{OLS}$ and $\widehat{\sigma} = \widehat{\sigma}_{OLS}$,the least square estimate obtained from the Linear model $\boldsymbol{y} = \boldsymbol{X}\beta + \boldsymbol{\epsilon}$. Then,

$$RSS(\boldsymbol{c}) = \sum_{i=1}^{m} \{y_i - \widehat{\pi}_i(\boldsymbol{c})\}^2.$$

We obtain optimal $\boldsymbol{c}$, denoted by $\boldsymbol{c}_{opt}$, for which $RSS(\boldsymbol{c})$ is minimum. Using that, as in the case of scalar c, we obtain $QAIC(\boldsymbol{c})$, $y_{pi}(\boldsymbol{c})$ and $Prop_p(\boldsymbol{c})$.

The case of more than two response categories can be tackled in the same way. For K categories of response $\boldsymbol{y}$, say, $C_0, C_1, \ldots, C_{K-1}$, the cutoffs $d_0, d_1, \ldots, d_{K-2}$ with $d_0 \leq d_1 \leq \cdots \leq d_{K-2}$, not depending on covariates, may be written as $\boldsymbol{d} = (d_0, d_1, \ldots, d_{K-2})'$. Then assuming normality of errors, we model the probabilities

$$\begin{aligned}
\pi_{0i}(\boldsymbol{d}) &= P(y_i \leq d_0|X_i) = \Phi\left(\frac{d_0 - X_i'\beta}{\sigma}\right),\\
\pi_{(K-1)i}(\boldsymbol{d}) &= P(y_i > d_{K-2}|X_i) = 1 - \Phi\left(\frac{d_{K-2} - X_i'\beta}{\sigma}\right),\\
\pi_{ki}(\boldsymbol{d}) &= P(d_{k-1} < y_i \leq d_k|X_i) = \Phi\left(\frac{d_k - X_i'\beta}{\sigma}\right) - \Phi\left(\frac{d_{k-1} - X_i'\beta}{\sigma}\right), \quad k = 1, \ldots, K-2,
\end{aligned}$$

which are estimated by replacing β by $\widehat{\beta} = \widehat{\beta}_{OLS}$ and σ by $\widehat{\sigma} = \widehat{\sigma}_{OLS}$ in the above expressions. Here also, assuming the value of C_k as k, for $k = 0, 1, \ldots, K-1$, we define

$$\widehat{E}(y_i|X_i)(\boldsymbol{d}) = \sum_{k=0}^{K-1} k\widehat{\pi}_{ki}(\boldsymbol{d}), \quad i = 1, \ldots, m.$$

Consequently,

$$RSS(\boldsymbol{d}) = \sum_{i=1}^{m} \left(y_i - \widehat{E}(y_i|X_i)(\boldsymbol{d})\right)^2.$$

As earlier, we find the optimal value of $\boldsymbol{d}$, given by $\boldsymbol{d}_{opt}$, for which $RSS(\boldsymbol{d})$ is minimum. Using that, we define

$$QAIC(\boldsymbol{d}) = -2\sum_{i=1}^{m}\sum_{k=0}^{K-1} I(y_i \in C_k)\log\widehat{\pi}_{ki}(\boldsymbol{d})) + 2(p+K).$$

For the purpose of prediction, for $k = 0, 1, \ldots, K-1$, we define $y_{pi}(\boldsymbol{d}) = k$ if $\widehat{\pi}_{ki}(\boldsymbol{d}) = \max_{0\le k\le K-1}\{\widehat{\pi}_{ki}(\boldsymbol{d})\}$. Proportion of correct prediction is given by

$$Prop_p(\boldsymbol{d}) = \frac{1}{n-m}\sum_{i=m+1}^{n} I(y_i = y_{pi}(\boldsymbol{d})).$$

If, for each i, the cutoff depends on the elements of Z_i, say $Z_i'\boldsymbol{f}$, where $\boldsymbol{f} = (f_1, f_2, ..., f_{p-1})'$ and $X_i = (1, Z_i')'$, then, as earlier, we can model the probabilities $\pi_{ki}(\boldsymbol{d}, \boldsymbol{f})$, $k = 0, 1, \ldots, K-1$. These are estimated by replacing β by $\widehat{\beta}$ and σ by $\widehat{\sigma}$. We can obtain $\widehat{E}(y_i|X_i)(\boldsymbol{d}, \boldsymbol{f})$ and hence $RSS(\boldsymbol{d}, \boldsymbol{f})$ in the same way, and obtain optimal values of $\boldsymbol{d}$ and $\boldsymbol{f}$, denoted by $\boldsymbol{d}_{opt}$ and $\boldsymbol{f}_{opt}$, by minimizing $RSS(\boldsymbol{d}, \boldsymbol{f})$. Using that we define $QAIC(\boldsymbol{d}, \boldsymbol{f})$ and $Prop_p(\boldsymbol{d}, \boldsymbol{f})$.

3 Simulations

We now discuss some simulation studies to understand the effectiveness of the new method. First, we generate n observations randomly from normal distribution to generate covariate(s). Then take distribution function of a specified distribution to transform the value of the drawn observations between [0, 1]. After that we take a random observation from Uniform(0, 1) distribution and compare the two values to define a new binary random variable. After generating the whole dataset, we take a part of the data as training sample of size m, and remaining part of the data is for testing.

For comparing the models for methods for two response categories with two covariates ($p = 2$), we consider $n = 1000$. For each $i = 1, 2, \ldots, 1000$, draw x_i from $N(-3, 4^2)$. Then we calculate $y_i{}^* = F(g(x_i))$, where $F(\cdot)$ is a distribution function of some random variable and $g(\cdot)$ is a linear function of x_i. Draw u_i from $U(0, 1)$, and define

$$y_i = \begin{cases} 0 & \text{if } 0 \le y_i{}^* < u_i; \\ 1 & \text{if } u_i \le y_i{}^* \le 1. \end{cases}$$

We call it *Type I model*. Taking $m = 800$, we compare logistic regression and linear regression with two types of cutoffs. The simulation is carried out with 10,000 replications.

In Table 1, we consider the following choices of $F(\cdot)$ (standard error (s.e.) of each measurement is given inside the braces):

1. F is the cumulative distribution function (cdf) of $-\chi^2$ distribution with degree of freedom (d.f.) 10 and noncentrality parameter (ncp) 0, and $g(x) = x - 2.5$. In this case, the simulated dataset contains about 80% 1's.
2. F is the cdf of Uniform $(-5, -1)$ distribution, and $g(x) = x$. In this case, the dataset contains about 50% 1's.
3. F is the cdf of t-distribution with d.f. 10 and ncp $= 0$, and $g(x) = x$. Here, the simulated dataset contains about 25% 1's.
4. F is the cdf of a logistic distribution with location $= 5$ and scale $= 4$, and $g(x) = x$. Here, the simulated dataset contains about 15% 1's.
5. F is the cdf of a normal $(0, 0.25)$ distribution and $g(x) = x$. Here, the dataset contains about 10% 1's.
6. F is the cdf of a Cauchy distribution with location parameter 0 and scale parameter 2, and $g(x) = x$. Here, the simulated dataset contains about 30% 1's.
7. F is the cdf of a log-normal distribution with location parameter 2 and scale parameter 5, and $g(x) = -x$. In this case, the dataset contains about 33% 1's.
8. F is the cdf of a Laplace distribution with location $= 3$ and scale $= 5$, and $g(x) = -3x$. The simulated dataset contains about 68% 1's.
9. F is the cdf of a Gamma distribution with shape parameter 1 and scale parameter 25, and $g(x) = 5 - 4x$. Here, the dataset contains about 30% 1's.

In a second simulation study, we consider $x_i \sim N(-3, 4^2)$, i.i.d., and $y^*_{1i} = F(g_1(x_i))$ and $y^*_{2i} = F(g_2(x_i))$ where $g_1(\cdot)$ is a linear function of x_i and $g_2(\cdot) = a + g_1(\cdot)$ for some $a > 0$. We draw u_i from $U(0, 1)$ and define

$$y_i = \begin{cases} 0 & \text{if } 0 \le u_i < y_{1i}{}^* \text{ or } y_{2i}{}^* \le u_i \le 1 \\ 1 & \text{if } y_{1i}{}^* \le u_i < y_{2i}{}^*. \end{cases}$$

This is *Type II model*. Different choices of $F(\cdot)$ studied under this model in Table 2 are:

1. F is the cdf of $-\chi^2$ distribution with d.f. $= 5$ and ncp $= 0$, and $g_1(x) = x - 6$ and $a = 3$. Here, the dataset contains about 80% 1's.
2. F is the cdf of a Gamma distribution with shape parameter $= 2$ and scale parameter $= 1$, and $g_1(x) = x + 4$ and $a = 6$. The dataset contains about 54% 1's.
3. F is the cdf of the logistic distribution with location $= 0$ and scale $= 1$, and $g_1(x) = x$ and $a = 5$. The dataset contains about 58% 1's.
4. F is the cdf of a Uniform$(-5, -1)$ distribution, and $g_1(x) = x - 1$ and $a = 2$. Simulated dataset contains about 80% 1's.
5. F is the cdf of the Cauchy distribution with location parameter 0 and scale parameter 1, and $g_1(x) = x - 2$ and $a = 12$. The dataset contains about 26% 1's.

For comparing the models for two response categories with three covariates ($p = 3$), we again take $n = 1000$, and draw x_{1i} from $N(-3, 4^2)$ and x_{2i} from $N(0, 2.5^2)$ independently. We then obtain $y_i^* = F(g(x_{1i}, x_{2i}))$, where $F(\cdot)$ is a distribution function of some random variable and $f(\cdot, \cdot)$ is a linear function of x_{1i} and x_{2i}. The remaining part is similar to the previous simulation exercise. With same m, we consider the following choices of $F(\cdot)$ in Table 3:

1. F is the cdf of $-\chi^2$ distribution with d.f. $= 10$ and ncp $= 0$, and $g(x_1, x_2) = x_1 + x_2$. The dataset contains about 85% 1's.
2. F is the cdf of Uniform$(-5, -1)$ distribution with $g(x_1, x_2) = x_1 + x_2$. The dataset contains about 50% 1's.
3. F is the cdf of a t-distribution with d.f. $= 10$ and ncp $= 0$, and $g(x_1, x_2) = x_2 - x_1$. Here the dataset contains about 73% 1's.
4. F is the cdf of a logistic distribution with location $= 2$ and scale $= 5$, and $g(x_1, x_2) = x_1 + x_2$. The simulated dataset contains about 30% 1's.
5. F is the cdf of a Cauchy distribution with location $= 0$ and scale $= 2$, and $g(x_1, x_2) = x_2 - 3x_1$. The dataset contains about 73% 1's.

In Table 4, we compare different methods for three response categories (i.e., $K = 3$) with two covariates ($p = 2$). We generate $x_i \sim N(-3, 4^2)$, i.i.d., $y_{1i}^* = F(g_1(x_i))$ and $y_{2i}^* = F(g_2(x_i))$ where $g_1(\cdot)$ is a linear function of x_i and $g_2(\cdot) = a + g_1(.)$ for some $a > 0$. We draw u_i from $U(0, 1)$ and define

$$y_i = \begin{cases} 0 & \text{if } 0 \le u_i < y_{1i}{}^* \\ 1 & \text{if } y_{1i}{}^* \le u_i < y_{2i}{}^* \\ 2 & \text{if } y_{2i}{}^* \le u_i \le 1. \end{cases}$$

With same m and n as earlier, the following choices of $F(\cdot)$ are taken:

1. F is the cdf of $N(-2, 1)$ distribution with $g_1(x) = x - 2$ and $a = 4$. Here, the dataset contains about 22% 0's, 36% 1's and 42% 2's.
2. F is the cdf of a t-distribution with d.f. $= 5$ and ncp $= 0$, and $g_1(x) = x$ and $a = 4$. The simulated dataset contains about 25% 0's, 35% 1's and 40% 2's.
3. F is the cdf of a Gamma distribution with scale $=$ 10 and shape $=$ 1, and $g_1(x) = x$ and $a = 4$. The dataset contains about 26% 0's, 20% 1's and 54% 2's.
4. F is the cdf of Uniform$(-6, 1)$ distribution, with $g_1(x) = x$ and $a = 2$. Simulated dataset contains about 45% 0's, 15% 1's and 40% 2's.
5. F is the cdf of a logistic distribution with location parameter -2 and scale parameter 1, and $g_1(x) = x$ and $a = 4$. Simulated dataset contains about 25% 0's, 35% 1's and 40% 2's.

From Table 1, we observe that for two response categories with two covariates, the method with cutoffs depending on covariates is really performing very well in terms of smaller RSS than the case of logistic regression for all the cases. The $Prop_p$ is same for both the cases. The performance of this method is competitive with respect to AIC/QAIC when the latent variable follows $-\chi^2$ or Uniform or logistic or log-normal or Gamma distribution; but the performance is not at all satisfactory for t_{10}

Table 1 RSS, AIC/QAIC and $Prop_p$ (s.e. in parentheses) for different modeling under different distributions of error for Type I binary model

Distribution	Models	RSS	AIC/QAIC	$Prop_p$
$-\chi^2_{10}$	Logistic	64.11 (0.05)	402.15 (0.27)	0.88 (0.0002)
	constant c	72.53 (0.04)	479.43 (0.23)	0.86 (0.0002)
	$c = (c_0, c_1)$	63.55 (0.04)	406.93 (0.27)	0.88 (0.0002)
Uniform(−5, −1)	Logistic	52.52 (0.04)	326.08 (0.25)	0.90 (0.0002)
	constant c	69.50 (0.04)	490.20 (0.20)	0.90 (0.0002)
	$c = (c_0, c_1)$	51.99 (0.04)	330.41 (0.25)	0.90 (0.0002)
t_{10}	Logistic	36.50 (0.04)	240.70 (0.24)	0.94 (0.0002)
	constant c	58.70 (0.03)	429.48 (0.19)	0.93 (0.0002)
	$c = (c_0, c_1)$	36.42 (0.04)	260.06 (0.25)	0.94 (0.0002)
Logistic(5, 4)	Logistic	92.30 (0.06)	606.52 (0.34)	0.85 (0.0003)
	constant c	94.17 (0.06)	624.93 (0.33)	0.85 (0.0003)
	$c = (c_0, c_1)$	92.28 (0.06)	613.79 (0.34)	0.85 (0.0003)
$N(0, 0.25)$	Logistic	8.55 (0.02)	58.69 (0.12)	0.98 (0.0000)
	constant c	34.74 (0.03)	285.97 (0.22)	0.94 (0.0002)
	$c = (c_0, c_1)$	8.44 (0.02)	67.35 (0.18)	0.98 (0.0000)
Cauchy(0, 2)	Logistic	118.37 (0.06)	748.39 (0.33)	0.80 (0.0003)
	constant c	124.02 (0.06)	777.79 (0.29)	0.79 (0.0003)
	$c = (c_0, c_1)$	118.15 (0.06)	778.73 (0.35)	0.80 (0.0003)
log-normal(2, 5)	Logistic	158.32 (0.05)	917.44 (0.23)	0.67 (0.0003)
	constant c	158.05 (0.05)	927.49 (0.22)	0.66 (0.0003)
	$c = (c_0, c_1)$	157.49 (0.05)	922.55 (0.23)	0.66 (0.0003)
Laplace(3, 5)	Logistic	77.35 (0.06)	505.15 (0.32)	0.87 (0.0002)
	constant c	88.81 (0.05)	589.65 (0.25)	0.86 (0.0002)
	$c = (c_0, c_1)$	77.30 (0.06)	525.80 (0.35)	0.87 (0.0002)
Gamma(1, 25)	Logistic	47.22 (0.04)	303.58 (0.26)	0.91 (0.0002)
	constant c	66.21 (0.04)	470.06 (0.20)	0.91 (0.0002)
	$c = (c_0, c_1)$	47.06 (0.04)	309.78 (0.26)	0.91 (0.0002)

(which is close to normal), normal, Cauchy or Laplace distribution. So, if there is any prior idea about the distribution of the latent variable, we may choose between logistic and linear regression model. However, the model with constant cutoff (i.e., the case where the cutoff does not depend on covariates) performs really bad.

From Table 2, we see that the, for Type II model, the proposed theory is doing better with respect to smaller RSS and higher $Prop_p$ with comparable AIC/QAIC values for $-\chi^2$ and logistic distributions. Though RSS is quite smaller and $Prop_p$ is quite higher, the proposed linear regression based model produces very high AIC/QAIC for Gamma and Cauchy distribution. And for Uniform distribution, it gives almost the same result in all the respects. So if we have any idea about the latent variable

Table 2 RSS, AIC/QAIC and $Prop_p$ (s.e. in parentheses) for different modeling under different distributions of error for Type II binary model

Distribution	Models	RSS	AIC/QAIC	$Prop_p$
$-\chi^2_5$	Logistic	121.30 (0.06)	749.89 (0.30)	0.79 (0.0003)
	constant c	120.51 (0.06)	757.63 (0.30)	0.80 (0.0003)
	$\boldsymbol{c} = (c_0, c_1)$	120.39 (0.06)	756.08 (0.31)	0.80 (0.0003)
Gamma(2, 1)	Logistic	169.41 (0.06)	990.60 (0.22)	0.70 (0.0005)
	constant c	171.79 (0.05)	1003.24 (0.20)	0.71 (0.0005)
	$\boldsymbol{c} = (c_0, c_1)$	162.44 (0.09)	2120.94 (0.58)	0.75 (0.0003)
Logistic(0, 1)	Logistic	194.56 (0.02)	1090.03 (0.09)	0.51 (0.0006)
	constant c	194.50 (0.02)	1094.18 (0.09)	0.53 (0.0005)
	$\boldsymbol{c} = (c_0, c_1)$	194.47 (0.02)	1095.96 (0.09)	0.52 (0.0006)
Uniform(−5, −1)	Logistic	122.77 (0.07)	779.77 (0.32)	0.81 (0.0003)
	constant c	122.75 (0.07)	783.78 (0.32)	0.81 (0.0003)
	$\boldsymbol{c} = (c_0, c_1)$	122.75 (0.07)	785.79 (0.32)	0.81 (0.0003)
Cauchy(0, 1)	Logistic	141.75 (0.07)	878.47 (0.29)	0.75 (0.0005)
	constant c	144.13 (0.06)	886.79 (0.28)	0.75 (0.0005)
	$\boldsymbol{c} = (c_0, c_1)$	133.15 (0.09)	1963.23 (0.54)	0.81 (0.0003)

Table 3 RSS, AIC/QAIC and $Prop_p$ (s.e. in parentheses) for different modeling under different distributions of error for Type I binary model with three covariates

Distribution	Models	RSS	AIC/QAIC	$Prop_p$
$-\chi^2_{10}$	Logistic	59.03 (0.05)	379.06 (0.28)	0.89 (0.0002)
	$\boldsymbol{c} = (c_0, c_1, c_2)$	58.58 (0.05)	383.07 (0.28)	0.89 (0.0002)
Uniform(−5, −1)	Logistic	44.63 (0.04)	280.23 (0.28)	0.92 (0.0002)
	$\boldsymbol{c} = (c_0, c_1, c_2)$	44.14 (0.04)	284.86 (0.28)	0.92 (0.0002)
t_{10}	Logistic	33.26 (0.04)	221.95 (0.24)	0.94 (0.0002)
	$\boldsymbol{c} = (c_0, c_1, c_2)$	33.29 (0.04)	237.59 (0.47)	0.94 (0.0002)
Logistic(2, 5)	Logistic	143.22 (0.06)	861.92 (0.27)	0.73 (0.0003)
	$\boldsymbol{c} = (c_0, c_1, c_2)$	143.25 (0.06)	868.40 (0.29)	0.73 (0.0003)
Cauchy(0, 2)	Logistic	67.14 (0.06)	485.79 (0.38)	0.90 (0.0002)
	$\boldsymbol{c} = (c_0, c_1, c_2)$	64.80 (0.06)	700.12 (1.27)	0.90 (0.0002)

in this non-monotone categorization, we may or may not use this new procedure accordingly.

From Table 3, i.e., for two response categories and three covariates, we did not report the case where the cutoffs do not depend on the covariates. From the five distributions that have been studied, the $-\chi^2$, Uniform, and logistic distributions favor our linear model based method in terms of smaller RSS, same $Prop_p$, and very close AIC/QAIC (though RSS for logistic regression in our method is slightly higher than that of the logistic regression). For Cauchy and t_{10} distributions, linear regression

Table 4 RSS, AIC/QAIC and $Prop_p$ (s.e. in parentheses) for different modeling under different distributions of error for three category model with two covariates

Distribution	Models	RSS	AIC/QAIC	$Prop_p$
$N(-2, 1)$	Logistic	74.69 (0.02)	482.63 (0.10)	0.41 (0.0002)
	$\boldsymbol{d} = (d_0, d_1)$	80.27 (0.02)	2556.42 (1.13)	0.41 (0.0002)
	$(\boldsymbol{d}, f_1) = (d_0, d_1, f_1)$	74.56 (0.02)	486.96 (0.10)	0.41 (0.0002)
t_5	Logistic	93.54 (0.07)	604.66 (0.38)	0.41 (0.0003)
	$\boldsymbol{d} = (d_0, d_1)$	97.81 (0.06)	2003.86 (2.84)	0.41 (0.0003)
	$(\boldsymbol{d}, f_1) = (d_0, d_1, f_1)$	93.41 (0.07)	633.64 (0.47)	0.41 (0.0003)
Gamma(10, 1)	Logistic	453.62 (0.06)	1404.62 (0.09)	0.54 (0.0002)
	$\boldsymbol{d} = (d_0, d_1)$	448.87 (0.06)	6614.58 (1.13)	0.54 (0.0002)
	$(\boldsymbol{d}, f_1) = (d_0, d_1, f_1)$	445.15 (0.06)	2663.33 (0.81)	0.54 (0.0002)
Uniform(−6, 1)	Logistic	234.78 (0.05)	925.84 (0.10)	0.37 (0.0003)
	$\boldsymbol{d} = (d_0, d_1)$	244.29 (0.05)	5893.23 (1.36)	0.37 (0.0003)
	$(\boldsymbol{d}, f_1) = (d_0, d_1, f_1)$	232.40 (0.05)	1271.24 (1.07)	0.37 (0.0003)
Logistic(−2, 1)	Logistic	140.49 (0.03)	830.02 (0.10)	0.41 (0.0002)
	$\boldsymbol{d} = (d_0, d_1)$	141.49 (0.03)	5128.44 (1.58)	0.41 (0.0003)
	$(\boldsymbol{d}, f_1) = (d_0, d_1, f_1)$	140.23 (0.03)	854.92 (0.10)	0.42 (0.0002)

based method is of no use (though there is a gain in RSS for Cauchy distribution but QAIC is too high). So, if there is any indication that the latent variable follows either $-\chi^2$ or Uniform or logistic distribution, we may proceed our analysis with the proposed linear regression based method.

From Table 4, i.e., for three response categories with two covariates, our method gives less RSS and same $Prop_p$ for all the cases. It performs well in terms of AIC/QAIC if the latent variable follows Normal distribution; for all the other studied cases, i.e., t_5, Gamma, Uniform or logistic distribution, the QAIC is too high. So depending on any prior knowledge of underlying the latent variable, we may choose the model.

4 Relative Entropy Based Assessment

In Sects. 2 and 3, we used RSS as a measure of goodness of fit of different models. An alternative measure of goodness of fit is used in this section, the *relative entropy* (RE), which is basically Kullback–Leibler divergence. Also, RE has the advantage that it does not take care of the values of the variable, and it depends on the probabilities only. Thus, for ordinal categorical random variables, the RE will be invariant of the values of the categories, as long as the ordinal structure is maintained. Relative entropy is a widely used information-theoretic measure of stochastic dependency in case of discrete random variables (see Kullback 1968; Cover and Thomas 1991).

The RE for a K-category ordinal random variable for the fit by logistic, with scalar c or vector $\boldsymbol{c}$, can be defined as

$$RE(logistic) = \sum_{k=0}^{K-1} p_k \log\left(\frac{p_k}{\pi_k(logistic)}\right),$$

$$RE(c) = \sum_{k=0}^{K-1} p_k \log\left(\frac{p_k}{\pi_k(c)}\right),$$

$$RE(\boldsymbol{c}) = \sum_{k=0}^{K-1} p_k \log\left(\frac{p_k}{\pi_k(\boldsymbol{c})}\right),$$

where p_k is the sample proportion for category k, $k = 0, 1, \ldots, K-1$. RE takes always nonnegative values and lower the value, better is the fitting through the particular model.

Here, we study RE only for binary data; however, the same approach can be used for $K > 2$. To use RE, we divide the whole training data (i.e., 800 data points) into 16 blocks of size 50 each. In each block, we draw a random observation from $N(-3, 4^2)$ and then replicate it 50 times. For two response categories, we thus get 16 p_1's, the

Table 5 RE, AIC/QAIC and $Prop_p$ (s.e. in parentheses) for different modeling under different distributions of error for Type I binary model

Distribution	Models	RE	AIC/QAIC	$Prop_p$
$-\chi^2_{10}$	Logistic	0.0108 (0.0000)	471.81 (0.93)	0.82 (0.0003)
	constant c	0.0349 (0.0000)	515.48 (0.99)	0.80 (0.0002)
	$\boldsymbol{c} = (c_0, c_1)$	0.0091 (0.0000)	475.14 (0.95)	0.82 (0.0003)
Uniform$(-5, -1)$	Logistic	0.0096 (0.0000)	321.22 (1.10)	0.90 (0.0002)
	constant c	0.0906 (0.0000)	454.74 (1.19)	0.90 (0.0003)
	$\boldsymbol{c} = (c_0, c_1)$	0.0092 (0.0003)	326.56 (1.10)	0.90 (0.0002)
Logistic(5, 4)	Logistic	0.0100 (0.0000)	612.43 (0.83)	0.85 (0.0003)
	constant c	0.0161 (0.0000)	626.12 (0.83)	0.85 (0.0003)
	$\boldsymbol{c} = (c_0, c_1)$	0.0100 (0.0000)	618.33 (0.83)	0.85 (0.0003)
Cauchy(0, 2)	Logistic	0.0172 (0.0000)	707.53 (0.64)	0.80 (0.0003)
	constant c	0.0332 (0.0000)	737.04 (0.63)	0.79 (0.0004)
	$\boldsymbol{c} = (c_0, c_1)$	0.0217 (0.0000)	720.64 (0.63)	0.78 (0.0004)
Gamma(1, 25)	Logistic	0.0245 (0.0001)	807.32 (0.78)	0.73 (0.0003)
	constant c	0.0347 (0.0001)	827.70 (0.77)	0.72 (0.0003)
	$\boldsymbol{c} = (c_0, c_1)$	0.0237 (0.0001)	812.04 (0.80)	0.73 (0.0003)

Table 6 RE, AIC/QAIC and $Prop_p$ (s.e. in parentheses) for different modeling under different distributions of error for Type II binary model

Distribution	models	RE	AIC/QAIC	$Prop_p$
$-\chi_5^2$	Logistic	0.0310 (0.0002)	756.74 (1.10)	0.78 (0.0004)
	constant c	0.0338 (0.0002)	765.22 (1.09)	0.80 (0.0003)
	$\boldsymbol{c} = (c_0, c_1)$	0.0297 (0.0002)	760.74 (1.11)	0.78 (0.0004)
Gamma(2, 1)	Logistic	0.1881 (0.0011)	783.92 (1.75)	0.64 (0.0013)
	constant c	0.1976 (0.0010)	803.06 (1.63)	0.64 (0.0013)
	$\boldsymbol{c} = (c_0, c_1)$	0.1897 (0.0011)	792.50 (1.71)	0.64 (0.0013)
Logistic(0, 1)	Logistic	0.1909 (0.0006)	1044.31 (0.40)	0.54 (0.0007)
	constant c	0.1921 (0.0006)	1050.16 (0.35)	0.55 (0.0006)
	$\boldsymbol{c} = (c_0, c_1)$	0.1909 (0.0006)	1050.24 (0.40)	0.54 (0.0007)
Uniform(−5, −1)	Logistic	0.1402 (0.0004)	713.64 (2.55)	0.76 (0.0011)
	constant c	0.1424 ((0.0004)	721.27 (2.48)	0.80 (0.0006)
	$\boldsymbol{c} = (c_0, c_1)$	0.1424 ((0.0004)	723.67 (2.42)	0.77 (0.0011)
Cauchy(0, 1)	Logistic	0.1214 (0.0006)	892.58 (1.47)	0.81 (0.0003)
	constant c	0.1270 (0.0006)	905.60 (1.42)	0.79 (0.0003)
	$\boldsymbol{c} = (c_0, c_1)$	0.1249 (0.0006)	904.20 (1.44)	0.81 (0.0003)

proportions of 1 in the jth block and also 16 values of relative entropy denoted by $RE_j(\cdot)$, $j = 1, \ldots, 16$. We finally obtain the RE as the simple average of these 16 RE-values:

$$RE(\cdot) = \frac{1}{16} \sum_{j=1}^{16} RE_j(\cdot).$$

For obtaining $RE(c)$ and $RE(\boldsymbol{c})$, we optimize with respect to c and $\boldsymbol{c}$, respectively.

The computational results are given in Tables 5 and 6, for Type I and Type II binary models, respectively. Once again we observe that for Type I binary data, the proposed model for vector $\boldsymbol{c}$ is working very well for almost all the situations except Cauchy; the RE-values are low and the AIC/QAIC are not too large, keeping the $Prop_p$-values intact. For Type II model, the proposed method works well except for Gamma and Cauchy. In both the cases, our proposed method is doing very good for $-\chi^2$ distribution.

5 Concluding Remarks

The present paper illustrates that the OLS can be successfully applied in several situations, instead of painstaking binary modeling by logistic model. Similar situation is expected if we implement probit model, instead of logit one.

The main advantage of the new process is that there is no convergence issue present for estimation of the model parameters unlike logistic regression. The main disadvantage is that the constrained optimization is really time consuming and complicated to implement as the number of covariates as well as the number of response categories increases. With the increase in the number of parameters, the optimization may become inaccurate also, which increases the chance of getting inaccurate optimal c or $\boldsymbol{c}$ at some local mode of the likelihood. Thus, the proposed method is recommended when the number of components in $\boldsymbol{c}$ is not too large.

References

Agresti, A. (2012). *Categorical data analysis* (3rd ed.). New York: Wiley.

Cover, T. M., & Thomas, J. A. (1991). *Elements of information theory*. Wiley series in telecommunications. New York, NY, USA: Wiley.

King, E. N., & Ryan, T. P. (2002). A preliminary investigation of maximum likelihood logistic regression versus exact logistic regression. *The American Statistician, 56*, 163–170.

Kullback, S. (1968). *Information theory and statistics*. Dover.

Peng, C. Y. J., T. H., So, S. Stage, F. K., & St. John, E. P., (2002). The use and interpretation of logistic regression in higher education: 1988–1999. *Research in Higher Education, 43*, 259–294.

Stigler, S. M. (1981). Gauss and the invention of least squares. *Annals of Statistics*, *9*, 465–474.

Estimation of Parameters of Misclassified Size Biased Borel Tanner Distribution

B. S. Trivedi and M. N. Patel

Abstract Different types of statistical methods are useful in data analysis in the field of science, engineering, medical, etc. In this paper, we have considered a statistical data analysis and estimation of the data by using size biased Borel–Tanner distribution. At the time of classification and analysis, there may arise error, like an observation may be misclassified into different classes or groups. Such type of data is known as misclassified data. Also, when sample units are selected with a probability proportional to the size of the units, the resultant distribution is known as a size biased distribution or weighted distributions. In this paper, we have studied misclassified size biased Borel–Tanner distribution and estimated its parameters by applying the method of maximum likelihood, method of moment, and Bayes' estimation method. Simulation study has been carried out for comparing the three methods of estimation.

Keywords Maximum likelihood estimation · Moments of the distribution
Lindley's approximation · Simulation

1 Introduction

All fields of Science and Economics are interested in statistical data analysis as a part of their study but have substantial problems due to the error in the observed data. There are several probable sources through which data error might occur. Due to error in the data, sampling process may not suggest an appropriate probability distribution and hence inference will also get affected. When these types of data

B. S. Trivedi (✉)
Amrut Mody School of Management, Ahmedabad University, Navrangpura, Ahmedabad 380009, Gujarat, India
e-mail: bhaktida.trivedi@ahduni.edu.in

M. N. Patel
Department of Statistics, School of Sciences, Gujarat University, Ahmedabad 380009, Gujarat, India
e-mail: mnpatel.stat@gmail.com

A. K. Laha (ed.), *Advances in Analytics and Applications*, Springer Proceedings in Business and Economics, https://doi.org/10.1007/978-981-13-1208-3_19

errors are identified in variables, it is expected to think about the solution of the problem in terms of classification errors.

Sometimes, regular method of estimation may not reduce bias or error in the estimate. When the data contains some kind of error like misclassification of the observations, the appropriate method also gives unreliable results. In case of discrete variables, the measurement error may be due to misclassification of the observations. Thus in the estimation of parameters and statistical inference, it is important to check the effect of misclassification of outcomes. In Medical science, the effect of misclassification is investigated by numerous Epidemiologists like Diamond and Lilienfeld (1962) and Shy et al. (1978). Effect of misclassification on binomial probabilities was first investigated by Bross (1954).

Sometimes, the sample units may be wrongly classified into the improper class instead of the correct class. Such type of classification is known as misclassification of the data. In data analysis, several statistical samplings are applicable. In this paper, the concept of weighted distribution, considered by Fisher (1934), is used in case of misclassified size biased discrete distribution.

Estimation under misclassified size biased distributions for various models like generalized negative binomial, log series distribution, Borel distribution, discrete Lindley distribution, and Poisson–Lindley distribution have been studied by Trivedi and Patel (2013, 2015a, b, 2017).

The paper is developed for estimation of the parameters of misclassified size biased Borel–Tanner distribution. Borel–Tanner distribution was investigated by a French mathematician Felix Edouard Justin Emile Borel in (1942). Nowadays, Borel–Tanner distribution has so many applications to model a variety of real-world phenomena, including highway traffic flows, online server traffic, and various investment behaviors relative to the existing financial portfolios. In addition to this, Borel–Tanner distribution is related to a number of other discrete statistical distributions like Poisson–Consul distribution, binomial distribution, negative binomial distribution, and log series distribution. Height and Breuer studied Borel–Tanner distribution and some of its important properties.

The Borel–Tanner distribution is a unimodal distribution also known as Tanner Borel distribution. This distribution is also used in finance (Nirei et al. 2012). The p.m.f. of Borel–Tanner distribution is given by

$$P_{BT}(x;\beta,a) = \begin{cases} \frac{a e^{-x\beta} x^{x-a} \beta^{x-a}}{x(x-a)!}, & x = a+1, a+2, \ldots, \infty \\ 0, & x = 1, 2, \ldots, a \end{cases}$$
$$0 < \beta < 1, \text{ and } a \text{ any positive integer} \tag{1.1}$$

A Borel–Tanner distribution is used in a queuing theory having single server with Poisson arrival when service times are fixed.

In (1.1) by taking $a = 1$, Borel–Tanner simply reduces to Borel distribution. Borel–Tanner is a discrete statistical distribution is based on the concept of the queuing theory that the customers arrive at random and that all customers spend the same amount of time using the cash machine.

Initially, Borel defined this distribution for the case $a = 1$ using geometric argument which was later on adapted by Tanner (1953), who showed that the Borel's argument is in fact valid for any positive integers a.

Borel–Tanner distribution is related to a number of other discrete statistical distributions like Poisson–Consul distribution, binomial distribution, negative binomial distribution, and log series distribution (Height and Breuer 1960).

First three-row moments and variance of the Borel–Tanner distribution are as follows:

$$\mu'_{1(BT)} = \frac{a}{1-\beta} \tag{1.2}$$

$$\mu'_{2(BT)} = \frac{a}{(1-\beta)^3}[\beta + a - a\beta] \tag{1.3}$$

$$\mu'_{3(BT)} = \frac{-a\left(a^2 + \beta + 3a\beta - 2a^2\beta + 2\beta^2 - 3a\beta^2 + a^2\beta^2\right)}{(-1+\beta)^5} \tag{1.4}$$

$$V(x)_{(BT)} = \frac{a\beta}{(1-\beta)^3} \tag{1.5}$$

Section 2 of this paper presented size biased Borel–Tanner distribution. Section 3 covered the study of the effect of the situation of misclassification under the size biased Borel–Tanner distribution. Different methods of estimation to estimate the parameter of the misclassified size biased Borel–Tanner distribution are considered in Sects. 4 and 5, and by using the simulation, we derive some conclusions regarding the performance of the methods of estimation, which will be very helpful to reduce the effect of misclassification in future.

2 Size Biased Borel–Tanner Distribution (SBBTD)

Size biased sampling is one of the most important nonrandom sampling methods. Size biased distribution arises when the observations of the sampling distributions have the same chance of being selected but each one is selected on the basis of its length or weight, e.g., weight $w(x) = x$ or x^2 (Fisher 1934). In such situation, the resulting distribution is denoted by size biased or length biased distribution.

To obtain the new p.m.f. under size biased distribution, consider a mechanism producing nonnegative random variable with a probability density function $P(x) = P(x;\theta)$, and let $w(x)$ be a nonnegative weight function under the assumption that $E[w(X)]$ exists. The p.m.f. of size biased distribution can be written by (Patil 2002)

$$P_{SB}(x;\theta) = \frac{w(x)\cdot P(x;\theta)}{E[w(X)]} \tag{2.1}$$

Here $P(x;\theta)$ and $P_{SB}(x;\theta)$ are stated to be the original probability distribution and weighted probability distribution. Also, the random variables of respective distributions are known as original and weighted random variables.

By using (1.1), (1.2) and taking weight $w(x) = x$, the p.m.f. of size biased Borel–Tanner distribution can be obtained as

$$
\begin{aligned}
P_{SB}(x;\beta,a) &= \frac{x P_{BT}(x;\beta,a)}{\mu'_{1(BT)}} \\
&= \frac{x\left[\frac{ae^{-x\beta}x^{x-a}\beta^{x-a}}{x(x-a)!}\right]}{\frac{a}{1-\beta}} \\
&= \frac{e^{-x\beta}x^{x-a}\beta^{x-a}(1-\beta)}{(x-a)!}
\end{aligned}
\tag{2.2}
$$

- **Mean and Variance of SBBTD**

Mean of the size biased Borel–Tanner distribution:

$$
\begin{aligned}
\mu'_{1(SB)} &= \sum_{x=a+1}^{\infty} x P_{SB}(x;\beta,a) \\
&= \sum_{x=a+1}^{\infty} x \frac{x P_{BT}(x;\beta,a)}{\mu'_{1(BT)}} \\
&= \frac{\mu'_{2(SB)}}{\mu'_{1(BT)}}
\end{aligned}
$$

by using (1.2) and (1.3),

$$
\mu'_{1(SB)} = \frac{\beta + a - a\beta}{(1-\beta)^2}
\tag{2.3}
$$

Variance of size biased Borel–Tanner distribution:

$$
\begin{aligned}
\mu'_{2(SB)} &= \sum_{x=a+1}^{\infty} x^2 P_{SB}(x;\beta,a) \\
&= \sum_{x=a+1}^{\infty} \frac{x^3 P_{BT}(x;\beta,a)}{\mu'_{1(BT)}} \\
&= \frac{\mu'_{3(BT)}}{\mu'_{1(BT)}}
\end{aligned}
$$

by using (1.2) and (1.4),

$$\mu'_{2(SB)} = \frac{\left(a^2 + \beta + 3a\beta - 2a^2\beta + 2\beta^2 - 3a\beta^2 + a^2\beta^2\right)}{(1-\beta)^4} \tag{2.4}$$

From (2.3) and (2.4), the variance of the size biased Borel–Tanner distribution can be obtained as

$$\begin{aligned} V(x)_{(SB)} &= \mu'_{2(SB)} - \left(\mu'_{1(SB)}\right)^2 \\ &= \frac{\left(a^2 + \beta + 3a\beta - 2a^2\beta + 2\beta^2 - 3a\beta^2 + a^2\beta^2\right)}{(1-\beta)^4} - \left(\frac{\beta + a - a\beta}{(1-\beta)^2}\right)^2 \\ &= \frac{\beta(1 + a + \beta - a\beta)}{(-1+\beta)^4} \end{aligned} \tag{2.5}$$

3 Misclassified Size Biased Borel–Tanner Distribution (MSBBTD)

As described in Sect. 1, misclassification means wrongly reporting or recording the responses. During the classification of data, it is possible to happen by an interviewer to record some observations into class two when actually they are from class three. It may happen when the respondent misunderstands the question or the interviewer simply checks the wrong class. Such type of measurement error can be observed in historical data also.

Suppose X is a random variable of the random sample having size biased Borel—Tanner distribution (SBBTD). Let some of the observations of SBBTD corresponding to class $b+1$ be wrongly reported as the observations of class b, then the resultant SBBTD of X is known as misclassified size biased Borel–Tanner distribution with the misclassifying probability α. By using (2.2), and the concept given by Parikh and Shah (1969), p.m.f. of misclassified size biased Borel–Tanner distribution for $x = b$ and $x = b+1$ can be written as

$$\begin{aligned} P(b;\alpha,\beta,a) &= P_{SB}(b;\beta,a) + \alpha P_{SB}(b+1;\beta,a) \\ &= \frac{e^{-b\beta} b^{b-a} \beta^{b-a}(1-\beta)}{(b-a)!} + \alpha \frac{e^{-(b+1)\beta}(b+1)^{(b+1)-a}\beta^{(b+1)-a}(1-\beta)}{((b+1)-a)!} \\ &= e^{-b\beta}\beta^{b-a}(1-\beta)\left[\frac{b^{b-a}}{(b-a)!} + \alpha \frac{e^{-\beta}(b+1)^{b-a+1}\beta}{(b-a+1)!}\right] \end{aligned} \tag{3.1}$$

$$\begin{aligned} P(b+1;\alpha,\beta,a) &= P_{SB}(b+1;\beta,a) - \alpha P_{SB}(b+1;\beta,a) \\ &= (1-\alpha)\frac{e^{-(b+1)\beta}(b+1)^{(b+1)-a}\beta^{(b+1)-a}(1-\beta)}{(b-a+1)!} \end{aligned} \tag{3.2}$$

By using (2.2), (3.1) and (3.2), the p.m.f. of misclassified size biased Borel–Tanner distribution can be obtained as

$$P(x;\alpha,\beta,a)=\begin{cases} e^{-b\beta}\beta^{-a+b}(1-\beta)\left[\frac{b^{b-a}}{(b-a)!}+\alpha\frac{e^{-\beta}(b+1)^{b-a+1}\beta}{(b-a+1)!}\right], & x=b\\ (1-\alpha)\frac{e^{-(b+1)\beta}(b+1)^{(b+1)-a}\beta^{(b+1)-a}(1-\beta)}{(b-a+1)!}, & x=b+1\\ \frac{e^{-x\beta}x^{x-a}\beta^{x-a}(1-\beta)}{(x-a)!}, & x\in S \end{cases} \tag{3.3}$$

$S=\{a+1,a+2,\ldots,\infty\}-\{b,b+1\}$, where N is a set of natural numbers, a is any positive integer, $0<\beta<1, 0\le\alpha\le 1, 0<\beta<1, 0\le\alpha\le 1$.

- Moments of the MSBBTD

Let the nth row moment of MSBBTD be denoted by μ'_n.

$$\mu'_n=\sum_{x=a+1}^{\infty}x^nP(x;\alpha,\beta,a) \tag{3.4}$$

From the p.m.f. (3.3), the relation between MSBBTD and SBBTD can be obtained as

$$\mu'_n=\mu'_{n(SB)}+\alpha P_{SB}(b+1;\beta,a)\left[b^n-(b+1)^n\right] \tag{3.5}$$

where $\mu'_{n(SB)}=n$th row moment of SBBTD.

By taking $n=1$, and using (2.2) and (2.3), the first row moment (μ'_1) of MSBBTD (Mean of MSBBTD) can be obtained as

$$\begin{aligned}\mu'_1&=\sum_{i=a+1}^{\infty}xP_{SB}(x;\beta,a)-\alpha P_{SB}(b+1;\beta,a)\\ &=\frac{\beta+a-a\beta}{(1-\beta)^2}-\alpha\left[\frac{e^{-(b+1)\beta}(c+1)^{-a+b+1}\beta^{-a+b+1}(1-\beta)}{(-a+b+1)!}\right]\end{aligned} \tag{3.6}$$

Similarly, by taking $n=2$, and using (2.2) and (2.4), the second row moment (μ'_2) of MSBBTD can be obtained.

$$\begin{aligned}\mu'_2&=\sum_{i=a+1}^{\infty}x^2P_{SB}(x;\beta,a)-\alpha(2b+1)P_{SB}(b+1;\beta,a)\\ &=\frac{\left(a^2+\beta+3a\beta-2a^2\beta+2\beta^2-3a\beta^2+a^2\beta^2\right)}{(-1+\beta)^4}\\ &\quad-\alpha(2b+1)\left[\frac{e^{-(b+1)\beta}(b+1)^{-a+b+1}\beta^{-a+b+1}(1-\beta)}{(-a+b+1)!}\right]\end{aligned} \tag{3.7}$$

From (3.6) and (3.7), the variance of the MSBBTD can be obtained as

$$
\begin{aligned}
\mathrm{V(X)} = \mu_2 = & \frac{\left(a^2 + \beta + 3a\beta - 2a^2\beta + 2\beta^2 - 3a\beta^2 + a^2\beta^2\right)}{(1-\beta)^4} \\
& - \alpha(2b+1)\left[\frac{e^{-(b+1)\beta}(b+1)^{b-a+1}\beta^{b-a+1}(1-\beta)}{(-a+b+1)!}\right] \\
& - \left(\frac{\beta + a - a\beta}{(1-\beta)^2} - \alpha\left[\frac{e^{-(b+1)\beta}(b+1)^{b-a+1}\beta^{b-a+1}(1-\beta)}{(b-a+1)!}\right]\right)^2
\end{aligned}
\tag{3.8}
$$

4 Methods of Estimation of the Parameters of MSBBTD

The p.m.f. of MSBBTD consists of three unknown parameters: $\beta, a,$ and α, where the random variable $X > a$.

We consider three methods to estimate the parameters by considering r known.

(i) Method of Maximum Likelihood
(ii) Method of Moment
(iii) Bayes Estimation Method.

(i) Method of Maximum Likelihood

$x_1, x_2, \ldots, x_i, \ldots, x_k$ are the observed values of random variable X in a random sample, where $k \in \{1, 2, 3, \ldots, \infty\}$. Such that

$$\sum_{i=1}^{k} n_i = n, \text{ the number of observations in the sample}$$

From (3.3), the likelihood function L of the sample of n observations can be written as

$$L \propto \prod_{i=1}^{k} P_i^{n_i}, \quad \text{where } P_i = P(X = X_i) \tag{4.1}$$

By using (3.3), and taking natural logarithm on both sides, the log likelihood function of MSBBTD can be written as

$$
\begin{aligned}
\ln L &= \text{Constant} + n_b \ln P_b + n_{b+1} \ln P_{b+1} + \sum_{i \neq b, b+1}^{k} n_i \ln P_i \\
&= \text{Constant} + n_b \ln\left\{e^{-b\beta}\beta^{b-a}(1-\beta)\left[\frac{b^{b-a}}{(b-a)!} + \alpha\frac{e^{-\beta}(b+1)^{b-a+1}\beta}{(b-a+1)!}\right]\right\} \\
&+ n_{b+1} \ln\left\{(1-\alpha)\frac{e^{-(b+1)\beta}(b+1)^{b-a+1}\beta^{b-a+1}(1-\beta)}{(b-a+1)!}\right\} \\
&+ \sum_{i \neq b, b+1}^{k} n_i \ln\left\{\frac{e^{-i\beta} i^{i-a}\beta^{i-a}(1-\beta)}{(i-a)!}\right\} \\
&= \text{Constant} - n\,\bar{x}\,\beta + \sum_{i=1}^{k} n_i (i-a)\ln i - n_b(b-a)\ln b + (-an + n\bar{x})\ln\beta \\
&+ n\,\ln(1-\beta) - \sum_{i=1}^{k} n_i \ln(i-a)! - n_b \ln(b-a)! \\
&+ n_b \ln\left[\frac{b^{b-a}}{(b-a)!} + \alpha\frac{e^{-\beta}(b+1)^{b-a+1}\beta}{(b-a+1)!}\right] + n_{b+1}\ln(1-\alpha)
\end{aligned}
\tag{4.2}
$$

To obtain the maximum likelihood (ML) estimates of the parameters α and β of MSBBTD, we set the partial derivative of (4.2) with respect to α and β to be 0. That is,

$$
\frac{\partial \ln L}{\partial \alpha} = \frac{n_b \frac{e^{-\beta}(b+1)^{b-a+1}\beta}{(b-a+1)!}}{\frac{b^{b-a}}{(b-a)!} + \alpha\frac{e^{-\beta}(b+1)^{b-a+1}\beta}{(b-a+1)!}} + \frac{n_{b+1}(-1)}{(1-\alpha)} = 0 \tag{4.3}
$$

$$
\begin{aligned}
\frac{\partial \ln L}{\partial \beta} &= -n\bar{x} + \frac{n(\bar{x}-a)}{\beta} - \frac{n}{1-\beta} \\
&+ \frac{n_c\alpha}{\left[\frac{b^{b-a}}{(b-a)!} + \alpha\frac{e^{-\beta}(b+1)^{b-a+1}\beta}{(b-a+1)!}\right]}\left[\frac{(b+1)^{b-a+1}e^{-\beta}(1-\beta)}{(b-a+1)!}\right] = 0
\end{aligned}
\tag{4.4}
$$

Solving Eqs. (4.3) and (4.4), we get the ML estimators of α and β as follows:

$$
\alpha = \frac{n_b}{(n_b + n_{b+1})} - \frac{n_{b+1}\left[\frac{b^{b-a}}{(-a+b)!}\right]}{\left[\frac{e^{-\beta}(b+1)^{b-a+1}\beta}{(b-a+1)!}\right](n_b + n_{b+1})} \tag{4.5}
$$

$$
\beta = \frac{n\left[\frac{b^{b-a}}{(b-a)!} + \alpha\frac{e^{-\beta}(b+1)^{b-a+1}\beta}{(b-a+1)!}\right](\bar{x}-a)}{n\left[\frac{b^{b-a}}{(b-a)!} + \alpha\frac{e^{-\beta}(b+1)^{b-a+1}\beta}{(b-a+1)!}\right]\left[\bar{x} + \frac{1}{1-\beta}\right] - n_b\alpha\left[\frac{(b+1)^{b-a+1}e^{-\beta}(1-\beta)}{(b-a+1)!}\right]} \tag{4.6}
$$

Substituting an expression of α from (4.5) in (4.4), we get an equation with only parameter β. Solving this equation by any iterative procedure (like Newton–Raphson method of Iteration), we can get MLE of β. Finally substitute this value of β in (4.5), we can obtain the MLE of α.

- Asymptotic Variance–Covariance Matrix for ML Estimators

It is essential to obtain the second-order partial derivative of (4.2) with respect to α and β, to find the asymptotic variances and covariance of α and β.

$$\frac{\partial^2 \ln L}{\partial \alpha^2} = -n_b \frac{\left[\frac{e^{-\beta}(b+1)^{b-a+1}\beta}{(b-a+1)!}\right]^2}{\left[\frac{b^{b-a}}{(b-a)!} + \alpha \frac{e^{-\beta}(b+1)^{b-a+1}\beta}{(b-a+1)!}\right]^2} - \frac{n_{b+1}}{(1-\alpha)^2} \tag{4.7}$$

$$\begin{aligned}
\frac{\partial^2 \ln L}{\partial \beta^2} &= \frac{n(r-\bar{x})}{\beta^2} - \frac{n}{(1-\beta)^2} \\
&\quad + \frac{n_b\alpha(b+1)^{b-a+1}}{(b-a+1)!}\left\{\frac{e^{-\beta}[\beta-2]}{\left[\frac{b^{b-a}}{(b-a)!} + \alpha\frac{e^{-\beta}(b+1)^{b-a+1}\beta}{(b-a+1)!}\right]} - \frac{\left[e^{-\beta}(1-\beta)\right]^2 \alpha \frac{(b+1)^{b-a+1}}{(b-a+1)!}}{\left[\frac{b-a+b}{(b-a)!} + \alpha\frac{e^{-\beta}(b+1)^{b-a+1}\beta}{(b-a+1)!}\right]^2}\right\} \\
&= \frac{n(a-\bar{x})}{\beta^2} - \frac{n}{(1-\beta)^2} \\
&\quad + \frac{n_b(b+1)^{b-a+1}e^{-\beta}(\beta-2)}{(b-a+1)!}\left[\frac{\alpha}{\left[\frac{b^{b-a}}{(b-a)!} + \alpha\frac{e^{-\beta}(b+1)^{b-a+1}\beta}{(b-a+1)!}\right]}\right] \\
&\quad - \left[\frac{(b+1)^{b-a+1}}{(b-a+1)!}\right]^2 n_b\left[e^{-\beta}(1-\beta)\right]^2\left[\frac{\alpha^2}{\left[\frac{b^{b-a}}{(b-a)!} + \alpha\frac{e^{-\beta}(b+1)^{b-a+1}\beta}{(b-a+1)!}\right]^2}\right]
\end{aligned} \tag{4.8}$$

$$\begin{aligned}
\frac{\partial^2 \ln L}{\partial \beta \partial \alpha} &= \frac{n_b \frac{(b+1)^{b-a+1}}{(b-a+1)!}}{\left(\frac{b^{b-a}}{(b-a)!} + \alpha\frac{e^{-\beta}(b+1)^{b-a+1}\beta}{(b-a+1)!}\right)^2} \\
&\quad \left[\left(\frac{b^{b-a}}{(b-a)!} + \alpha\frac{e^{-\beta}(b+1)^{b-a+1}\beta}{(b-a+1)!}\right)\left(e^{-\beta}(1-\beta) - e^{-\beta}\beta\alpha\frac{(b+1)^{b-a+1}}{(b-a+1)!}e^{-\beta}(1-\beta)\right)\right] \\
&= n_c\frac{(b+1)^{b-a+1}}{(b-a+1)!}\left[\frac{e^{-\beta}(1-\beta)}{\left(\frac{b^{b-a}}{(b-a)!} + \alpha\frac{e^{-\beta}(b+1)^{b-a+1}\beta}{(b-a+1)!}\right)} - \frac{\alpha\frac{(b+1)^{b-a+1}}{(b-a+1)!}e^{-2\beta}\beta(1-\beta)}{\left(\frac{b^{b-a}}{(b-a)!} + \alpha\frac{e^{-\beta}(b+1)^{b-a+1}\beta}{(b-a+1)!}\right)^2}\right]
\end{aligned} \tag{4.9}$$

By using (4.7)–(4.9), Fisher information matrix $J(\beta, \alpha)$ can be obtained as

$$J(\beta,\alpha) = \begin{bmatrix} -E\left(\frac{\partial^2 \ln L}{\partial \beta^2}\right) & -E\left(\frac{\partial^2 \ln L}{\partial \beta \partial \alpha}\right) \\ -E\left(\frac{\partial^2 \ln L}{\partial \beta \partial \alpha}\right) & -E\left(\frac{\partial^2 \ln L}{\partial \alpha^2}\right) \end{bmatrix} \tag{4.10}$$

By taking the inverse of the Fisher information matrix, asymptotic variance–covariance matrix of MLE (Σ) can be found as

$$\Sigma = \begin{bmatrix} V(\beta) & Cov(\beta,\alpha) \\ Cov(\beta,\alpha) & V(\alpha) \end{bmatrix} \text{and } SE\left(\hat{\beta}\right) = \sqrt{V(\beta)},\ SE(\hat{\alpha}) = \sqrt{V(\alpha)} \tag{4.11}$$

(ii) Method of Moment

By using the first and second moments of MSBBTD, moment estimator of the parameters β and α can be obtained.

From the first row moment of MSBBTD given in the result (3.6), the moment estimator of the parameter α can be obtained as

$$\alpha = \frac{\frac{\beta+a-a\beta}{(1-\beta)^2} - \mu_1'}{\frac{e^{-(b+1)\beta}(b+1)^{b-a+1}\beta^{b-a+1}(1-\beta)}{(b-a+1)!}} = \frac{(b-a+1)!\left[(\beta+a-a\beta) - \mu_1'(1-\beta)^2\right]}{(1-\beta)^3 e^{-(b+1)\beta}[(b+1)\beta]^{b-a+1}} \tag{4.12}$$

From the second row moment of MSBBTD given in the result (3.7), the moment estimator of the parameter β can be obtained as

$$\beta = 1 + (b-a+1)!\left\{\frac{\mu_2' - \frac{(a^2+\beta+3a\beta-2a^2\beta+2\beta^2-3a\beta^2+a^2\beta^2)}{(1-\beta)^4}}{\alpha(2b+1)e^{-(b+1)\beta}[(b+1)\beta]^{b-a+1}}\right\} \tag{4.13}$$

Substitute α from (4.12) in (4.13) and replacing μ_1' by m_1' and μ_2' by m_2', the equation of the moment estimator of parameter β can be obtained which can be solved by applying any method of iteration like Newton–Raphson. To get the exact value of the moment estimator of parameter α, again substitute moment estimator of parameter β in (4.12).

- Asymptotic Variance–Covariance Matrix for Moment Estimators

The first two moments of MSBBTD obtained in Eqs. (3.6) and (3.7) are

$$\mu_1' = m_1' = H_1(\beta, \alpha, a) = \frac{\beta+a-a\beta}{(1-\beta)^2} - \alpha\left[\frac{e^{-(b+1)\beta}(b+1)^{b-a+1}\beta^{b-a+1}(1-\beta)}{(b-a+1)!}\right] \tag{4.14}$$

and

$$\mu_2' = m_2' = H_2(\beta, \alpha, a) = \frac{(a^2+\beta+3a\beta-2a^2\beta+2\beta^2-3a\beta^2+a^2\beta^2)}{(-1+\beta)^4} - \alpha(2b+1)\left[\frac{e^{-(b+1)\beta}(b+1)^{b-a+1}\beta^{b-a+1}(1-\beta)}{(b-a+1)!}\right] \tag{4.15}$$

To obtain the asymptotic variance–covariance matrix for moment estimators, by using (4.14) and (4.15), define a matrix A as

$$A = \begin{bmatrix} \frac{\partial H_1}{\partial \beta} & \frac{\partial H_1}{\partial \alpha} \\ \frac{\partial H_2}{\partial \beta} & \frac{\partial H_2}{\partial \alpha} \end{bmatrix} \tag{4.16}$$

and the variance–covariance matrix of sample moments m_1' and m_2' as

$$\Sigma = \begin{bmatrix} V(m_1') & Cov(m_1', m_2') \\ Cov(m_1', m_2') & V(m_2') \end{bmatrix} \tag{4.17}$$

where

$$V(m_a') = \frac{\mu_{2a}' - (\mu_a')^2}{n}; \quad Cov(m_a', m_s') = \frac{\mu_{a+s}' - \mu_a'\mu_s'}{n}; \quad a \neq s = 1, 2$$

and m_a' is the ath sample raw moment of MSBBTD

$$\text{i.e., } m_a' = \frac{1}{n}\sum_{i=1}^{k} f_i x_i^a$$

by using the results obtained in (4.16) and (4.17), the variance–covariance matrix of moment estimators $\tilde{\beta}$ and $\tilde{\alpha}$ of MSBBTD is given by

$$V = A^{-1}\Sigma(A^{-1})' = \begin{bmatrix} V(\tilde{\beta}) & Cov(\tilde{\beta}, \tilde{\alpha}) \\ Cov(\tilde{\beta}, \tilde{\alpha}) & V(\tilde{\alpha}) \end{bmatrix} \tag{4.18}$$

(iii) Bayes Estimation Method

In Bayesian estimation, it is required to define the prior distributions for the parameters β and α. For both the parameters, we consider the piecewise independent priors, namely the uniform prior for α and exponential prior for β.

$$\text{i.e., } g_1(\beta) = \gamma e^{-\gamma\beta}, \quad \gamma > 0; \beta > 0 \tag{4.19}$$

$$g_2(\alpha) = 1, \quad 0 \leq \alpha \leq 1 \tag{4.20}$$

Thus, the joint prior for β and α is

$$g(\beta, \alpha) = g_1(\beta)g_2(\alpha) = \gamma e^{-\gamma\beta}, \quad \gamma > 0; \beta > 0 \tag{4.21}$$

From (4.1) and (4.21), the posterior distribution of β and α can be obtained. In Bayes estimation method, posterior distribution takes a ratio form which involves an integration in the denominator. To estimate the Bayes estimators, it is required to get the closed form of this integration which is tedious for the MSBBTD. In this paper, we have adopted Lindley's approximation method to obtain the Bayes estimator of the parameters of MSBBTD in the next section.

- Lindley's Approximation

In this section, Bayes estimators of the parameters β and α of MSBBTD are obtained according to the concept of Lindley's approximation method (see: Box and Tiao (1973): Bayesian Inference in Statistical Analysis).

Press (1989) has suggested the use of Lindley's (1980) approximation method to obtain Bayes estimators of the parameters when the number of unknown parameters is not more than 5.

MSBBTD is a three-parameter discrete distribution. So in this section, we have used Lindley's approach to derive estimates of the parameters β and α. This method provides an approximation for

$$I(\underline{x}) = \frac{\int u(\Theta)L(\Theta)g(\Theta)d\Theta}{\int L(\Theta)g(\Theta)d\Theta} \tag{4.22}$$

Lindley's approximation of (4.22) reduces to

$$\begin{aligned} I(\underline{x}) \cong u\left(\tilde{\Theta}\right) &+ \frac{1}{2}\left\{\left[\frac{\partial^2 u}{\partial\theta^2} + 2\left(\frac{\partial u(\Theta)}{\partial\theta}\right)\left(\frac{\partial \ln g(\theta,\alpha)}{\partial\theta}\right)\right]\hat{\sigma}_{11}\right. \\ &+ \left[\frac{\partial^2 u}{\partial\theta\partial\alpha} + 2\left(\frac{\partial u(\Theta)}{\partial\theta}\right)\left(\frac{\partial \ln g(\theta,\alpha)}{\partial\alpha}\right)\right]\hat{\sigma}_{12} \\ &+ \left[\frac{\partial^2 u}{\partial\alpha\partial\theta} + 2\left(\frac{\partial u(\Theta)}{\partial\alpha}\right)\left(\frac{\partial \ln g(\theta,\alpha)}{\partial\theta}\right)\right]\hat{\sigma}_{21} \\ &+ \left.\left[\frac{\partial^2 u}{\partial\alpha^2} + 2\left(\frac{\partial u(\Theta)}{\partial\alpha}\right)\left(\frac{\partial \ln g(\theta,\alpha)}{\partial\alpha}\right)\right]\hat{\sigma}_{22}\right\} \\ &+ \frac{1}{2}\left\{\frac{\partial^3 \ln L}{\partial\theta^3}\left[\frac{\partial u(\Theta)}{\partial\theta}\hat{\sigma}_{11}^2 + \frac{\partial u(\Theta)}{\partial\alpha}\hat{\sigma}_{11}\hat{\sigma}_{21}\right]\right. \\ &+ \frac{\partial^3 \ln L}{\partial\theta^2\partial\alpha}\left[3\frac{\partial u(\Theta)}{\partial\theta}\hat{\sigma}_{11}\hat{\sigma}_{12} + \frac{\partial u(\Theta)}{\partial\alpha}\left(\hat{\sigma}_{11}\hat{\sigma}_{21} + 2\hat{\sigma}_{12}^2\right)\right] \\ &+ \frac{\partial^3 \ln L}{\partial\theta\partial\alpha^2}\left[3\frac{\partial u(\Theta)}{\partial\alpha}\hat{\sigma}_{12}\hat{\sigma}_{22} + \frac{\partial u(\Theta)}{\partial\theta}\left(\hat{\sigma}_{11}\hat{\sigma}_{22} + 2\hat{\sigma}_{12}^2\right)\right] \\ &+ \left.\frac{\partial^3 \ln L}{\partial\alpha^3}\left[\frac{\partial u(\Theta)}{\partial\alpha}\hat{\sigma}_{22}^2 + \frac{\partial u(\Theta)}{\partial\theta}\hat{\sigma}_{12}\hat{\sigma}_{22}\right]\right\}_{\Theta_1=\hat{\Theta}_1} \end{aligned} \tag{4.23}$$

where $\Theta = (\beta, \alpha)$ and so $u(\Theta) = u(\beta, \alpha)$ is a function of β and α only and $L(\Theta)$ is a log likelihood of β and α.

Here, to obtain the Bayes estimate of α, we use $u(\beta, \alpha) = \alpha$ and for β, we use

$$u(\beta, \alpha) = \beta. \tag{4.24}$$

Moreover to estimate Bayes estimators of parameters β and α, second- and third-order partial derivatives of the parameters β and α are required. Second-order partial derivatives are obtained in earlier part of Sect. 4 in (4.7)–(4.9), and the third-order partial derivatives of the parameters β and α are as follows:

$$\begin{aligned}
\frac{\partial^3 \ln L}{\partial \beta^3} &= -\frac{2n(a-\bar{x})}{\beta^3} - \frac{2n}{(1-\beta)^3} + \frac{n_b \alpha (b+1)^{b-a+1}}{(b-a+1)!}\left\{\left[\left[\frac{e^{-\beta}[-(\beta-2)+1]}{\left[\frac{b^{b-a}}{(b-a)!} + \alpha \frac{e^{-\beta}(b+1)^{b-a+1}\beta}{(b-a+1)!}\right]}\right.\right.\right. \\
&\quad \left. - \frac{\left[e^{-\beta}(\beta-2)\right]e^{-\beta}\left[\alpha \frac{(b+1)^{b-a+1}}{(b-a+1)!}\right](-\beta+1)}{\left[\frac{b^{b-a}}{(b-a)!} + \alpha \frac{e^{-\beta}(b+1)^{b-a+1}\beta}{(b-a+1)!}\right]^2}\right] \\
&\quad - \alpha \frac{(b+1)^{b-a+1}}{(b-a+1)!}\left[\frac{2e^{-\beta}\left[e^{-\beta}(1-\beta)\right](-(1-\beta)-1)}{\left[\frac{b^{b-a}}{(b-a)!} + \alpha \frac{e^{-\beta}(b+1)^{b-a+1}\beta}{(b-a+1)!}\right]^2}\right. \\
&\quad \left.\left. - \frac{\left[e^{-\beta}(1-\beta)\right]^2 2e^{-\beta}\alpha \frac{(b+1)^{b-a+1}}{(b-a+1)!}(-\beta+1)}{\left[\frac{b^{b-a}}{(b-a)!} + \alpha \frac{e^{-\beta}(b+1)^{b-a+1}\beta}{(b-a+1)!}\right]^3}\right]\right\} \\
&= -\frac{2n(a-\bar{x})}{\beta^3} - \frac{2n}{(1-\beta)^3} + \frac{n_b \alpha (b+1)^{b-a+1}}{(b-a+1)!}\left\{\frac{e^{-\beta}(3-\beta)}{\left[\frac{b^{b-a}}{(b-a)!} + \alpha \frac{e^{-\beta}(b+1)^{b-a+1}\beta}{(b-a+1)!}\right]}\right. \\
&\quad - \frac{2\left[e^{-\beta}(\beta-2)\right]e^{-\beta}\left[\alpha \frac{(b+1)^{b-a+1}}{(b-a+1)!}\right](1-\beta)}{\left[\frac{b^{b-a}}{(b-a)!} + \alpha \frac{e^{-\beta}(b+1)^{b-a+1}\beta}{(b-a+1)!}\right]^2} \\
&\quad \left. - \alpha \frac{(b+1)^{b-a+1}}{(b-a+1)!} \frac{\left[e^{-\beta}(1-\beta)\right]^3 2\alpha \frac{(b+1)^{b-a+1}}{(b-a+1)!}}{\left[\frac{b^{b-a}}{(b-a)!} + \alpha \frac{e^{-\beta}(b+1)^{b-a+1}\beta}{(b-a+1)!}\right]^3}\right\}
\end{aligned} \tag{4.25}$$

$$\begin{aligned}
\frac{\partial^3 \ln L}{\partial \alpha^3} &= \frac{-n_b\left[\frac{e^{-\beta}(b+1)^{b-a+1}\beta}{(b-a+1)!}\right]^2(-2)\left[\frac{e^{-\beta}(b+1)^{b-a+1}\beta}{(b-a+1)!}\right]}{\left[\frac{b^{b-a}}{(b-a)!} + \alpha \frac{e^{-\beta}(b+1)^{b-a+1}\beta}{(b-a+1)!}\right]^3} - \frac{2n_{b+1}}{(1-\alpha)^3} \\
&= \frac{2n_b\left[\frac{e^{-\beta}(b+1)^{b-a+1}\beta}{(b-a+1)!}\right]^3}{\left[\frac{b^{b-a}}{(b-a)!} + \alpha \frac{e^{-\beta}(b+1)^{b-a+1}\beta}{(b-a+1)!}\right]^3} - \frac{2n_{b+1}}{(1-\alpha)^3}
\end{aligned} \tag{4.26}$$

$$\frac{\partial^3 \ln L}{\partial\beta\partial\alpha^2} = -n_b \frac{\left[\frac{(b+1)^{b-a+1}}{(b-a+1)!}\right]^2}{\left[\frac{b^{b-a}}{(b-a)!} + \alpha\frac{e^{-\beta}(b+1)^{b-a+1}\beta}{(b-a+1)!}\right]^4}\left\{\left[\frac{b^{-a+b}}{(-a+b)!}\right.\right.$$

$$\left. + \alpha\frac{e^{-\beta}(b+1)^{b-a+1}\beta}{(b-a+1)!}\right]^2 2\left(e^{-\beta}\beta\right)\left(-e^{-\beta}\beta + e^{-\beta}\right)$$

$$\left. - \left(e^{-\beta}\beta\right)^2 2\left[\frac{b^{b-a}}{(b-a)!} + \alpha\frac{e^{-\beta}(b+1)^{b-a+1}\beta}{(b-a+1)!}\right]\left[\alpha\frac{(b+1)^{b-a+1}}{(b-a+1)!}\right]\left(-e^{-\beta}\beta + e^{-\beta}\right)\right\}$$

$$= -2n_b\left(e^{-2\beta}\beta\right)(1-\beta)\left\{\frac{\left[\frac{(b+1)^{b-a+1}}{(b-a+1)!}\right]^2}{\left[\frac{b^{b-a}}{(b-a)!} + \alpha\frac{e^{-\beta}(b+1)^{b-a+1}\beta}{(b-a+1)!}\right]^2}\right.$$

$$\left. - \frac{\alpha\left(e^{-\beta}\beta\right)\left[\frac{(b+1)^{b-a+1}}{(b-a+1)!}\right]^3}{\left[\frac{b^{b-a}}{(b-a)!} + \alpha\frac{e^{-\beta}(b+1)^{b-a+1}\beta}{(b-a+1)!}\right]^3}\right\} \tag{4.27}$$

$$\frac{\partial^3 \ln L}{\partial\beta^2\partial\alpha} = n_b\frac{(b+1)^{b-a+1}}{(b-a+1)!}\left\{\left[\frac{\left[\frac{b^{b-a}}{(b-a)!} + \alpha\frac{e^{-\beta}(b+1)^{b-a+1}\beta}{(b-a+1)!}\right]\left(-e^{-\beta}(1-\beta) - e^{-\beta}\right)}{\left[\frac{b^{b-a}}{(b-a)!} + \alpha\frac{e^{-\beta}(b+1)^{b-a+1}\beta}{(b-a+1)!}\right]^2}\right.\right.$$

$$\left. - \frac{\alpha e^{-\beta}(1-\beta)\frac{(b+1)^{b-a+1}}{(b-a+1)!}\left(e^{-\beta}(1-\beta)\right)}{\left[\frac{b^{b-a}}{(b-a)!} + \alpha\frac{e^{-\beta}(b+1)^{b-a+1}\beta}{(b-a+1)!}\right]^2}\right]$$

$$- \alpha\frac{(b+1)^{b-a+1}}{(b-a+1)!}\left[\frac{\left[-2e^{-2\beta}\beta(1-\beta) + e^{-2\beta}(1-\beta) - e^{-2\beta}\beta\right]}{\left[\frac{b^{b-a}}{(b-a)!} + \alpha\frac{e^{-\beta}(b+1)^{b-a+1}\beta}{(b-a+1)!}\right]^2}\right.$$

$$\left.\left. - \frac{2\alpha\frac{(b+1)^{b-a+1}}{(b-a+1)!}\left(e^{-\beta}(1-\beta)\right)e^{-2\beta}\beta(1-\beta)}{\left[\frac{b^{b-a}}{(b-a)!} + \alpha\frac{e^{-\beta}(b+1)^{b-a+1}\beta}{(b-a+1)!}\right]^3}\right]\right\}$$

$$= n_b\frac{(b+1)^{-a+b+1}e^{-\beta}}{(-a+b+1)!\left(\frac{b^{-a+b}}{(-a+b)!} + \alpha\frac{e^{-\beta}(b+1)^{-a+b+1}\beta}{(-a+b+1)!}\right)}$$

$$\left\{(\beta-2) - \frac{\alpha\frac{(b+1)^{-a+b+1}}{(-a+b+1)!}e^{-\beta}\left[\alpha(1-\beta)^2 + \left(2\beta^2 - 4\beta + 1\right)\right]}{\left(\frac{b^{-a+b}}{(-a+b)!} + \alpha\frac{e^{-\beta}(b+1)^{-a+b+1}\beta}{(-a+b+1)!}\right)}\right.$$

$$\left. - \frac{2\alpha^2\left[\frac{(b+1)^{-a+b+1}}{(-a+b+1)!}\right]^2\left[e^{-2\beta}\beta(1-\beta)^2\right]}{\left(\frac{b^{-a+b}}{(-a+b)!} + \alpha\frac{e^{-\beta}(b+1)^{-a+b+1}\beta}{(-a+b+1)!}\right)^2}\right\} \tag{4.28}$$

By using the second- and third-order partial derivatives with respect to β and α, the Bayes estimates of parameters β and α are given by

$$\hat{\alpha}_{BL} \cong \hat{\alpha} - \gamma\,\hat{\sigma}_{21} + \frac{1}{2}\left[\frac{\partial^3 \ln L}{\partial\beta^3}\left(\hat{\sigma}_{11}\hat{\sigma}_{21}\right) + \frac{\partial^3 \ln L}{\partial\beta^2\partial\alpha}\left(\hat{\sigma}_{11}\hat{\sigma}_{21} + 2\hat{\sigma}_{12}^2\right) + \frac{\partial^3 \ln L}{\partial\beta\partial\alpha^2}\left(3\hat{\sigma}_{12}\hat{\sigma}_{22}\right) + \frac{\partial^3 \ln L}{\partial\alpha^3}\left(\hat{\sigma}_{22}^2\right)\right]\Bigg|_{\alpha=\hat{\alpha}} \tag{4.29}$$

$$\hat{\beta}_{BL} \cong \hat{\beta} - \gamma\hat{\sigma}_{11} + \frac{1}{2}\left[\frac{\partial^3 \ln L}{\partial\beta^3}\left(\hat{\sigma}_{11}^2\right) + \frac{\partial^3 \ln L}{\partial\beta^2\partial\alpha}\left(3\hat{\sigma}_{11}\hat{\sigma}_{12}\right) + \frac{\partial^3 \ln L}{\partial\beta\partial\alpha^2}\left(\hat{\sigma}_{11}\hat{\sigma}_{22} + 2\hat{\sigma}_{12}^2\right) + \frac{\partial^3 \ln L}{\partial\alpha^3}\left(\hat{\sigma}_{12}\hat{\sigma}_{22}\right)\right]\Bigg|_{\gamma=\hat{\gamma}} \tag{4.30}$$

Here, $\hat{\alpha}$ and $\hat{\beta}$ are the MLEs of α and β, respectively. Equations (4.29) and (4.30) give the Bayes estimates of α and β.

5 Simulation Study

Once we estimate the parameters of the MSBBTD using ML, MOM, and Bayes estimation method, it is advisable to study its efficiency. By considering the different values of the parameters of distribution, simulated risk (SR) is calculated for both the parameters considering r known for all the three methods of estimation.

To study the effect of a varied collection of values of the parameters, 1000 random samples are generated, each of size $n = 20, 40$ and 100.

The values of the parameters of the MSBBTD considered are $a = 2, \alpha = 0.05, 0.1, 0.15$, and $\beta = 0.2, 0.4, 0.5$ and prior parameter $(\gamma) = 0.5$ and 2.

The simulated risk of the estimates in the three methods is calculated using the formula:

$$SR = \sqrt{\frac{\sum_{i=1}^{1000}\left(\hat{\beta}_i - \beta\right)^2}{1000}}$$

The table shows the results of simulation for various combinations of values of α, β and γ and by fixing $a = 2$.

From Tables 1, 2, 3, and 4, it can be concluded that

Out the three methods of estimations, for any of the abovementioned pairs of values of parameters β and α, as the sample size n increases, $SR(\beta)$ and $SR(\alpha)$ decrease. For a fixed value of β, as the probability of misclassification α increases, $SR(\beta)$ and $SR(\alpha)$ decrease in ML and Bayes estimation method, whereas the method of moment estimation is not giving any noteworthy conclusion. In the comparison of Bayes estimators for different values of prior parameter γ, it can be observed that for almost all the pairs of β and α, as the value of prior parameter increases, the simulated risk of β and α decreases. Moment estimators are not comparable with ML estimators as well as Bayes estimators, as it gives an erratic behavior for almost all

Table 1 Simulated risk of β and α by ML estimation method for different values of β, α, and a different sample size (n)

		$n=20$		$n=40$		$n=100$	
β	α	$SR(\beta)$	$SR(\alpha)$	$SR(\beta)$	$SR(\alpha)$	$SR(\beta)$	$SR(\alpha)$
0.2	0.05	0.3827	0.3622	0.1514	0.2559	0.0217	0.1456
	0.1	0.3719	0.3405	0.1469	0.2280	0.0213	0.1349
	0.15	0.3429	0.3115	0.1272	0.2223	0.0225	0.1263
0.4	0.05	0.0490	0.3326	0.0369	0.2483	0.0252	0.1409
	0.1	0.0475	0.3088	0.0352	0.2175	0.0236	0.1338
	0.15	0.0468	0.2725	0.0345	0.2131	0.0228	0.1316
0.5	0.05	0.0593	0.3306	0.0511	0.2598	0.0417	0.1637
	0.1	0.0591	0.2982	0.0490	0.2374	0.0408	0.1400
	0.15	0.0502	0.2885	0.0476	0.2344	0.0401	0.1314

Table 2 Simulated risk of β and α by Bayes estimation method for different values of β, α and a sample size (n) with prior $\gamma = 0.5$

		$n=20$		$n=40$		$n=100$	
β	α	$SR(\beta)$	$SR(\alpha)$	$SR(\beta)$	$SR(\alpha)$	$SR(\beta)$	$SR(\alpha)$
0.2	0.05	0.4013	0.2760	0.0915	0.2108	0.0242	0.1362
	0.1	0.3725	0.2359	0.0813	0.1929	0.0238	0.1247
	0.15	0.3706	0.2202	0.0782	0.1751	0.0234	0.1134
0.4	0.05	0.0509	0.2369	0.0385	0.2030	0.0247	0.1326
	0.1	0.0482	0.2111	0.0376	0.1857	0.0240	0.1231
	0.15	0.0452	0.1863	0.0367	0.1791	0.0238	0.1194
0.5	0.05	0.0600	0.2135	0.0501	0.2060	0.0428	0.1366
	0.1	0.0609	0.1899	0.0495	0.1851	0.0420	0.1278
	0.15	0.0615	0.1627	0.0490	0.1737	0.0418	0.1179

Table 3 Simulated risk of β and α by Bayes estimation method for different values of β, α and a sample size (n) with prior $\gamma = 2$

		$n=20$		$n=40$		$n=100$	
β	α	$SR(\beta)$	$SR(\alpha)$	$SR(\beta)$	$SR(\alpha)$	$SR(\beta)$	$SR(\alpha)$
0.2	0.05	0.3842	0.2526	0.1316	0.2072	0.0217	0.1293
	0.1	0.3733	0.2345	0.1473	0.1835	0.0214	0.1213
	0.15	0.3442	0.2138	0.1275	0.1824	0.0208	0.1174
0.4	0.05	0.0509	0.2259	0.0381	0.2065	0.0237	0.1295
	0.1	0.0508	0.2055	0.0369	0.1775	0.0241	0.1231
	0.15	0.0507	0.1738	0.0366	0.1737	0.0253	0.1229
0.5	0.05	0.0653	0.2089	0.0531	0.2090	0.0419	0.1492
	0.1	0.0635	0.1765	0.0510	0.1854	0.0417	0.1271
	0.15	0.0630	0.1660	0.0496	0.1830	0.0410	0.1191

Table 4 Simulated risk of β and α by moment estimation method for different values of β, α and a sample size (n)

		n=20		n =40		n = 100	
β	α	$SR(\beta)$	$SR(\alpha)$	$SR(\beta)$	$SR(\alpha)$	$SR(\beta)$	$SR(\alpha)$
0.2	0.05	0.4013	0.2760	0.0915	0.2108	0.0242	0.1362
	0.1	0.3725	0.2359	0.0813	0.1929	0.0238	0.1247
	0.15	0.3706	0.2202	0.0782	0.1751	0.0234	0.1134
0.4	0.05	0.0509	0.2369	0.0385	0.2030	0.0247	0.1326
	0.1	0.0482	0.2111	0.0376	0.1857	0.0240	0.1231
	0.15	0.0452	0.1863	0.0367	0.1791	0.0238	0.1194
0.5	0.05	0.0600	0.2135	0.0501	0.2060	0.0428	0.1366
	0.1	0.0609	0.1899	0.0495	0.1851	0.0420	0.1278
	0.15	0.0615	0.1627	0.0490	0.1737	0.0418	0.1179

the pairs of the parameters β and α. It is observed that ML estimators give minimum simulated risk for β and α compared to Bayes and moment estimators.

References

Borel, E. (1942). *Sur* l'emploi du theoreme de Bernoulli pour faciliter le calcul d'un infinite de coefficients. Application au probleme de l' attente a un guichet, comptes rendus. *Acedemic des Sciences, Paris, Series A, 214,* 452–456.

Box, G. E. P., & Tioa, G. C. (1973). *Bayesian inference in statistical analysis.* Wiley.

Bross, I. D. J. (1954). Misclassification in 2 × 2 tables. *Biometrics, 10,* 478–486.

Diamond, E. L., & Lilienfeld, A. M. (1962). Effect of errors in classification and diagnosis in various types of epidemiological studies. *American Journal of Public Health, 52*(7), 1137–1144.

Fisher, R. A. (1934). The effects of methods of ascertainments upon the estimation of frequencies. *Annals of Eugenics, 6,* 13–25.

Haight, F. A., & Breuer, M. A. (1960). The Borel–Tanner distribution. *Biometrika, 47,* 143–150.

Lindley, D. V. (1980). Approximate Bayes method. *Trabajos de Estadistica, 31,* 223–237.

Nirei, M., Stamatiou, T., & Sushko, V. (2012). Stochastic herding in financial markets evidence from institutional investor equity portfolios. BIS Working Papers, No. 371.

Parikh, N. T., & Shah, S. M. (1969). Misclassification in power series distribution in which the value one is sometimes reported as zero. *Journal of Indian Statistical Association, 7,* 11–19.

Patil, G. P. (2002). Weighted distributions. *Encyclopedia of Environmetrics, 4,* 2369–2377.

Press, S. J. (1989). *Bayesian statistics: Principles, models and applications.* Wiley.

Shy, C. M., Kleinbaum, D. G., & Morgenstern, H. (1978). The effect of misclassification of exposure status in epidemiological studies of air pollution health effects. *Journal of Urban Health, Springer, 54*(11), 1155–1165.

Tanner, J. C. (1953). A problem of interference between two queues. *Biometrika, 40,* 58–69.

Trivedi, B. S., & Patel, M. N. (2013). Estimation in misclassified size-biased generalized negative binomial distribution. *Mathematics and Statistics*, Horizon Research (Publishing Corporation), CA, USA, *1*(2), 74–85.

Trivedi, B. S., & Patel, M. N. (2015). Estimation in misclassified size-biased log series distribution. *Mathematical Theory and Modelling*, IISTE—International Knowledge sharing Platform, USA, *5*(2), 128–135.

Trivedi, B. S., & Patel, M. N. (2015). Estimation of parameters of misclassified size biased Borel distribution. *Journal of Modern Applied Statistical Methods (JMASM)*, Wayne State University, Detroit, Michigan, USA, *15*(2), 475–494.

Trivedi, B. S., & Patel, M. N. (2017). Estimation of parameters of misclassified size biased discrete lindley distribution. *Communications in Statistics—Simulation and Computation*, Taylor and Francis Group, UK, *45*(7), 5541–5552.

A Stochastic Feedback Queuing Model with Encouraged Arrivals and Retention of Impatient Customers

Bhupender Kumar Som

Abstract Globalization has introduced never ending competition in business. This competition empowers customers and somewhat ensures quality at reduced cost. High competition along with uncertain customer behavior complicates the situation further for organizations. In order to stay ahead in the competition, organizations introduce various discounts and offers to attract customers. These discounts and offers encourage customers to visit the particular firm (online or offline). Encouraged arrivals result in heavy rush at times. Due to this, customers have to wait longer in queues before they can be serviced. Long waiting times results in customer impatience and a customer may decide to abandon the facility without completion of service, known as reneging. Reneging results in loss of goodwill and revenue both. Further, heavy rush and critical occupation of service counters may lead to unsatisfactory service and some customers may remain unsatisfied with the service. These customers (known as feedback customers) may rejoin the facility rather than leaving satisfactorily. Unsatisfactory service in these situations may cause harm to the brand image and business of the firm. If the performance of the system undergoing such pattern can be measured in advance with some probability, an effective management policy can be designed and implemented. A concrete platform for measuring performance of the system can be produced by developing a stochastic mathematical model. Hence, in this paper, a stochastic model addressing all practically valid and contemporary challenges mentioned above is developed by classical queuing theory model development approach. The model is solved for steady-state solution iteratively. Economic analysis of the model is also performed by introduction of cost model. The necessary measures of performance are derived, and numerical illustrations are presented. MATLAB is used for analysis as and when needed.

Keywords Stochastic models · Queuing theory · Globalization · Governance

---

B. K. Som (✉)
JIMS, Rohini, Sec-5, New Delhi 110085, India
e-mail: bksoam@live.com

A. K. Laha (ed.), *Advances in Analytics and Applications*, Springer Proceedings in Business and Economics, https://doi.org/10.1007/978-981-13-1208-3_20

1 Introduction

Contemporary business environment is highly challenging due to factors like uncertainty, globalization, and competition. Engaging new customers and retaining existing customers require a full proof strategy. The margin of error in given times is so low that a slight compromise with the strategy may result in loss to the business.

In order to engage new customers, firms often release offers and discounts. These discounts can be observed very frequently. Let it be online stores or offline stores, every firm comes up with discounts and offers every now and then. These offers and discounts attract customers to visit the stores or online web portals. These mobilized customers are termed as *encouraged arrivals* in this paper. At times customers also get attracted toward a firm once they see that lot of people are engaged in service with the particular firm already. Since a high volume of customers ensures a better quality, affordable price, or service, therefore, an arriving customer may get attracted toward a firm where there are customers already exist in large volumes. For example, in a healthcare facility, a large patient base ensures better doctors, better facilities, better treatment at affordable cost, or better service. This phenomenon was termed as *reverse balking* by Jain et al. (2014) which is contrary to classical queuing phenomenon named *balking* mentioned by Ancker and Gaffarian (1963a, b). Encouraged arrivals are different from reverse balking in the sense that reverse balking deals with probability calculated from existing system size and system capacity, while encouraged arrivals are directly related to the system size at given time or percentage increase in customers' arrival due to offers and discounts. The phenomenon of encouraged arrivals is contrary to discouraged arrivals discussed by Kumar and Sharma (2012).

Once the customers are encouraged, they result in higher volumes to the system. High volumes result in longer queues. These queues can be physical or digital. Service facility experiences increased load and smooth functioning becomes a tedious task. A dissatisfactory service at this point may result in customer impatience and a customer may decide to abandon the facility. This phenomenon is termed as customer *reneging* and discussed first by Haight (1957). Reneging is a loss to business, goodwill of the company, and revenue. A number of papers emerged on impatient customers since the inception of concept. Bareer (1957) studied phenomenon of balking and reneging in many ways. Natvig (1974) derived transient probabilities of queueing system with arrivals discouraged by queue length. Dooran (1981) further discussed the notion of discouraged arrivals in his work. Xiong and Altiok (2009) discussed impatience of customers. Kumar and Sharma (2013, 2014) studied queuing systems with retention and discouraged arrivals. Kumar and Som (2014) study single-server stochastic queuing model with reverse balking and reverse reneging. They mentioned that customer impatience will also decrease with increase in volume of customers. Kumar and Som (2015a, b) further present a queuing model with reverse balking, reneging, and retention. Kumar and Som (2015a, b) further added feedback customers to their previous model. Recently, Som (2016) presented a queueing system with reverse balking, reneging, retention of impatient customers, and feedback with heterogeneous service. He discussed the case of healthcare facility going through mentioned contemporary challenges.

There is a need to have a strategy that can help in smooth functioning of the system for better result. Strategies driven by scientific methods in such cases result in better output. Owing to this practically valid aspect of business, we develop a single-server feedback stochastic queuing model addressing contemporary challenges in this paper. Cost–profit analysis of the model is discussed later.

2 Formulation of Stochastic Model

A Markovian single-server queuing model is formulated under the following assumptions:

(i) The customers are arriving at the facility in accordance with Poisson process one by one with a mean arrival rate of $\lambda + n$, where "n" represents the number of customers exists in the system.
(ii) Customers are provided service exponentially with parameter μ.
(iii) An FCFS discipline is followed.
(iv) Service is provided through single server and the capacity of system is finite (say, N).
(v) The reneging times are independently and exponentially distributed with parameter ξ.
(vi) Unsatisfied serviced customer may retire to the system with probability q and may decide to abandon the facility $p = (1 - q)$.
(vii) A reneging customer may be retained with probability $q_1 = (1 - p_1)$.

In order to construct a stochastic model based on assumptions (i)–(viii), the following differential-difference equations are derived. These equations govern exhaustive stages the system can experience. Equation (1) governs initial stage of the system, Eq. (2) governs the functioning stage, and Eq. (3) governs the stage when system is full. The three differential-difference equations governing the system are given by

$$P'_0(t) = -\lambda P_0(t) + \mu p P_1(t) \tag{1}$$

$$P'_n(t) = \{\lambda + (n-1)\}P_{n-1}(t) + \{(-\lambda + n) - \mu p - (n-1)\xi p_1\}P_n(t) + (\mu p + n\xi p_1)P_{n+1}(t) \tag{2}$$

$$P'_N(t) = \{\lambda + (N-1)\}P_{N-1}(t) - \{\mu p + (N-1)\xi p_1\}P_N(t) \tag{3}$$

3 Steady-State Equations

In steady-state equation, the system of equations becomes

$$0 = -\lambda P_0 + \mu p P_1 \tag{4}$$

$$0 = \{\lambda + (n-1)\}P_{n-1} + \{(-\lambda + n) - \mu p - (n-1)\xi p_1\}P_n + (\mu p + n\xi p_1)P_{n+1} \tag{5}$$

$$0 = \{\lambda + (N-1)\}P_{N-1} - \{\mu p + (N-1)\xi p_1\}P_N \tag{6}$$

4 Steady-State Solution

On solving above equations iteratively, we get

$$P_n = \prod_{i=0}^{n-1} \frac{\lambda + i}{\mu p + i\xi p_1} P_0$$

As normality condition explains $\sum_{n=0}^{N} P_n = 1$

$$P_0 = \left\{ 1 + \sum_{n=1}^{N} \prod_{i=0}^{n-1} \frac{\lambda + i}{\mu p + i\xi p_1} \right\}^{-1}$$

And the probability that system is full is given by

$$P_N = \prod_{i=0}^{N-1} \frac{\lambda + i}{\mu p + i\xi p_1} P_0$$

5 Measures of Performance

5.1 Expected System Size (L_s)

$$L_s = \sum_{n=0}^{N} nP_n$$

$$L_s = \sum_{n=0}^{N} n \left\{ \prod_{i=0}^{n-1} \frac{\lambda + i}{\mu p + i\xi p_1} P_0 \right\}$$

5.2 Expected Queue Length (L_q)

$$L_q = \sum_{n=0}^{N} (n-1)P_n$$

$$L_q = \sum_{n=0}^{N} (n-1)\left\{\prod_{i=0}^{n-1} \frac{\lambda + i}{\mu p + i\xi p_1} P_0\right\}$$

5.3 Average Rate of Reneging Is Given by (R_r)

$$R_r = \sum_{n=1}^{N} (n-1)\xi p_1 P_n$$

$$R_r = \sum_{n=1}^{N} (n-1)\xi\left\{\prod_{i=0}^{n-1} \frac{\lambda + i}{\mu p + i\xi p_1} P_0\right\}$$

5.4 Average Rate of Retention Is Given by (R_R)

$$R_R = \sum_{n=1}^{N} (n-1)\xi q_1 P_n$$

$$R_R = \sum_{n=1}^{N} (n-1)\xi\left\{\prod_{i=0}^{n-1} \frac{\lambda + i}{\mu p + i\xi q_1} P_0\right\}$$

6 Numerical Illustration

Numerical validity of the model is tested by performing numerical analysis. The analysis is performed by varying parameters arbitrarily. Sensitivity of appropriate measures of performance is observed with respect to varying arbitrary values of parameters (Table 1).

We can observe that an increasing average rate of arrival leaves a positive impact on average system size; as a result queue length also increases, while increase in rate of reneging means that high volume of customers put service under pressure and lot of customers fail to wait after a threshold value of time and renege thereafter, i.e.,

Table 1 Variation in $L_s, L_q, and\ R_r$ and R_R with respect to λ. We take $N = 10, \mu = 3, \xi = 0.2, q = 0.2, q_1 = 0.3$

(λ)	(Ls)	(L_q)	(R_r)	(R_R)
2	9.511	8.5118	1.1917	0.5107
2.5	9.553	8.5528	1.1974	0.5132
3	9.585	8.5854	1.2020	0.5151
3.5	9.613	8.6127	1.2058	0.5168
4	9.636	8.6361	1.2090	0.5182
4.5	9.656	8.6564	1.2119	0.5194
5	9.674	8.6744	1.2144	0.5205
5.5	9.690	8.6904	1.2167	0.5214
6	9.705	8.7049	1.2187	0.5223
6.5	9.718	8.7179	1.2205	0.5231
7	9.730	8.7298	1.2222	0.5238
7.5	9.741	8.7407	1.2237	0.5244
8	9.751	8.7507	1.2251	0.5250
8.5	9.760	8.7599	1.2264	0.5256
9	9.768	8.7685	1.2276	0.5261
9.5	9.776	8.7764	1.2287	0.5266
10	9.784	8.7838	1.2297	0.5270

Source simulated data

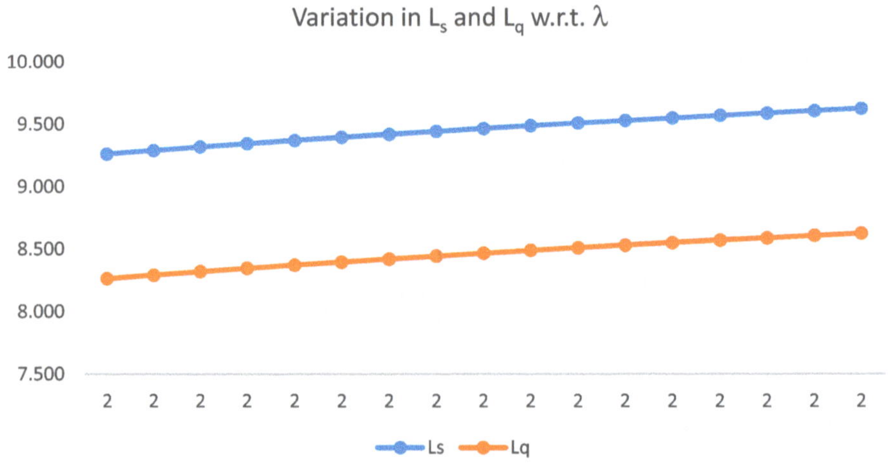

Fig. 1 Variation in system size and length of queue with respect to arrival rate

abandon the facility without completion of service. The following graphs explain the phenomenon (Fig. 1).

Similarly, the numerical results are obtained by varying service rate and rate of reneging. The following tables explain the same (Table 2).

Table 2 Variation in $L_s, L_q, R_R and\ R_r$ with respect to μ. We take $N = 10, \lambda = 2, \xi = 0.2, q = 0.2, q_1 = 0.3$

(μ)	(Ls)	(L_q)	(R_r)	(R_R)
3	9.511	8.5118	1.1917	0.5107
3.1	9.492	8.4925	1.1890	0.5096
3.2	9.472	8.4722	1.1861	0.5083
3.3	9.450	8.4509	1.1831	0.5071
3.4	9.427	8.4283	1.1800	0.5057
3.5	9.403	8.4045	1.1766	0.5043
3.6	9.378	8.3792	1.1731	0.5028
3.7	9.351	8.3523	1.1693	0.5011
3.8	9.322	8.3237	1.1653	0.4994
3.9	9.291	8.2932	1.1611	0.4976
4	9.257	8.2606	1.1565	0.4956
4.1	9.222	8.2257	1.1516	0.4935
4.2	9.184	8.1882	1.1464	0.4913
4.3	9.142	8.1480	1.1407	0.4889
4.4	9.098	8.1048	1.1347	0.4863
4.5	9.050	8.0583	1.1282	0.4835

Source simulated data

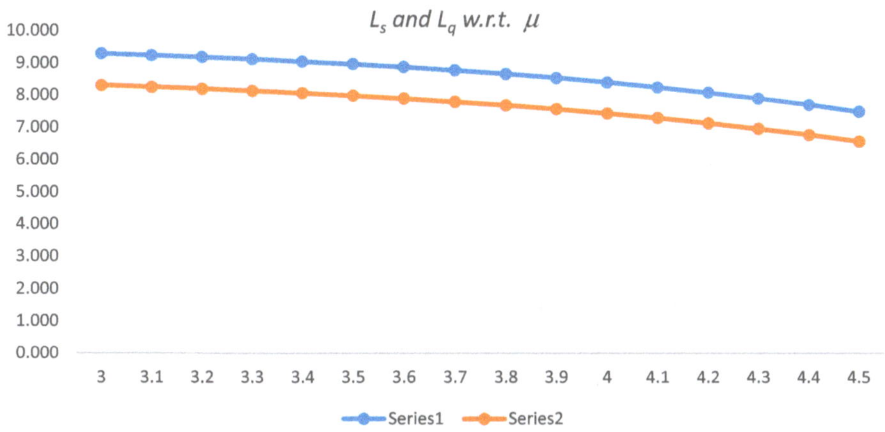

Fig. 2 System size and queue length with respect to service rate

We can observe as service rate increases, the expected system size decreases and so does expected length of queue, while decrease in rate of reneging, it means that decreasing volume of customers eases out the pressure on service and high number customers choose to wait rather than leaving the facility. The following graphs explain the phenomenon (Fig. 2).

From Table 3, it can be observed that feedback results in increasing expected system size and queue length. The rate of reneging goes high as more and more

Table 3 Variation in $L_s, L_q, R_R and\ R_r$ with respect to q. We take $N = 10, \lambda = 2, \xi = 0.1, \mu = 3, q_1 = 0.3$

(q)	(Ls)	(L_q)	(R_r)	(R_R)
0	9.148	8.1539	1.1415	0.4892
0.05	9.232	8.2359	1.1530	0.4942
0.1	9.304	8.3065	1.1629	0.4984
0.15	9.366	8.3679	1.1715	0.5021
0.2	9.421	8.4217	1.1790	0.5053
0.25	9.469	8.4693	1.1857	0.5082
0.3	9.511	8.5118	1.1917	0.5107
0.35	9.550	8.5503	1.1970	0.5130
0.4	9.585	8.5855	1.2020	0.5151
0.45	9.618	8.6178	1.2065	0.5171
0.5	9.648	8.6478	1.2107	0.5189
0.55	9.676	8.6759	1.2146	0.5206
0.6	9.702	8.7023	1.2183	0.5221
0.65	9.727	8.7272	1.2218	0.5236
0.7	9.751	8.7508	1.2251	0.5250
0.75	9.773	8.7732	1.2283	0.5264

Source simulated data

customers in the system cause high level of delays in service and cause high level of impatience.

From Table 4, it can be observed that retention leaves a positive impact on system size. Though the queue length increases, retaining impatient customers is an important phenomenon that leads to increased revenue at the end.

7 Economic Analysis of the System

This section discusses the cost–profit analysis of the model. An algorithm is written for newly formulated functions of total expected revenue, total expected cost, and total expected profit for obtaining numerical outputs.

Total expected cost (TEC) of the model is given by

$$TEC = C_s\mu + C_hL_s + C_rR_r + C_L\lambda P_N + +C_fqL_S + C_RR_R$$

$$TEC = C_s\mu + C_h\sum_{n=0}^{N} n\left\{\prod_{i=0}^{n-1}\frac{\lambda + i}{\mu p + i\xi p_1}P_0\right\}$$

$$+ C_r\sum_{n=1}^{N}(n-1)\xi p_1\left\{\prod_{i=0}^{n-1}\frac{\lambda + i}{\mu p + i\xi p_1}P_0\right\}$$

Table 4 Variation in $L_s, L_q, R_R and\ R_r$ with respect to q_1. We take $N = 10, \lambda = 2, \xi = 0.1, \mu = 3, q = 0.2$

Probability of retention (q_1)	Expected system size (Ls)	Expected queue length (L_q)	Average rate of reneging (R_r)	Average rate of retention (R_R)
0	9.263	8.2645	1.6529	0.0000
0.05	9.291	8.2929	1.5756	0.0829
0.1	9.319	8.3203	1.4977	0.1664
0.15	9.345	8.3469	1.4190	0.2504
0.2	9.371	8.3726	1.3396	0.3349
0.25	9.396	8.3975	1.2596	0.4199
0.3	9.421	8.4217	1.1790	0.5053
0.35	9.444	8.4450	1.0979	0.5912
0.4	9.467	8.4677	1.0161	0.6774
0.45	9.489	8.4896	0.9339	0.7641
0.5	9.510	8.5109	0.8511	0.8511
0.55	9.531	8.5315	0.7678	0.9385
0.6	9.551	8.5515	0.6841	1.0262
0.65	9.570	8.5709	0.6000	1.1142
0.7	9.589	8.5897	0.5154	1.2026
0.75	9.608	8.6079	0.4304	1.2912

Source simulated data

$$+ C_L\lambda \prod_{i=0}^{N-1} \frac{\lambda + i}{\mu p + i\xi p_1} P_0 + +C_f \times q \sum_{n=0}^{N} n\left\{\prod_{i=0}^{n-1} \frac{\lambda + i}{\mu p + i\xi p_1} P_0\right\}$$
$$+ C_R \sum_{n=1}^{N} (n-1)\xi q_1 \left\{\prod_{i=0}^{n-1} \frac{\lambda + i}{\mu p + i\xi p_1} P_0\right\}$$

Total expected revenue (TER) of the model is given by

$$TER = R \times \mu \times (1 - P_0) + R_f \times q \times L_s$$

Total expected profit (TEP) of the model is given by

$$TEP = R \times \mu \times (1 - P_0) + R_f \times q \times L_s - C_s\mu + C_h \sum_{n=0}^{N} n\left\{\prod_{i=0}^{n-1} \frac{\lambda + i}{\mu p + i\xi p_1} P_0\right\}$$
$$+ C_r \sum_{n=1}^{N} (n-1)\xi p_1 \left\{\prod_{i=0}^{n-1} \frac{\lambda + i}{\mu p + i\xi p_1} P_0\right\} + C_L\lambda \prod_{i=0}^{N-1} \frac{\lambda + i}{\mu p + i\xi p_1} P_0 + +C_f$$
$$\times q \sum_{n=0}^{N} n\left\{\prod_{i=0}^{n-1} \frac{\lambda + i}{\mu p + i\xi p_1} P_0\right\} + C_R \sum_{n=1}^{N} (n-1)\xi q_1 \left\{\prod_{i=0}^{n-1} \frac{\lambda + i}{\mu p + i\xi p_1} P_0\right\}$$

Table 5 Variation in *TEC*, *TER* and *TEP* with respect to μ. We take $N = 10, \lambda = 2, \xi = 0.1, C_S = 10, C_l = 15, C_h = 2, C_f = 50, C_r = 2, R = 500$

(μ)	(TEC)	(TER)	(TEP)
3	83.995	1589.6643	1505.6697
3.1	84.883	1638.4003	1553.5177
3.2	85.760	1686.8742	1601.1145
3.3	86.624	1735.0321	1648.4078
3.4	87.475	1782.8114	1695.3363
3.5	88.310	1830.1395	1741.8294
3.6	89.128	1876.9333	1787.8058
3.7	89.925	1923.0979	1833.1726
3.8	90.701	1968.5263	1877.8251
3.9	91.453	2013.0986	1921.6456
4	92.178	2056.6818	1964.5036
4.1	92.874	2099.1296	2006.2553
4.2	93.539	2140.2829	2046.7443
4.3	94.169	2179.9708	2085.8021
4.4	94.762	2218.0114	2123.2494
4.5	95.316	2254.2144	2158.8984

Source simulated data

where

C_s Cost of service,
C_h Holding cost of a customer,
C_r Cost of reneging,
C_L Cost of a lost customer,
C_f Cost of feedback,
R Revenue earned from each customer,
R_f Revenue earned from each feedback customer, and
C_R Cost of retention of a customer.

The cost model formulated above is translated into MATLAB, and numerical results are obtained for varying rate of arrival and service (Table 5).

Though the cost increases with increase in the rate of service, as service cost increases but due to reduced rate of reneging, the revenue goes high and firms profit keeps on increasing with an improving rate of service (Table 6).

The table shows an increase in TEP with increase in average arrival rate which is obvious as increasing rate of arrival results in higher number of customers to the system and firm enjoys more revenue from each customer.

Figure 3 shows total expected profit is higher with improving service and fixed arrival (—). In comparison to the case when arrivals increase, the customers are serviced at a constant rate. This shows an improving service that ensures better profits and a firm shall focus more on providing better service rather than bringing more customers to the system.

Table 6 Variation in TEC, TERandTEP with respect to μ. We take $N = 10, \mu = 3, \xi = 0.1, C_S = 10, C_l = 15, C_h = 2, C_r = 2, R = 500$

(λ)	(TEC)	(TER)	(TEP)
2	83.995	1589.6643	1505.6697
2.5	91.747	1592.2844	1500.5369
3	99.422	1593.5561	1494.1341
3.5	107.054	1594.3050	1487.2510
4	114.660	1594.8157	1480.1558
4.5	122.248	1595.2010	1472.9531
5	129.823	1595.5115	1465.6884
5.5	137.389	1595.7726	1458.3841
6	144.946	1595.9983	1451.0521
6.5	152.498	1596.1970	1443.6994
7	160.044	1596.3744	1436.3306
7.5	167.586	1596.5342	1428.9486
8	175.124	1596.6793	1421.5556
8.5	182.659	1596.8119	1414.1533
9	190.191	1596.9337	1406.7430
9.5	197.720	1597.0461	1399.3258

Source simulated data

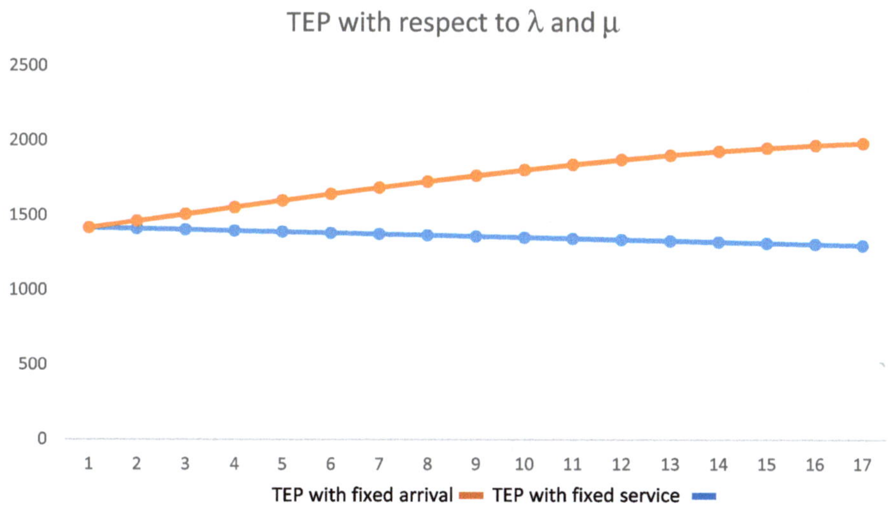

Fig. 3 TEP with respect to fixed and improving service rate

8 Conclusion and Future Scope

The results of the paper are of immense value for any firm encountering the phenomenon of encouraged customers and load on service. By knowing the measures of performance in advance, the overall performance of the system can be measured and an effective strategy can be planned for smooth functioning. By adopting and implementing this model, the economic analysis of the facility can also be measured and bird's eye view on financial aspect of the business can also be observed.

Further optimization of service rate and system size can be achieved while the system can be studied in transient state. The system can also be studied for heterogeneous service. A multi-server model can also be explored.

References

Ancker, C. J., & Gafarian, A. V. (1963a). Some queuing problems with balking and reneging I. *Operations Research, 11*(1), 88–100. https://doi.org/10.1287/opre.11.1.88.

Ancker, C. J., & Gafarian, A. V. (1963b). Some queuing problems with balking and reneging—II. *Operations Research, 11*(6), 928–937. https://doi.org/10.1287/opre.11.6.928.

Barrer, D. Y. (1957). Queuing with impatient customers and indifferent clerks. *Operations Research, 5*(5), 644–649. https://doi.org/10.1287/opre.5.5.644.

Haight, F. A. (1957). Queueing with balking. *Biometrika, 44*(3/4), 360. https://doi.org/10.2307/2332868.

Jain, N. K., Kumar, R., & Som, B. K. (2014). An M/M/1/N Queuing system with reverse balking. *American Journal of Operational Research, 4*(2), 17–20.

Kumar, R., & Sharma, S. K. (2012). A multi-server Markovian queuing system with discouraged arrivals and retention of reneged customers. *International Journal of Operational Research, 9*(4), 173–184.

Kumar, R., & Sharma, S. K. (2013). An M/M/c/N queuing system with reneging and retention of reneged customers. *International Journal of Operational Research, 17*(3), 333. https://doi.org/10.1504/ijor.2013.054439.

Kumar, R., & Sharma, K. (2014). A single-server Markovian queuing system with discouraged arrivals and retention of reneged customers. *Yugoslav Journal of Operations Research, 24*(1), 119–126. https://doi.org/10.2298/yjor120911019k.

Kumar, R., & Som, B. K. (2014). An M/M/1/N queuing system with reverse balking and reverse reneging. *Advance Modeling and Optimization, 16*(2), 339–353.

Kumar, R., & Som, B. K. (2015a). An M/M/1/N feedback queuing system with reverse balking, reverse reneging and retention of reneged customers. *Indian Journal of Industrial and Applied Mathematics, 6*(2), 173. https://doi.org/10.5958/1945-919x.2015.00013.4.

Kumar, R., & Som, B. K. (2015b). An M/M/1/N queuing system with reverse balking, reverse reneging, and retention of reneged customers. *Indian Journal of Industrial and Applied Mathematics, 6*(1), 73. https://doi.org/10.5958/1945-919x.2015.00006.7.

Natvig, B. (1974). On the transient state probabilities for a queueing model where potential customers are discouraged by queue length. *Journal of Applied Probability, 11*(02), 345–354. https://doi.org/10.1017/s0021900200036792.

Som, B. K. (2016). *A Markovian feedback queuing model for health care management with heterogeneous service*. Paper presented at 2nd International Conference on "Advances in Healthcare Management Services", IIM Ahmedabad.

Van Doorn, E. A. (1981). The transient state probabilities for a queueing model where potential customers are discouraged by queue length. *Journal of Applied Probability, 18*(02), 499–506. https://doi.org/10.1017/s0021900200098156.

Xiong, W., & Altiok, T. (2009). An approximation for multi-server queues with deterministic reneging times. *Annals of Operations Research, 172*(1), 143–151. https://doi.org/10.1007/s10479-009-0534-3.

Part VIII
Econometric Applications

Banking Competition and Banking Stability in SEM Countries: The Causal Nexus

Manju Jayakumar, Rudra P. Pradhan, Debaleena Chatterjee, Ajoy K. Sarangi and Saurav Dash

Abstract The study tries to evaluate whether banking competition impacts the banking stability in the case of 32 Single European Market countries, and whether a guiding principle focus on banking competition is appropriate as a loom to boost banking stability. This study finds the interactions between banking competition and banking stability in 32 European countries between 1996 and 2014 with a panel vector auto-regressive model. Our observed results show that there is a cointegrating relationship between the two. Moreover, banking competition is a causative factor in banking stability in the long run. Thus, a focus on banking competition will enhance the banking stability of these countries.

Keywords Banking stability · Banking competition · Granger causality European countries

1 Introduction

A long line of relationship was postulated between finance and economic by eminent scholars. It is widely accepted that efficient allocation of fund to projects which has high return, in turn, stimulates saving, investment, and economic growth (Levine

M. Jayakumar · R. P. Pradhan (✉) · D. Chatterjee · A. K. Sarangi · S. Dash
Vinod Gupta School of Management, Indian Institute of Technology Kharagpur, Kharagpur 721302, West Bengal, India
e-mail: rudrap@vgsom.iitkgp.ernet.in

M. Jayakumar
e-mail: manjhu_jk@yahoo.com

D. Chatterjee
e-mail: debaliena@gmail.com

A. K. Sarangi
e-mail: ajoyketan@yahoo.com

S. Dash
e-mail: saurav.stat@gmail.com

A. K. Laha (ed.), *Advances in Analytics and Applications*, Springer Proceedings in Business and Economics, https://doi.org/10.1007/978-981-13-1208-3_21

2003). Patrick (1966) states that if the financial sector is ill-organized, economic growth and development would have a discussion which was supported by Owusu and Odhiambo (2014). In the last couple of decades, there has been a rising flow of research that aims to study the relationship between financial development and economic growth (see, inter alia, Pradhan et al. 2015; Ngare et al. 2014; Wolde-Rufael 2009; Al-Yousif 2002).

As most transactions in the real economy happen through banks and other financial system, financial stability considers being a paramount for economic growth.[1] There are also studies which observed that bank competition and bank stability have influenced economic growth a large in both positive and negative ways (see, for instance, Fernandez et al. 2016; Soedarmono et al. 2011). However, in the financial industry, the importance for banking competition[2] has been given very small attention to analyze the growth of financial sector, especially in the banking sector. In spite of the reality that it can notably influence banks' efficiency, the excellence of the banking services and innovation presented to the customer. Though in history bank competition and financial stability have no strong connection, bank competition plays an important role. However, the financial crisis 2007–2009 brought this importance between the relationship of bank competition and bank stability (see, for instance, Fernandez et al. 2016).

After the crisis, many developed countries tightened the bank regulation so that the competition was greatly controlled. Compared to this, three decades prior to the financial crisis, many unbalanced reforms took place in the banking sector in which financial stability was surpassed by banking efficiency and competition (Vives 2016). Bank competition always has an impact on financial stability. So in this paper, an attempt is made to analyze the chance of relationship linking banking competition and banking stability in SEM[3] countries.

Mainly, we aim to concentrate on two issues on the connection between banking competition and banking stability. First, whether to consider these two variables together for study as per the literature available which deals with banking competition and banking stability (Vives 2016; Fu et al. 2014; Jeon and Lim 2013; Tabak et al.

[1]Theories of finance and economic growth nexus suggest that the financial purposes performed by banks (and other nonbanking financial intermediaries) are significant in encouraging economic growth (Levine 2005).

[2]Competition has both positive and negative impact on banking stability. There exist two strands of literature which argue that competition in banking industry endangers financial stability. Increased competition in bank deposits wears away the profitability in the banking market and thus lowers the market power of banks, this is due to the increase in bank interest (see, for instance, Matutes and Vives 2000; Schaeck et al. 2009). Second, positive relationship exists between the two; competition created a number of benefits such as reducing liquidity and giving them in the form of bank loans to the public and sells securities to the investors. Less concentration in banking industry leads to good stability by diversifying both interest and noninterest income activities (Amidu and Wolfe 2013; Hellmann et al. 2000; Berger et al. 2009).

[3]The SEM is meant for the benefit of all the European Union (EU) countries. Its mainstreams are the "four freedoms": free movement of people, goods, services, and capital between all the EU member countries. These can be enjoyed, with limited exemptions, by everyone living and working in the EU.

2012; Beck et al. 2006; Boyd and De Nicolo 2005; Claessens and Laeven 2003); second, to find any long-run stable relationship between banking competition and banking stability (Creel et al. 2015; Kasman and Carvallo 2014). The large literature on the relationship between banking competition and banking stability is mostly concerned with countrywise and bankwise but causality between the two has not been documented. This study intends to find the connection between banking competition and banking stability using dynamic panel Granger causality approach.

The remaining of the paper is planned as follows: Sect. 2 presents a literature review; Sect. 3 describes the methods of study; Sect. 4 discusses the observed results; and finally, we summarize the main findings in Sect. 5.

2 Review of Literature and Rationale of Analysis

Many researchers have studied the likely connection between bank stability and bank competition as competition plays an important role in determining the stability of any bank (Fiordelisi and Mare 2014; Liu et al. 2013; Berger et al. 2009; Boyd and De Nicolo 2005). Though a number of researcher have focused on the existence of relationship between the two, not enough investigation has been done to find the causal nexus between bank stability and bank competition. In this paper, we try to investigate whether there is a Granger causality link involving these two variables in our sample taken for study.

Several on-hand literature give an ample debate on the relationship between banking stability and economic growth (Hogart et al. 2002; Serwa 2010; Soedarmono et al. 2011; Louzis et al. 2012; Carby et al. 2012; Jokipii and Monnin 2013; Creel et al. 2015) and between banking competition and economic growth (Cetrolli and Gambere 2001; Coccorese 2008; De Guevara and Maudos 2011; Mitchener and Wheelock 2013; Gaffeo and Mazzocchi 2014). This section strictly highlights the literature connecting to the direction of relationship between banking competition and banking stability.

The relationship between the two variables namely banking competition and banking stability can be described in four forms of hypotheses. The first form shows that banking competition has strong impact on banking stability (*supply-leading hypothesis*: SLH), and in the second form, banking stability shows a positive impact on banking competition (*demand-following hypothesis*: DFH), whereas the third form shows that both have impact on each other (*feedback hypothesis*: FBH), and the fourth form shows that both banking competition and stability are independent of each other (*neutrality hypothesis*: NEH).

The SLH assumes that increase (decrease) in banking competition will result in increase (decrease) in banking stability in the economy. More competition is favorable to greater financial stability (Amidu and Wolfe 2013; Kasmanand Carvallo 2014; Fiordelisi and Mare 2014). Less competition in the bank can result in less credit rationing and larger loans, eventually increasing the probability of bank failure (Yeyati and Micco 2007; Fu et al. 2014). More competition in banking industry can lead to increased risk taking behaviour like inappropriate screening of borrowers, who approach for loans and lending in cheap interest rate (Allen and Gale 2004). Competition may make bigger financial instability due to bigger loan risk which may lead to further nonperforming loans (Boyd and De Nicolo 2005; Berger et al. 2009).

The DFH assumes that increase (decrease) in banking stability will result in increase (decrease) in banking competition in the economy. The nonperforming loans (act as a proxy indicator for bank stability) show a negative relationship with the market size, and competition in market increases when the NPL reduces, which was created by good stability in banks (Chang 2006). Bank stability competition in banking industry increases in the countries where the regulatory frameworks and monetary policies are strict. Banks' transparency decreases with the growth in bank instability, thus affecting the competition in the market (Fernandez et al. 2016; Andrievskaya and Semenova 2016).

The FBH viewpoint put forward that both the variables, i.e., banking competition and banking stability have an impact on one another. Martinez-Miera and Repullo (2010), Hakenes and Schnabel (2011), James et al. (2013), and Liu et al. (2013) examine the link between the bank's competition and stability and support the above hypothesis. Few findings from the former studies prove that stable banking system in a country attracts more funds in savings and investment. Due to this increase in investments and savings, competition in banking market increases to enhance a balance in the market. This competitive market can have a constructive impact on stability levels in the banking industry.

NEH describes that banking competition and banking stability do not have any relationship with each other. Few studies support this hypothesis, Lee and Hsieh (2014)[4] investigate the relationship between the two variables and find that there is no relationship between each other. Similarly, results of other studies suggest that there is no impact of concentration on bank stability but foreign penetration weakens the banking competition, and the latter is negatively related with bank risk and shows a nonlinear relationship with each other (Liu et al. 2013; Martinez-Miera and Repullo 2010). Table 1 provides a brief summary of these literatures.

It is clear that the relationship between the stability of the banking system and competition has been widely examined in the above-cited literature. It is a well-known factor that banking competition and banking stability trade off each other by few theories,[5] but the direction of causality is unknown due to its intricacy. These studies show the mixed relationship between banks' stability and banks' competition and leave the result inconclusive.

---

[4]Under different conditions of bank reforms in the host country, the degree of relationship between the two (banking stability and banking competition) varies.

[5]Competition-fragility view (Marcus 1984; Keeley 1990; Hellmann et al. 2000) and competition-stability view (Boyd and De Nicolo 2005).

Table 1 Summary of studies on the causal connection between banking competition and banking stability

Study	Study area	Data period	Hypothesis confirmed
Adjei-Frimpong (2013)	Ghana	2001–2010	SLH
Ferreira (2013)	European Union	1996–2008	DFH, FBH
Fiordelisi and Mare (2014)	Europe	1998–2009	DFH
Fu et al. (2014)	14 APCs	2003–2010	SLH
James et al. (2013)	10 African countries	2005–2010	SLH
Kasman and Carvallo (2014)	15 Latin American countries	2001–2008	DFH
Tabak et al. (2012)	10 LACs	2001–2008	SLH, NEH
Titko et al. (2015)	Latvia	2007–2013	NEH

Note SLH is supply-leading hypothesis, representing causality from banking competition to banking stability; DFH is demand-following hypothesis, representing causality from banking stability to banking competition; FBH is feedback hypothesis, representing bidirectional causality between banking competition and banking stability; and NEH is neutrality hypothesis, representing no causality between banking competition and banking stability

2.1 Hypotheses Tested

This study intends to test the following hypotheses:

H_{1A} Banking competition (BCP) Granger-causes banking stability (BSI). This is termed as BCP-led BSI hypothesis.

H_{1B} Banking stability Granger-causes banking competition. This is named as BSI-led BCP hypothesis.

3 Data, Variables, and Econometric Model

The analysis of the study is based on secondary data over the period 1996–2014[6] collected annually for 32 European countries.[7] The data have been collected from the following sources: World Development Indicators, International Financial Statistics by International Monetary Fund, and World Development Reports by World Bank.

[6]The panels used in this study are unbalanced due to data unavailability for some of these countries over the complete sample period.

[7]European Single Market Countries (see, for instance, Jayakumar et al. 2018). It can be noted that some of these countries are transcontinental.

A recognized difficulty in the banking industry is that competition cannot be measured in a straight line. So, based on the literature, six frequently used indicators of banking competition (BCP)[8] were carefully chosen for the study. They are as follows: Lerner index (LEI), Boone indicator (BOI), H-statistic (HST), CR-5 firm concentration ratio (CR5), CR-3 firm concentration ratio (CR3), and foreign ownership (FOW).

There is also a range of indicators to measure financial soundness and few studies attempt to measure the systemic stability using aggregate firm-level stability measures. In this study, five most commonly used proxy variables for banking stability (BSI)[9] are identified for the analysis of steadiness in banks. The indicators used are as follows bank capitalization (BCA), nonperforming loans (NPL), Z-index (ZSC), provision of nonperforming loans (PNL), and private credit by deposit money banks (PCD). Apart from these, the sixth indicator, composite banking stability index (CBS), was derived using principal component analysis (PCA) (Pradhan et al. 2014). Our composite index (CBS) is formed by exploiting four indicators such as BCA, return on assets (ROA), performing loan assets (PLA), and PCD which are clustered together as they all are expressed in the same units of measurement, i.e., percentage.

The variables integrated for the formation of CBS and the statistical values from the principal component analysis are mentioned in detail in Tables 6 and 7 (see Appendix 1). The summary statistics and correlation matrix of these variables are available in Table 2.

Considering six different banking competition indicators and six different banking stability indicators, we have six models and six cases. The panels used in this study are unbalanced due to data unavailability in some of these countries. The first model considers BCA and banking competition. The second model considers NPL and banking competition. The third model considers ZSC and banking competition. The fourth model considers PNL and banking competition. The fifth model considers PCD and banking competition. The sixth model considers CBS and banking competition. The individual model consists of six different cases, depending on which indicator of banking competition like LEI, BOI, HST, CR5, CR3, AND FOW.

---

[8]Some of these measures of banking competition have been used previously. (See, for instance, by Hamada et al. (2017), Fernandez et al. (2016), Kasman and Kasman (2015), Hakam et al. (2013), Bikker et al. (2007), Turk-Ariss (2009), and Boone (2008).)

[9]Some of these measures of banking stability have been used previously, for example, by Carretta et al. (2015), Creel et al. (2015), Kasman and Kasman (2015), Tabak et al. (2012), Berger et al. (2009), Cihak and Hesse (2010), and Chang et al. (2008).

Table 2 Descriptive statistics for the decision variables

Variables	LEI	BOI	HST	CR5	CR3	FOW	BCA	NPL	ZSC	PNL	PCD	CBS
Part 1: Summary statistics												
Mean	0.261	0.312	−0.945	1.868	1.789	1.458	0.831	0.488	1.451	1.759	1.889	0.782
Median	0.2666	0.311	−1.300	1.905	1.821	1.613	0.826	0.556	1.442	1.753	1.888	0.931
Maximum	0.304	0.366	0.116	2.000	2.000	2.000	1.153	1.391	1.848	2.442	2.358	1.128
Minimum	0.196	0.279	−2.010	1.509	1.336	0.001	0.432	−0.836	0.671	1.158	1.122	−1.288
Standard dev.	0.023	0.014	0.539	0.114	0.149	0.518	0.136	0.469	0.181	0.189	0.254	0.374
Skewness	−0.797	0.573	0.805	−0.932	−0.632	−0.850	0.113	−0.503	−0.381	0.278	−0.185	−2.458
Kurtosis	3.193	3.763	1.817	3.011	2.670	2.656	2.839	2.891	5.093	5.188	0.278	8.722
IQR	19.89	14.60	30.75	26.80	12.83	23.19	0.592	7.893	38.22	39.27	5.188	13.86
Part 2: Correlation matrix												
LEI	1.000											
BOI	0.053	1.000										
HST	0.015	0.139**	1.000									
CR5	0.161	0.048	−0.114	1.000								
CR3	0.160	0.009	−0.128**	0.920*	1.000							
FOW	0.086	−0.039	−0.110	0.302*	0.277*	1.000						
BCA	0.001	−0.081	−0.078	0.311*	0.3444*	0.557*	1.000					
NPL	−0.064	−0.007	0.282*	−0.047	−0.084	0.152	0.304*	1.000				
ZSC	0.272	0.119	−0.001	−0.110	−0.215*	0.274*	−0.193*	0.069	1.000			
PNL	−0.095	−0.018	−0.160	0.039	0.064	−0.129	0.022	−0.265	−0.092	1.000		
PCD	−0.069	0.146**	0.163**	0.161**	−0.084	−0.652*	−0.495*	−0.171	−0.201*	−0.075	1.000	
CBS	−0.087	−0.113	−0.041	−0.016	0.024	−0.357*	0.034	−0.161	−0.223*	0.184**	0.565*	1.000

Note 1 LEI is Lerner index; BOI is Boone indicator; HST is H-statistic; CR5 is firm concentration of five largest banks; CR3 is firm concentration of three largest banks; FOW is foreign ownership; BCA is bank capitalization; NPL is nonperforming loans; ZSC is bank-level Z-index; PNL is provision of nonperforming loans; PCD is private credit by deposit money banks; and CBS is composite index of banking stability
Note 2 Reported values in square brackets are the probability levels of significance
Note 3 * and ** indicate significance level at 1% and 5%, respectively

The probable direction of causality between banking competition and banking stability is examined using vector error correction modeling (VECM).

$$\begin{bmatrix} \Delta\, Banking\ Competition_{it} \\ \Delta\, Banking\ Stability_{it} \end{bmatrix} = \begin{bmatrix} \eta_{1j} \\ \eta_{2j} \end{bmatrix} + \sum_{k=1}^{n} \begin{bmatrix} \mu_{11ik}(L)\mu_{12ik}(L) \\ \mu_{21ik}(L)\mu_{22ik}(L) \end{bmatrix} \begin{bmatrix} \Delta\, Banking\ Competition_{it-k} \\ \Delta\, Banking\ Stability_{it-k} \end{bmatrix} + \begin{bmatrix} \delta_{1i} ECT_{it-1} \\ \delta_{2i} ECT_{it-1} \end{bmatrix} + \begin{bmatrix} \varepsilon_{1it} \\ \varepsilon_{2it} \end{bmatrix} \tag{1}$$

where Δ is the first difference filter, i is the country description in the panel, t is the time period, and ε is the error term. *Banking competition* is defined as LEI, BOI, HST, CR5, CR3, or FOW, while *banking stability* is defined as BCA, NPL, ZSC, PNL, or PCD, according to the number of case and condition.

ECT-1's is the lagged error correction terms. The particular representation provides strong results if the time series variables are of order one integration and have the presence of cointegration. While using the VECM technique, suppose, the variable does not show any cointegration, the lagged error correction terms are removed from the estimation process.

The hypotheses of this study are to test:

$$H_{1A}^{0} : \boldsymbol{\mu_{12ik} = 0;\ and\ \delta_{1i} = 0\ for\ k = 1, 2, 3, \ldots, n}$$
$$H_{1A}^{1} : \boldsymbol{\mu_{12ik} \neq 0;\ and\ \delta_{1i} \neq 0\ for\ k = 1, 2, 3, \ldots, n}$$
$$H_{1B}^{0} : \boldsymbol{\mu_{21ik} = 0;\ and\ \delta_{2i} = 0\ for\ k = 1, 2, 3, \ldots, n}$$
$$H_{1B}^{1} : \boldsymbol{\mu_{21ik} \neq 0;\ and\ \delta_{2i} \neq 0\ for\ k = 1, 2, 3, \ldots, n}$$

If we consider the direction of causality between banking competition and banking stability, there exist a number of possible situations. For example, in the first scenario, banking competition and banking stability can be said to be not causally related, if neither μ_{12ik} nor μ_{21ik} are significantly different from zero. Second, we can assume only banking stability Granger-causes banking competition, if only μ_{12ik} is statistically different from zero. Third, we can infer that only banking competition Granger-causes banking stability, if only μ_{21ik} is statistically different from zero. Finally, we can deduce that banking competition and banking stability reinforce each other if all μ_{12ik} and μ_{21ik} are statistically different from zero.

In VECM estimation process, the lag length is a key factor because the result of causality check totally depends on the lag specification. It is not possible to fix best lag lengths as there is no general method or rule to solve it. But, it may create huge complication in calculation process as we are handling a large panel of data with too many cases and specifications. To resolve this issue, in Eq. (1), the three variables are set to run in different maximum lag lengths, at the same time, it is kept under restriction to vary across countries. The optimum lag length is decided using AIC[10] statistics in this test.

[10]AIC-Akaike information criterion. A suitable statistical technique to fix optimum lag length (Brooks 2014).

4 Empirical Results

To initiate the discussion, we need to lay a hand on properties like integration and cointegration of our time series variables. In order to find out the integration and cointegration properties of the time series variables, it is necessary to make use of the three-panel unit root tests and Pedroni's panel cointegration test (Pedroni 2004). The outcome of these test shows that the variables are cointegrated and they are of I (1) (see Tables 3 and 4, respectively). This result signifies that there exists a long-run equilibrium relationship between banking competition and banking stability.

Moreover, in order to confirm the long-run relationship between banking competition and banking stability, we also apply a fully modified OLS (FMOLS[11]) and dynamic OLS (DOLS[12]) (see, Pedroni 2000). As a result, it confirms that banking stability is considerably correlated with banking competition.

This is accurate for all six forms (case 1: linking BCA, and BCP; case 2: linking NPL, and BCP; case 3: linking ZSC, and BCP; case 4: linking PNL, and BCP; case 5: linking PCD, and BCP; and case 6: linking CBS, and BCP) and the six cases within each model. Due to space limitations, the findings of FMOLS and DOLS are not presented here.

This is true for all six models (Model 1: between BCA, and BCP; Model 2: between NPL, and BCP; Model 3: between ZSC, and BCP; Model 4: between PNL, and BCP; Model 5: between PCD, and BCP; and Model 6: between CBS, and BCP) and the six cases within each model. The results of FMOLS and DOLS are not available here due to space constraints. From the above estimations, we can find a clear pattern between the banking competition and banking stability relationship. In addition, this connection is strong[13] to different measures of banking competition and banking stability, respectively.

As the three sets of variables hold a sturdy cointegrating relationship among them, it is prudent to apply a panel Granger causality test, based on a panel vector error correction model. This test facilitates to identify the direction of causality between the two. Based on the estimation of Eq. 1, Granger causal relationships among the variables are presented in Table 5. These results are summarized below.

4.1 Short-Run Causality Results Between Banking Competition and Banking Stability

In this scenario, the majority of the cases show a unidirectional causality running from banking competition to banking stability. This results are consistent with some of the findings of Kasman and Kasman (2015), Fu et al. (2014), Chang et al. (2008),

[11]See Maeso-Fernandez et al. (2006) and Pedroni (2001) for more details on FMOLS.

[12]See Kao and Chiang (2000) for detailed description on DOLS.

[13]The results are consistent with the findings of Fiordelisi and Mare (2014), Fu et al. (2014), Jeon and Lim (2013), Amidu and Wolfe (2013), Berger et al. (2009), and Allen and Gale (2004).

Table 3 Panel unit root test results

Unit root test statistics									
X	LLC			ADF			PP		
Variables	LD	FD	IN	LD	FD	IN	LD	FD	IN
Part A: Banking competition									
LEI	0.812	−19.46*	I [1]	38.91	387.88*	I [1]	50.22	530.71*	I [1]
BOI	−1.24	−17.90*	I [1]	45.71	346.6*	I [1]	52.28	532.5*	I [1]
HST	2.78	−12.36*	I [1]	49.4	247.3*	I [1]	43.4	381.1*	I [1]
CR5	1.042	−16.90*	I [1]	17.29	291.6*	I [1]	17.5	447.6*	I [1]
CR3	−0.956	−19.57*	I [1]	18.99	336.0*	I [1]	25.14	499.6*	I [1]
FOW	−0.262	−512.5*	I [1]	64.38	179.6*	I [1]	33.29	293.8*	I [1]
Part B: Banking stability									
BCA	0.912	−17.29*	I [1]	28.66	286.7*	I [1]	37.46	436.4*	I [1]
NPL	−0.219	−11.8*	I [1]	73.59	208.2*	I [1]	65.2	275.6*	I [1]
ZSC	0.902	−14.83*	I [1]	23.4	282.1*	I [1]	20.76	466.9*	I [1]
PNL	−0.177	−12.40*	I [1]	48.4	215.0*	I [1]	64.9	334.4*	I [1]
PCD	1.954	−9.379*	I [1]	26.23	190.1*	I [1]	16.14	232.76*	I [1]
CBS	−0.782	−216.0*	I [1]	34.49	496.2*	I [1]	37.85	507.8*	I [1]

Note 1 LEI is Lerner index; BOI is Boone indicator; HST is H-statistic; CR5 is firm concentration of five largest banks; CR3 is firm concentration of three largest banks; FOW is foreign ownership; BCA is bank capitalization; NPL is nonperforming loans; ZSC is bank-level Z-index; PNL is provision of nonperforming loans; PCD is private credit by deposit money banks; and CBS is composite index of banking stability

Note 2 LD stands for level data, FD stands for first difference data, LLC stands for Levine–Lin–Chu test, ADF stands for ADF-Fischer Chi-square test, and PP stands for PP-Fischer Chi-square test

Note 3 * denotes significance at 1% level

Note 4 I [1] indicates integrated of order one

Table 4 Results of panel cointegration test for various specifications and cases

Test statistics	Cases					
	Case 1: LEI	Case 2: BOI	Case 3: HST	Case 4: CR5	Case 5: CR3	Case 6: FOW
Specification1: BCA, BCP						
Panel υ	−2.277	−1.835	−5.729	1.007	1.176	−3.294
Panel ρ	−0.084	−1.165	−0.4222	−2.327	−2.255	−2.666*
Panel PP	−1.358***	−1.980**	−1.459***	−4.259*	−4.922*	−5.919*
Panel ADF	−0.287	−0.771	−2.306**	−3.669*	−4.790*	−3.071*
Group ρ	0.224	−0.589	0.896	0.001	0.281	1.289
Group PP	−3.464*	−4.819*	−2.891*	−3.833*	−4.691*	−5.770*
Group ADF	−2.993*	−3.086*	−3.533*	−1.324***	−2.619*	−1.907**
Inference	Cointegrated	Cointegrated	Cointegrated	Cointegrated	Cointegrated	Cointegrated
Specification 2: NPL, BCP						
Panel υ	0.054	−0.504	−6.107	0.561	0.577	−1.231
Panel ρ	1.059	1.688	0.313	1.104	1.096	2.478
Panel PP	0.942	1.284	0.882	1.431	1.523	2.927
Panel ADF	−0.633	0.243	0.042	0.581	0.441	0.680
Group ρ	3.515	3.724	4.110	3.293	3.025	4.041
Group PP	2.849	3.046	3.285	2.940	2.967	2.814
Group ADF	1.079	2.600	0.361	0.888	−1.249	0.427
Inference	Cointegrated	Cointegrated	Cointegrated	Cointegrated	Cointegrated	Cointegrated
Specification 3: ZSC, BCP						
Panel υ	10.36*	3.717*	10.96*	3.385*	5.142*	−12.33
Panel ρ	−1.958**	−2.314*	−1.918**	−5.734*	−4.221*	−3.330*
Panel PP	0.355	−1.618	−0.745	−11.63*	−6.221*	−6.794*
Panel ADF	−1.321	−1.307	−2.273	−7.19	−1.851**	−4.811*
Group ρ	0.411	−1.094	1.115	−3.380*	−3.281*	1.493
Group PP	−4.879*	−5.967*	−3.114*	−13.98*	−11.58*	−7.608*
Group ADF	−4.957*	−7.512*	−4.675*	−4.290*	−2.037*	−6.698*
Inference	Cointegrated	Cointegrated	Cointegrated	Cointegrated	Cointegrated	Cointegrated
Specification 4: PNL, BCP						
Panel υ	−3.754	−3.739	0.721	2.464*	2.436*	−8.504
Panel ρ	−1.788**	−3.123*	−2.432**	−1.834***	−2.437*	−3.227*
Panel PP	−1.639	−5.346*	−3.242*	−2.598*	−3.117*	−7.967*
Panel ADF	−4.878*	−4.945*	−3.667*	−2.247*	−2.337*	−3.777*
Group ρ	0.258	0.024	0.988	2.085	1.816	0.972
Group PP	−4.629*	−7.895*	−6.094*	−2.468*	−3.027*	−9.430*
Group ADF	−4.898*	−3.056*	−2.189*	−1.614***	−1.800***	−5.634*

(continued)

Table 4 (continued)

Inference	Cointegrated	Cointegrated	Cointegrated	Cointegrated	Cointegrated	Cointegrated
Specification 5: PCD, BCP						
Panel υ	−3.457	−3.319	−2.630	−0.877	−1.499	−9.068
Panel ρ	0.585	−1.367***	−1.700***	−1.806****	−1.738***	−3.337*
Panel PP	−0.814	−2.529	−0.364	−3.155	−3.707	−6.642*
Panel ADF	0.366	−1.843***	−1.986***	−2.437**	−2.841**	−4.669*
Group ρ	−0.602	1.898	3.159	2.766	2.352	1.849
Group PP	−4.327*	−3.847*	−2.125**	−1.566***	−2.192**	−9.302*
Group ADF	−3.106*	−3.118*	−2.147**	−2.102***	−2.244***	−11.38*
Inference	Cointegrated	Cointegrated	Cointegrated	Cointegrated	Cointegrated	Cointegrated
Specification 6: CBS, BCP						
Panel υ	5.09*	4.47*	−1.537	3.457*	3.045*	−14.7*
Panel ρ	−5.07*	−5.18*	−2.496*	−7.595*	−7.624*	−2.89*
Panel PP	−9.49*	−9.44*	−3.835*	−10.4*	−9.89*	−4.69*
Panel ADF	−12.9*	−12.8*	−3.91*	−7.97*	−6.82*	−3.99*
Group ρ	−2.26	−2.37*	1.717	−2.884*	−3.286*	1.016
Group PP	−13.5*	−13.9*	−2.743*	−9.878*	−10.74*	−6.954*
Group ADF	−24.4*	−23.7*	−3.668*	−6.713*	−7.182*	−9.015*
Inference	Cointegrated	Cointegrated	Cointegrated	Cointegrated	Cointegrated	Cointegrated

Note 1 Case 1: banking stability and LEI; Case 2: banking stability and BOI; Case 3: banking stability and HST; Case 4: banking stability and CR5; Case 5: banking stability and CR3; and Case 6: banking stability and FOW
Note 2 LEI is Lerner index; BOI is Boone indicator; HST is H-statistic; CR5 is firm concentration of five largest banks; CR3 is firm concentration of three largest banks; FOW is foreign ownership; BCA is bank capitalization; NPL is nonperforming loans; ZSC is bank-level Z-index; PNL is provision of nonperforming loans; PCD is private credit by deposit money banks; and CBS is composite index of banking stability
Note 3 BCP stands for banking competition and is used to indicate LEI, BOI, HST, CR5, CR3, or FOW
Note 4 *, ** and *** represent the parameter estimates are significant at the 1%, 5% and 10% levels, respectively

Yeyati and Micco (2007), Beck et al. (2006), Allen and Gale (2004), Caminal and Matutes (2002) and Mishkin (1999). We also find the support of bidirectional causality between the two in few cases by supporting the feedback hypothesis (FBH) of the BCP-led BSI nexus. Previously, few researchers namely Martinez-Miera and Repullo (2010), Hakenes and Schnabel (2011), James et al. (2013), and Liu et al. (2013) report that the two variables supports each other supporting the feedback hypothesis. These findings provide some support for the other two hypothesis SLH and DFH, i.e., BCP-led BSI nexus and the results are constant with a few studies. (Andrievskaya and Semenova 2016; Titko et al. 2015; Kasman and Kasman 2015; Fu et al. 2014; James et al. 2013; Tabak et al. 2012; Soedarmono et al. 2011). However, there is no proof for demand-following hypothesis in few cases. This study also supports the neutrality hypothesis of BCP-led BSI nexus and it is consistence with Lee and Hsieh (2014).

Table. 5 Results of panel Granger causality test for various specifications and cases

Dependent variables	Independent variables and ECT-1								
Specification1: BCA, BCP									
	Case 1: LEIx			Case 2: BOI			Case 3: HST		
	ΔBCA	ΔLEI	ECT-1	ΔBCA	ΔBOI	ECT-1	ΔBCA	ΔHST	ECT-1
ΔBCA	–	9.73*	−0.01*	–	1.817	−0.02	–	11.1*	−0.06*
ΔBCP	0.944	–	−0.04	14.54	–	−0.03*	10.15*	–	−0.10
	Case 4: CR5			Case 5: CR3			Case 6: FOW		
	ΔBCA	ΔCR5	ECT-1	ΔBCA	ΔCR3	ECT-1	ΔBCA	ΔFOW	ECT-1
ΔBCA	–	3.715***	−0.01***	–	5.147**	−0.05**	–	3.911**	−0.12**
ΔBCP	1.244	–	−0.032	0.819	–	−0.07	4.27**	–	−0.15**
Specification 2: NPL, BCP									
	Case 1: LEI			Case 2: BOI			Case 3: HST		
	ΔNPL	ΔLEI	ECT-1	ΔNPL	ΔBOI	ECT-1	ΔNPL	ΔHST	ECT-1
ΔBNPL	–	15.14*	–	–	8.891*	–	–	4.054**	–
ΔBCP	4.035**	–	–	9.906*	–	–	38.5*	–	–
	Case 4: CR5			Case 5: CR3			Case 6: FOW		
	ΔNPL	ΔCR5	ECT-1	ΔNPL	ΔCR3	ECT-1	ΔNPL	ΔFOW	ECT-1
ΔNPL	–	9.868*	–	–	9.25*	–	–	30.23*	–
ΔBCP	1.104	–	–	1.868	–	–	0.158	–	–
Specification 3: ZSC, BCP									
	Case 1: LEI			Case 2: BOI			Case 3: HST		
	ΔZSC	ΔLEI	ECT-1	ΔZSC	ΔBOI	ECT-1	ΔZSC	ΔHST	ECT-1
ΔZSC	–	4.177**	−0.01**	–	6.207*	−0.01*	–	19.2*	−0.24*
ΔBCP	0.139	–	−0.03	0.211	–	−0.03	8.70*	–	−0.04

(continued)

Table. 5 (continued)

Dependent variables	Independent variables and ECT-1								
	Case 4: CR5			Case 5: CR3			Case 6: FOW		
	ΔZSC	ΔCR5	ECT-1	ΔZSC	ΔCR3	ECT-1	ΔZSC	ΔFOW	ECT-1
ΔZSC	–	6.02*	−0.03*	–	21.8*	−0.04*	–	4.977*	−0.16*
ΔBCP	6.26*	–	−0.04*	6.09*	–	−0.03*	5.094*	–	−0.95*
Specification 4: PNL, BCP									
	Case 1: LEI			Case 2: BOI			Case 3: HST		
	ΔPNL	ΔLEI	ECT-1	ΔPNL	ΔBOI	ECT-1	ΔPNL	ΔHST	ECT-1
ΔPNL	–	24.0*	−0.12*	–	12.3*	−0.09*	–	2.79***	−0.03***
ΔBCP	18.5*	–	−0.04	3.304***	–	−0.11***	9.15*	–	−0.17
	Case 4: CR5			Case 5: CR3			Case 6: FOW		
	ΔPNL	ΔCR5	ECT-1	ΔPNL	ΔCR3	ECT-1	ΔPNL	ΔFOW	ECT-1
ΔPNL	–	16.1*	−0.32*	–	5.067**	−0.26**	–	4.33**	−0.19*
ΔBCP	3.306***	–	−0.05***	1.552	–	−0.07	3.894***	–	−0.023
Specification 5: PCD, BCP									
	Case 1: LEI			Case 2: BOI			Case 3: HST		
	ΔPCD	ΔLEI	ECT-1	ΔPCD	ΔBOI	ECT-1	ΔPCD	ΔHST	ECT-1
ΔPCD	–	6.863*	−0.053	–	6.863*	−0.053	–	6.863*	−0.053
ΔBCP	7.068*	–	−0.521	7.068*	–	−0.521	7.068*	–	−0.521
	Case 4: CR5			Case 5: CR3			Case 6: FOW		
	ΔPCD	ΔCR5	ECT-1	ΔPCD	ΔCR3	ECT-1	ΔPCD	ΔFOW	ECT-1
ΔPCD	–	3.546***	−0.73*	–	2.478	−0.847*	–	5.079**	−1.209*
ΔBCP	1.606	–	−0.03	2.132	–	−0.099	1.813	–	−0.05

(continued)

Table. 5 (continued)

Dependent variables	Independent variables and ECT-1								
Specification 5: CBS, BCP									
	Case 1: LEI			Case 2: BOI			Case 3: HST		
	ΔCBS	ΔLEI	ECT-1	ΔCBS	ΔBOI	ECT-1	ΔCBS	ΔHST	ECT-1
ΔCBS	–	4.89*	−0.46*	–	4.07*	−0.02*	–	4.97*	−0.03*
ΔBCP	0.19	–	−0.01	1.854	–	−0.07	4.04*	–	−0.16*
	Case 4: CR5			Case 5: CR3			Case 6: FOW		
	ΔBCA	ΔCR5	ECT-1	ΔCBS	ΔCR3	ECT-1	ΔCBS	ΔFOW	ECT-1
ΔCBS	–	7.21*	−0.48*	–	5.02*	−0.47*	–	24.1*	−0.83*
ΔBCP	0.276	–	−0.01	1.996	–	−0.02	0.87	–	−0.14

Note 1 Case 1: banking stability and LEI; Case 2: banking stability and BOI; Case 3: banking stability and HST; Case 4: banking stability and CR5; Case 5: banking stability and CR3; and Case 6: banking stability and FOW
Note 2 LEI is Lerner index; BOI is Boone indicator; HST is H-statistic; CR5 is firm concentration of five largest banks; CR3 is firm concentration of three largest banks; FOW is foreign ownership; BCA is bank capitalization; NPL is nonperforming loans; ZSC is bank-level Z-index; PNL is provision of nonperforming loans; PCD is private credit by deposit money banks; CBS is the composite index of banking stability; and ECT-1 is lagged error correction term
Note 3 BCP stands for banking competition and is used to indicate LEI, BOI, HST, CR5, CR3, or FOW
Note 4 *, ** and *** represent the parameter estimates are significant at the 1%, 5%, and 10% levels, respectively

4.2 Long-Run Causality Results Between the Variables

In this segment, we explicitly highlight the significance of ECTs in the VECM estimation (see Table 4). From this result, we can find that ΔBSI acts as a reliant variable; the lagged error correction term generally is statistically acceptable at a 1% level.[14] This implies that banking stability tends to converge to its long-run equilibrium path in response to change in banking competition. The general significance of the ECT-1 coefficient in the ΔBSI equation suggests the continuation of a long-run equilibrium between banking competition (using six different indicators, namely, LEI, BOI, HST, CR5, CR3, and FOW) and banking stability (using six different indicators, namely BCA, NPL, ZSC, PNL, PCD, and CBS).[15] In addition to this, we can generally say that banking competition Granger-causes banking stability in the long run in few SEM countries. In all the combination, the projected lagged ECT coefficients are negatively signed. This implies that, in reality, the change in the level of banking stability (ΔBSI) quickly makes any variation in the long-run equilibrium, or short-run disequilibrium, for the t-1 period. Whereas, in the long run, the consequence of an immediate shock to BCP on BSI will be totally adjusted. But as reported in Table 5, the pace of modification between these three differs from case to case. Conversely, there is no evidence of significant results, when we consider ΔBCP as the dependent variable in all six models and six cases. Here, in this case, it is clear that the lagged error correction terms (ECTs) are also not statistically significant.

4.3 Results from Innovation Accounting

In conclusion, to find the direction of the relationship between banking competition and banking stability in the study area, we have utilized the generalized forecast error variance decomposition method (using VAR system). The study is in line with the work of Pesaran and Shin (1998) and Engle and Granger (1987), this system also implies that the magnitude of the predicted error variance for a sequence accounted for by innovations from each of the independent variable over different time periods beyond the selected time periods. Due to the innovative stemming effect in one variable, the other variables may also mimic the proportional contribution. Like orthogonalized forecast error variance decomposition approach, it is one of the most important gains of this approach that it is insensitive with the ordering of the variables because the ordering of the variables is uniquely determined by the VECM system. Moreover, the abovementioned approach predicts the real-time changes that banking competition can cause to banking stability. The results are unavailable here in order to conserve space and can be availed on demand.

---

[14]As it is apparent and is true in a most of the cases that we consider.

[15]This is line with the study of Jayakumar et al. (2018).

5 Concluding Comments and Policy Implications

As mentioned earlier, our study tries to verify the possible connection the between banking competition and banking stability in 32 Single European Market countries using the data span from 1996 to 2014 collected annually. We use six different indicators for both banking competition and banking stability, including our own calculation (CBS), in this study.

Our above-given models make a clear attempt to show that there is a long-run connection between banking competition and banking stability. As the two variables are cointegrated, we confirm the existence of a long-run equilibrium relationship between them. By employing the fully modified OLS and dynamic OLS procedures, we have a substantial evidence to prove banking competition is correlated with banking stability.

On confirmation of long-run equilibrium relationship between the two variables, we tend to study the direction of causality among the associations further by investigating into the issue of causality among the variables using Granger causality method (using a panel VAR method). Our findings direct many short-run results, indicating the existence of unidirectional or bidirectional causal associations among the variables in few cases. But, our remarkable uniform outcome is that banking competition plays an important role in driving banking stability especially in Single European Market countries in the long run. So the question remains how one can exploit banking competition to achieve the stability[16] in the banking industry. The solution totally depends on the financial sector restructuring, because these reforms create financial innovation and elevate performance in the financial system. It is suggested that banking competition matter—positively and negatively—for managing a well-organized financial market[17] and their guidelines has become one of the key objectives of financial policy to reach good stability in the banking system.

Acknowledgements We are thankful to anonymous reviewers and session chair (5th IIMA International Conference of Advanced Data Analysis, Business Analytics, and Intelligence (ICADABAI), April 8–9, 2017) for detailed suggestions that significantly improved the paper. We are also grateful to Professor Arnab K. Laha (the conference convenor, ICADABAI) for his valuable direction and kind support.

[16]To promote the banking stability in the economy, the banking authority should implement better policies and reforms supporting the banking competition in the market in a well-refined form.

[17]It is widely regarded that a good banking system is as an important drivers of economic growth; while any failure in the banking system will have direct negative effect on any country's economy (see, inter alia, Louzis et al. 2012; Dell' Ariccia et al. 2008; Hogart et al. 2002; Levine and Zervous 1998).

Appendix 1: Description of Variables

Table 6 Definition of variables

Code	Definitions	Source	Expected effect
A. The six banking stability indicators:			
BAC	**The bank-level capitalization ratio**: Measured as the ratio of equity to total assets. A higher value indicates a good stability. Values are average over time [measured in percentage]	World Bank: The Global Financial Development Database	+
NPL	**Nonperforming loans**: The bank-level ratio of nonperforming loans to total loans; a higher value indicates a riskier loan portfolio. Values are averaged over time [measured in percentage]	World Bank: The Global Financial Development Database	–
ZSC	**Z-score** = (ROAA + CAR)/SROAA, where CAR represents capital assets ratio, and SROAA stands for standard deviation of return on assets. A higher value indicates a good stability [measured in percentage]	World Bank: The Global Financial Development Database	+
PNL	**Provision of nonperforming loans**: It is the NPL ratio being overdue for more than a certain number of days, usually more than 90 days [expressed in percentage]	World Bank: The Global Financial Development Database	+
PCD	**Private credit by deposit money banks**: Private credit by deposit money banks and other financial institutions to GDP [expressed as a percentage]	World Bank: The Global Financial Development Database	+
CBS	**Composite index of banking stability**: It is constructed using four banking sector indicators, namely BCA, ROA, PLA, and PCD. A higher value indicates a good stability [expressed in percentage]	Own calculation	+
B. The six banking competition indicators:			
LEI	**Lerner index**: A measure of market competition in the banking market. It compares output pricing and marginal costs. An increase in the Lerner index indicates a decline in competition among the banks	World Bank: The Global Financial Development Database	–

(continued)

Table 6 (continued)

Code	Definitions	Source	Expected effect
BOI	**Boone indicator**: The indicator is calculated based on the relationship between performance, in terms of profits, and efficiency, measured as marginal costs. An increase in the Boone indicator indicates a decline in competition among the banks	World Bank: The Global Financial Development Database	–
HST	**H-Statistics**: A measure of the degree of competition in the banking market. It measures the elasticity of banks revenues relative to input prices. Under perfect competition, an increase in input prices raises both marginal costs and total revenues by the same amount, and hence the H-statistic equals 1	World Bank: The Global Financial Development Database	?
CR5	**CR-5 firm concentration**: A country-level structural indicator of bank concentration, measured by the concentration of assets held by the five largest banks in each country	World Bank: The Global Financial Development Database	–
CR3	**CR-3 firm concentration**: A country-level structural indicator of bank concentration, measured by the concentration of assets held by the three largest banks in each country	World Bank: The Global Financial Development Database	–
FOW	Percentage of the total banking assets that are held by foreign banks. A foreign bank is a bank where 50% or more of its shares are owned by foreigners	World Bank: The Global Financial Development Database	+

Note 1 Definitions are adopting from World Development Indicators of World Bank. The monetary values of variables are in real US dollars
Note 2 ROA is return on assets (in percentage); and PLA is performing loan assets (in percentage)
Note 3 Six individual indicators are employed in this study for both banking competition and banking stability

Appendix 2: Devising of Composite Index of Financial Stability by Using PCA

The study constructs a composite index of financial stability (CBS) using principal component analysis (PCA). We deploy the subsequent steps to obtain the CBS: (a) data are arranged in order to create an input matrix for principal components (PCs), subsequently the matrix is normalized, based on the min-max criteria; (b) expending PCA, eigenvalues, factor loadings, and PCs are derived; and (c) PCs are used to construct the "CBS" for each country and for each year. The detailed discussion of these steps[18] is not available here due to conserve space.

[18] See Jayakumar et al. (2018) for more details.

Table 7 Instantaneous of PCA-related information for composite index of banking stability

Part A: Eigen analysis of correlation matrix

PCs	Eigen value	Proportion	Cumulative
1	1.41	0.35	0.35
2	1.23	0.31	0.66
3	0.72	0.18	0.84
4	0.64	0.16	1.00

Part B: Eigen vectors (component loadings)

Variables	PC1	PC2	PC3	PC4
BCA	0.65	−0.22	−0.33	0.65
PCD	0.57	−0.33	−0.75	0.08
PLA	0.39	0.64	0.09	0.66
ROA	0.31	0.66	−0.57	−0.38

Note 1 PCs is principal components; PC1 is the first principal component; PC2 is the second principal component; PC3 is the third principal component; and PC4 is the fourth principal component
Note 2 BCA is bank capitalization; PCD is private credit by deposit money banks; PLA is performing loan assets; and ROA is return on assets

To have the "CBS", combination of few important variables relating to banking stability was utilized. The following four variables are used for CBS: bank capitalization (BCA), private credit by deposit money banks (PCD), performing loan assets (PLA), and return on assets (ROA). These four indicators are selected to calculate CBS, as they have a common unit of measurement, i.e., in percentage. Table 7 offers the statistical analysis of PCA.

References

Adjei-Frimpong, K. (2013). Efficiency and competition in the Ghanaian banking industry: A panel granger causality approach. *Annals of Financial Economics, 8*(1), 1–16.

Allen, F., & Gale, D. (2004). Competition and financial stability. *Journal of Money, Credit, and Banking, 36*(3), 453–480.

Al-Yousif, Y. K. (2002). Financial development and economic growth: Another look at the evidence from developing countries. *Review of Financial Economics, 11*(2), 131–150.

Amidu, M., & Wolfe, S. (2013). Does bank competition and diversification lead to greater stability? Evidence from emerging markets. *Review of Development Finance, 3*(3), 152–166.

Andrievskaya, I., & Semenova, M. (2016). Does banking system transparency enhance bank competition? Cross-country evidence. *Journal of Financial Stability, 23,* 33–50.

Beck, T., Demirguc-Kunt, A., & Levine, R. (2006). Bank concentration, competition, and crises: First results. *Journal of Banking & Finance, 30*(5), 1581–1603.

Berger, A. N., Klapper, L. F., & Turk-Ariss, R. (2009). Bank competition and financial stability. *Journal of Financial Services Research, 35*(2), 99–118.

Bikker, J. A., Spierdijk, L., & Finnie, P. (2007). Market structure, contestability and institutional environment: The determinants of banking competition. DNB Working Paper, no. 156.

Boone, J. (2008). A new way to measure competition. *Economic Journal, 118,* 1245–1261.

Boyd, H., & Nicolo, G. (2005). The theory of bank risk taking and competition revisited. *Journal of Finance, 3,* 1329–1343.

Brooks, C. (2014). *Introductory econometrics for finance*. Cambridge: Cambridge University Press.

Caminal, R., & Matutes, C. (2002). Market power and banking failures. *International Journal of Industrial Organization, 20,* 1341–1361.

Carby, Y., Craigwell, R., Wright, A., & Wood, A. (2012). Finance and growth causality: A test of the Patrick's stage-of-development hypothesis. *International Journal of Business and Social Science, 3*(21), 129–139.

Carretta, A., Farina, V., Fiordelisi, F., Schwizer, P., & Lopes, F. S. S. (2015). Don't stand so close to me: The role of supervisory style in banking stability. *Journal of Banking & Finance, 52,* 180–188.

Cetorelli, N., & Gambera, M. (2001). Banking market structure, financial dependence and growth: International evidence from industry data. *The Journal of Finance, 56*(2), 617–648.

Chang, E. J., Guerra, S. M., Lima, E. J. A., & Tabak, B. M. (2008). The stability-concentration relationship in the Brazilian banking system. *Journal of International Financial Markets, Institutions and Money, 18*(4), 388–397.

Chang, Y. T. (2006). Role of non-performing loans (NPLs) and capital adequacy in banking structure and competition. University of Bath School of Management Working Paper, no. 2006.16.

Cihak, M., & Hesse, H. (2010). Islamic banks and financial stability: An empirical analysis. *Journal of Financial Services Research, 38*(2/3), 95–113.

Claessens, S., & Laeven, L. (2003). Financial development, property rights and growth. *Journal of Finance, 58*(6), 24–36.

Coccorese, P. (2008). Bank competition and regional differences. *Economics Letters, 101*(1), 13–16.

Creel, J., Hubert, P., & Labondance, F. (2015). Financial stability and economic performance. *Economic, 48,* 25–40.

De Guevara, F. H., & Maudos, J. (2011). Banking competition and economic growth: cross-country evidence. *The European Journal of Finance, 17*(8), 739–776.

Dell'Ariccia, G., Detragiache, E., & Rajan, R. (2008). The real effect of banking crises. *Journal of Financial Intermediation, 17*(1), 89–112.

Engle, R. F., & Granger, C. W. J. (1987). Cointegration and error correction: Representation estimation and testing. *Econometrica, 55*(2), 251–276.

Fernandez, A., González, A., & Suarez, N. (2016). Banking stability, competition, and economic volatility. *Journal of Financial Stability, 22,* 101–120.

Ferreira, C. (2013). Bank market concentration and bank efficiency in the European Union: A panel Granger causality approach. *International Economics and Economic Policy, 10*(3), 365–390.

Fiordelisi, F., & Mare, D. S. (2014). Competition and financial stability in European cooperative banks. *Journal of International Money and Finance, 45,* 1–16.

Fu, X., Lin, Y., Molyneux, P. (2014). Bank competition and financial stability in Asia Pacific. *Journal of Banking & Finance, 38,* 64–77.

Gaffeo, E., & Mazzocchi, R. (2014). Competition in the banking sector and economic growth: Panel-based international evidence. *SSRN*. https://doi.org/10.2139/ssrn.2416379.

Hakam, A., Fatine, F. A., & Zakaria, F. (2013). Determinants of banking competition in Morocco and evaluation of the structural reforms. *International Journal of Economics and Financial Issues, 3*(2), 447–465.

Hakenes, H., & Schnabel, I. (2011). Capital regulation, bank competition, and financial stability. *Economics Letters, 113*(3), 256–258.

Hamada, K., Kaneko, A., & Yanagihara, M. (2017). Oligopolistic competition in the banking market and economic growth. *Economic Modelling, 68*(C), 239–248.

Hellmann, T. F., Murdock, K. C., & Stiglitz, J. E. (2000). Liberalization, moral hazard in banking, and prudential regulation: Are capital requirements enough. *American Economic Review, 90*(1), 147–165.

Hogart, G., Reis, R., & Saporta, V. (2002). Costs of banking system instability: Some empirical evidence. *Journal of Banking & Finance, 26,* 825–855.

James H. C., Gwatidzo, T., & Ntuli, M. (2013). Investigating the effect of bank competition on financial stability in ten African countries. *International Business & Economics Research Journal, 12*(7), 755–76.

Jayakumar, M., Pradhan, R. P., Dash, S., Maradana, R. P. & Gaurav, K. (2018). Banking competition, banking stability, and economic growth: Are feedback effects at work? *Journal of Economics and Business, 96*, 15–41.

Jeon, J. Q., & Lim, K. K. (2013). Bank competition and financial stability: A comparison of commercial banks and mutual savings banks in Korea. *Pacific-Basin Finance Journal, 25,* 253–272.

Jokipii, T., & Monnin, P. (2013). The impact of banking sector stability on the real economy. *Journal of International Money and Finance, 32,* 1–16.

Kao, C., & Chiang, M. H. (2000). On the estimation and inference of a co-integrated regression in panel data. In B. H. Baltagi (ed.), *Advances in econometrics: Nonstationary panels, panel cointegration and dynamic panels*, Vol. 15, pp. 179–222.

Kasman, A., & Carvallo, O. (2014). Financial stability, competition and efficiency in Latin American and Caribbean banking. *Journal of Applied Economics, 17*(2), 301–324.

Kasman, S., & Kasman, A. (2015). Bank competition concentration and financial stability in the Turkish banking industry. *Economic Systems, 39*(3), 502–517.

Keeley, M. C. (1990). Deposit insurance, risk and market power in banking. *American Economic Review, 80*(5), 1183–1200.

Lee, C., & Hsieh, M. (2014). Bank reforms, foreign ownership, and financial stability. *Journal of International Money and Finance, 40*(3), 204–224.

Levine, R. (2003). More on finance and growth: More finance, more growth? *Federal Reserve Bank of St. Louis Review, 85*(6), 31–46.

Levine, R. (2005). Finance and growth: Theory and evidence. In P. Aghion & S. Durlauf (Eds.), *Handbook of economic growth*. Amsterdam: Elsevier Science.

Levine, R., & Zervos, S. (1998). Stocks market and economic growth. *American Economic Review, 88*(3), 537–558.

Liu, H., Molyneux, P., Wilson, J., & John, O. S. (2013). Competition and stability in European banking: A regional analysis* competition and stability in European banking: A regional analysis. *Academic Journal, Manchester School., 81*(2), 176–201.

Louzis, D. P., Vouldis, A. T., & Metaxas, V. L. (2012). Macroeconomic and bank-specific determinants of non-performing loans in Greece: A comparative study of mortgage, business and consumer loan portfolios. *Journal of Banking & Finance, 36,* 1012–1027.

Maeso-Fernandez, F., Osbat, C., & Schnatz, B. (2006). Towards the estimation of equilibrium exchange rates for CEE acceding countries: Methodological issues and a panel cointegration perspective. *Journal of Computational Economics, 34*(3), 499–517.

Marcus, A. J. (1984). Deregulation and bank financial policy. *Journal of Banking & Finance, 8*(4), 557–565.

Martinez-Miera, D., & Repullo, R. (2010). Does competition reduce the risk of bank failure? *Review of Financial Studies, 23*(10), 3638–3664.

Matutes, C., & Vives, X. (2000). Imperfect competition, risk taking, and regulation in banking. *European Economic Review, 44,* 1–34.

Mishkin, F. S. (1999). Financial consolidation: Dangers and opportunities. *Journal of Banking & Finance, 23,* 675–691.

Mitchener, K. J., & Wheelock, D. C. (2013). Does the structure of banking markets affect economic growth? Evidence from U.S. state banking markets. *Explorations in Economic History, 50,* 161–178.

Ngare, E., Nyamongo, M. M., & Misati, R. N. (2014). Stock market development and economic growth in Africa. *Journal of Economics and Business, 74,* 24–39.

Owusu, E. L., & Odhiambo, N. M. (2014). Financial liberalisation and economic growth in Nigeria: An ARDL-bounds testing approach. *Journal of Economic Policy Reform, 17*(2), 164–177.

Patrick, H. T. (1966). Financial development and economic growth in underdeveloped countries. *Economic Development and Cultural Change, 14*(2), 174–189.

Pedroni, P. (2000). Fully modified OLS for heterogeneous cointegrated panels. *Advances in Econometrics, 15*(1), 93–130.
Pedroni, P. (2001). Purchasing power parity tests in cointegrated panels. *Review of Economics and Statistics, 83*(4), 727–731.
Pedroni, P. (2004). Panel cointegration: Asymptotic and finite sample properties of pooled time series tests with an application to the PPP hypothesis. *Econometric Theory, 20*(3), 597–625.
Pesaran, M., & Shin, Y. (1998). Generalized impulse response analysis in linear multivariate models. *Economics Letters, 58*(1), 17–29.
Pradhan, R. P., Arvin, B. M., Norman, N. R., & Nishigaki, Y. (2014). Does banking sector development affect economic growth and inflation? A panel co-integration and causality approach. *Applied Financial Economics, 24*(7), 465–480.
Pradhan, R. P., Arvin, M. B., & Bahmani, S. (2015). Causal Nexus between economic growth, inflation, and stock market development: The case of OECD countries. *Global Finance Journal, 27,* 98–111.
Schaeck, K., Cihak, M. & Wolfe, S. (2009). Are competitive banking systems more stable? *Journal of Money, Credit and Banking*, *41*(4), 712–734.
Serwa, D. (2010). Larger crises cost more: Impact of banking sector instability on output growth. *Journal of International Money and Finance, 29*(8), 1463–1481.
Soedarmono, W., Machrouh, F., & Tarazi, A. (2011). Bank market power, economic growth and financial stability: Evidence from Asian banks. *Journal of Asian Economic, 22,* 460–470.
Tabak, B., Fazio, D., & Cajueiro, D. (2012). The relationship between banking market competition and risk-taking: Do size and capitalization matter? *Journal of Banking & Finance, 36*(12), 3366–3381.
Titko, J., Skvarciany, V., & Jurevičienė, D. (2015). Drivers of bank profitability: Case of Latvia and Lithuania. *Intellectual Economics, 9*(2), 120–129.
Turk-Ariss, R. (2009). Competitive behaviour in middle East and North Africa banking systems. *Quarterly Review of Economics and Finance, 49,* 693–710.
Vives, X. (2016). *Competition and stability in banking: The role of competition policy and regulation*. Princeton: Princeton University Press.
Wolde-Rufael, Y. (2009). Re-examining the financial development and economic growth Nexus in Kenya. *Economic Modelling, 26*(6), 1140–1146.
Yeyati, E. L., & Micco, A. (2007). Concentration and foreign penetration in Latin American banking sectors: Impact on competition and risk. *Journal of Banking & Finance, 31,* 1633–1647.

MIX
Papier aus verantwortungsvollen Quellen
Paper from responsible sources
FSC® C105338

If you have any concerns about our products,
you can contact us on
ProductSafety@springernature.com

In case Publisher is established outside the EU,
the EU authorized representative is:
Springer Nature Customer Service Center GmbH
Europaplatz 3, 69115 Heidelberg, Germany

Printed by Libri Plureos GmbH
in Hamburg, Germany